# BUSINESS STATISTICS
## *A First Course*

**JOHN E. FREUND**
*Professor of Mathematics*
*Arizona State University*

**BENJAMIN M. PERLES**
*Professor of Economics*
*Old Dominion University*

**PRENTICE-HALL, INC.,** Englewood Cliffs, New Jersey

Library of Congress Cataloguing in Publication Data

Freund, John E.
  Business statistics

  Bibliography: p. 321
  1. Statistics. I. Perles, Benjamin, joint author.
II. Title.
HA 29. F 6828        519.5        73-21565
ISBN 0-13-107714-7

©1974 by PRENTICE-HALL, INC., Englewood Cliffs, N.J.

All rights reserved. No part of this book
may be reproduced in any form or by any means
without permission in writing from the publishers.

10  9  8  7  6  5  4

Printed in the United States of America

PRENTICE-HALL INTERNATIONAL, INC., *London*
PRENTICE-HALL OF AUSTRALIA, PTY. LTD., *Sydney*
PRENTICE-HALL OF CANADA, LTD., *Toronto*
PRENTICE-HALL OF INDIA PRIVATE LIMITED, *New Delhi*
PRENTICE-HALL OF JAPAN, INC., *Tokyo*

# CONTENTS

**PREFACE**

# 1

**INTRODUCTION, 1**

    Statistics, Past and Present, 1
    Sources of Data, 4
    Statistics, What Lies Ahead, 5

# 2

**SUMMARIZING DATA, 9**

    Introduction, 9
    Frequency Distributions, 9
    Graphs and Charts, 17
    Statistics, 29
    The Mean, 29
    The Weighted Mean, 33
    The Median, 39
    Further Measures of Location, 45
    Index Numbers, 48
    Unweighted Index Numbers, 48
    Weighted Index Numbers, 54

# 3
## POSSIBILITIES AND PROBABILITIES, 59

    Introduction, 59
    The Sample Space, 59
    Outcomes and Events, 62
    Listing Outcomes and Counting Events, 69
    Probabilities and Odds, 80
    Expectations and Decisions, 85

# 4
## SOME RULES OF PROBABILITY, 93

    Introduction, 93
    Some Basic Rules, 93
    Further Addition Rules, 96
    Conditional Probabilities, 103
    Bayes' Rule, 115

# 5
## CHANCE VARIATION: PROBABILITY FUNCTIONS, 120

    Introduction, 120
    Probability Functions, 121
    The Binomial Distribution, 123
    The Hypergeometric Distribution, 126
    The Mean of a Probability Distribution, 133
    Measuring Chance Variation, 135
    Chebyshev's Theorem, 139

# 6
## THE NORMAL DISTRIBUTION, 144

    Introduction, 144
    The Standard Normal Distribution, 147
    Some Applications, 157
    Probability Graph Paper, 162
    Approximating the Binomial Distribution, 166

## 7

### CHANCE VARIATION: SAMPLING, 171

Introduction, 171
Random Sampling, 176
Further Problems of Sampling, 181
Chance Fluctuations of Means, 186
The Sample Standard Deviation, 190
The Sampling Distribution of the Mean, 201

## 8

### THE ANALYSIS OF MEASUREMENTS, 209

Introduction, 209
The Estimation of Means, 210
The Estimation of $\sigma$, 214
Tests Concerning Means, 220
Differences Between Means, 235
Differences Among $k$ Means (Analysis of Variance), 240

## 9

### THE ANALYSIS OF COUNT DATA, 248

Introduction, 248
The Estimation of Proportions, 248
Tests Concerning Proportions, 257
Differences Among Proportions, 260
Contingency Tables, 267
Goodness of Fit, 267

## 10

### THE ANALYSIS OF PAIRED DATA, 278

Introduction, 278
The Method of Least Squares, 281
The Coefficient of Correlation, 290

## 11

### THE ANALYSIS OF TIME SERIES, 300

> Introduction, 300
> Linear Trends, 304
> Moving Averages, 307
> Seasonal Variation, 312
> Some Further Considerations, 320

### BIBLIOGRAPHY, 321

> A. Statistics for the Layman, 321
> B. Some Books on the Theory of Probability and Statistics, 321
> C. Some General Books on Business Statistics, 321
> D. Some Books Dealing with Special Topics, 322
> E. Some General Reference Works and Tables, 322

### STATISTICAL TABLES, 323

### ANSWERS TO ODD-NUMBERED EXERCISES, 342

### INDEX, 353

# PREFACE

In recent years, courses in business statistics have taken many different directions. Although one can defend specialization in selected areas or preoccupation with logical and mathematical details, *the approach of this book emphasizes instruction in the understanding, use, and appreciation of meaningful modern and well established statistical techniques.*

In today's courses in statistics, the emphasis has shifted from descriptive methods to more and more inference, and there has also been a change in the level at which first courses in statistics are being taught. Whereas these courses used to be taught mostly to college juniors and seniors, they are now also taught to freshmen and sophomores, and attempts have been made to introduce probability and statistics into advanced high school programs. *Business Statistics: A First Course* is designed to reach the student at this early level, as will be apparent from the organization of the material, the language of the book, its format, its notation, and above all the exercises and the illustrations.

Nevertheless, controversial material (say, about the meaning of probability) has not been avoided, the reader is exposed to the strengths of statistical techniques as well as their weaknesses, and it is hoped that this honest approach will provide a challenge for further study in the field.

To attain a measure of flexibility and to accommodate some of the differences of opinion as to what should be taught in a first course in statistics, a number of topics are included among the exercises *with detailed explanations*. These exercises, which are clearly marked, have made it possible to present special material, extra details, and some more advanced work without cluttering up the main body of the text.

The authors would like to express their gratitude to Mrs. Linda Henry for typing the manuscript and assisting in the many tasks required for its preparation, to Suzanne Perles and Ray Ewer, students at Princeton University and Tufts University, respectively, for their many suggestions and comments which focused attention on student needs and attitudes, and to Doug Freund for his help with the proofreading. Sincere thanks are also extended to Burton Gabriel, Editor, Prentice-Hall, Inc., for his cooperation and encouragement in the preparation of the manuscript.

Finally, the authors are indebted to the Literary Executor of the late Sir Ronald Fisher, F.R.S., Cambridge, to Dr. Frank Yates, F.R.S., Rothamsted, and also to Oliver and Boyd, Edinburgh, for permission to reprint parts of Tables IV and VI, respectively, from their books *Statistical Methods for Research Workers* and *Statistical Tables for Biological, Agricultural, and Medical Research;* and to Professor E. S. Pearson and the Biometrika trustees for permission to reproduce parts of Tables 8, 18, and 41 from their *Biometrika Tables for Statisticians.*

JOHN E. FREUND
BENJAMIN M. PERLES

*Scottsdale, Arizona*
*Norfolk, Virginia*

# 1

# INTRODUCTION

## STATISTICS, PAST AND PRESENT

The origin of modern statistics can be traced to two areas which, on the surface, have very little in common: *government* (political science) and *games of chance*.

Governments have long used *censuses* to count persons and property. The ancient Romans used this technique to assist in the taxation of their subjects; indeed, the Bible tells how Mary and Joseph, subjects of Rome, went to Bethlehem to have their names listed in a census. Another famous census is reported in the *Domesday Book* of William of Normandy, completed in the year 1086. This census covered most of England, listing its economic resources, including property owners and the land which they owned. The U.S. census of 1790 was the first "modern" census, but government agents merely counted the population. More recent U.S. censuses have become much wider in scope, providing a wealth of information about the population and the economy, and they are conducted every ten years. The most recent one, the 19th decennial census, was conducted in 1970.

The problem of describing, summarizing, and analyzing census data led to the development of methods which, until recently, constituted almost all that there was to the subject of statistics. These methods, which originally consisted mainly of presenting the most important features of data by means of tables and charts, constitute what is now referred to as **descriptive statistics**. To be more specific, this term applies to *anything done to data that does not infer anything which goes (generalizes) beyond the data themselves.* Thus, if the government reports on the basis of census counts that the population of the United States was 179,323,175 in 1960 and 203,184,722 in 1970, this belongs to the field of descriptive statistics. This would also be the case if we calculated the corresponding percentage growth, which, as can easily be verified, was 13.3 percent, but *not* if we used these data to predict, say, the population of the United States in the year 2000.

There are essentially two reasons why the scope of statistics and the need to study statistics has grown enormously in the last few decades. One reason is that the amount of data that is collected, processed, and disseminated to the public for one reason or another has increased almost beyond comprehension. To act as watchdogs, more and more persons with some knowledge of statistics are needed to take an active part in the collection of the data, in the analysis of the data, and, what is equally important, in all of the preliminary planning. It is really frightening when one starts to think of all the things that can go wrong in the compilation of statistical data. The results of very costly surveys can be completely useless if questions are ambiguous or asked in the wrong way, if they are asked of the wrong persons, in the wrong place, or at the wrong time. The following illustrates how questions may inadvertently be asked of the wrong persons: In a survey conducted by a manufacturer of an "instant" dessert, the interviewers, who worked from 9 A.M. to 5 P.M. on weekdays, got only the opinions of persons who happened to be at home during that time. Unfortunately, this did not include working wives and other persons who may have been especially interested in the manufacturer's product, and as a result the whole survey was worthless.

There are many subtle reasons for getting **biased data.** For one thing, many persons are reluctant to give honest answers to questions about their sanitary habits, say, how often they use a deodorant, bathe, or brush their teeth; they may be reluctant to return a mail questionnaire inquiring about their success in life unless they happen to be doing rather well; and they may talk about their "Xerox copies" even though the copies were produced on a copying machine made by Smith-Corona-Marchant or Pitney-Bowes. A factor, too, is the human element which may lead a poll taker to reduce the difficulty or expense of his assignment by interviewing the first persons who happen to come along, by interviewing persons who all live in the same apartment house or neighborhood, by falsely reporting interviews which were never held, or by spending an inordinate amount of time interviewing members of the opposite sex.

Then there are those annoying hidden biases which may be very difficult to foresee or detect. For instance, if the registrar of a large university, wanting to select a "typical" group of students, happened to take a sample from the files of students whose last name begins with M, N, and O, he may well get a disproportionate number of students of Irish or Scottish descent: MacLeod, MacPherson, McDonald, O'Brien, O'Toole, and so forth. Similarly, if a quality control inspection procedure calls for a check of every tenth jar of jelly filled and sealed by a machine, the results would be disastrously misleading if it so happened that due to a defect in the machine, every fifth jar is improperly sealed. Needless to say, it can also happen that computers are programmed incorrectly or that information is transmitted or recorded incorrectly; *there is literally no end to what can conceivably cause trouble when it comes to the collecting, recording, and processing of data.* Of course, there are also such things as *intentional biases* and outright fraudulent misrepresentations (as in false advertising or when using "loaded" questions), but this is a matter of ethics rather than statistics.

The second, and even more important, reason why the scope of statistics and the need to study statistics have grown so tremendously in recent years is the increasingly *quantitative approach* employed in business and economics, and also, of course, in the sciences and many other activities which directly affect our lives. Since most of the information required by this approach comes from *samples* (namely, from observations made on only part of a large set of items), its analysis requires generalizations which go beyond the data, and this is why there has been a pronounced shift in emphasis from descriptive statistics to **statistical inference,** or **inductive statistics.** In other words,

> Statistics has grown from the art of constructing charts and tables to the science of basing decisions on numerical data, or even more generally the science of decision making in the face of uncertainty.

To mention a few examples, generalizations (that is, methods of statistical inference) are needed to estimate the number of long distance telephone calls which will be made in the United States ten years hence (on the basis of business trends and population projections); to determine the effect of a new trading stamp program on the sales of supermarkets (on the basis of experiments conducted at a few markets); to evaluate conflicting and uncertain legal evidence; to predict the extent to which elevators will be used in a large office building which has not yet been built; to estimate the demand on a new computer which is being developed; to rate the efficiency of salesmen (on the basis of partial information about their performance); and so forth. In each of these examples there are uncertainties, only partial or incomplete information, and it is here that we must use statistical methods which find their origin in *games of chance.*

Games of chance date back thousands of years, as is evidenced, for example, by the use of *astragali* (the forerunners of dice) in Egypt about 3500 B.C., but the mathematical study of such games began less than 400 years ago. This was due mainly to the fact that until then chance was looked upon as an expression of *divine intent,* and it would have been impious, or even sacrilegious, to analyze the "mechanics" of the supernatural through mathematics.

Although the mathematical study of games of chance, called **probability theory,** dates back to the seventeenth century, it was not until the early part of the nineteenth century that the theory developed for "heads or tails," for example, or "red or black" or "even or odd," was applied also to real life situations where the outcomes were "boy or girl," "life or death," "pass or fail," and so forth. Thus, probability theory was applied to many problems in the social as well as the natural sciences, and nowadays it provides an important tool for the analysis of *any* situation (in business, in science, or in everyday life) which in some way involves an element of uncertainty or risk. In particular, it provides the basis for the methods which we use when we generalize from observed data, namely, when we use the methods of statistical inference.

## SOURCES OF DATA

Statistical data come not only from censuses, but also from the day-to-day operation of businesses, from institutions (say, hospitals or universities), from government agencies, from experiments conducted by individuals or organizations, from market analyses, from opinion polls, and so forth.

Primarily in connection with business statistics, it is customary to make the following distinction: Data taken by a business firm or some other organization from its *own* accounting records, payrolls, production records, inventories, sales vouchers, and the like, are called **internal data,** while data coming from outside the particular firm or organization, such as data from federal, state, or local government agencies, trade associations, private reporting organizations, or perhaps other firms, are called **external data.**

External data may be classified further as **primary data** or **secondary data,** depending on whether the data are collected and released (reported, or published) by the same organization. To be more specific, primary data are collected and released by the same organization, while secondary data are released by an organization other than the one by which they are collected. For instance, the Bureau of Labor Statistics collects the data required for its Consumer Price Index and publishes it in its *Monthly Labor Review,* which is thus the primary source for this important index. On the other hand, when the values of the Consumer Price Index are given in the *Federal Reserve Bulletin* or in the financial pages of newspapers or magazines, we refer to these publications as secondary sources of the index. Generally, primary sources are preferred to secondary sources because the possibility of errors of transcription is reduced, and also because primary sources are often accompanied by documentation and precise definitions.

The U.S. government is undoubtedly the largest publisher of both primary and secondary statistical data. For instance, the *Statistical Abstract of the United States,* published annually by the Bureau of the Census, contains a wide variety of information from various sources on the life of the nation; the *Survey of Current Business,* issued monthly by the Department of Commerce, contains a wealth of information about prices, production, inventories, income, sales, employment, wages, . . ., as well as important indicators of business conditions in general; and the *Monthly Labor Review,* published monthly by the Bureau of Labor Statistics, provides data concerning the labor force, employment, hours, wages, work stoppages, labor turnover, and other related activities. In view of the vast scope of the statistical activities of the Federal government, the book by Hauser and Leonard (listed in the Bibliography at the end of the book) is a valuable aid in locating statistical data published by the various government agencies and departments.

Examples of non-governmental sources of business data are *Moody's*

*Investor Service*, which annually publishes financial material concerning corporations and government, and the *Economic Almanac*, published annually by the National Industrial Conference Board, which contains information about wages, salaries, and living costs. Then there are many periodicals such as the *Commercial and Financial Chronicle*, the *Wall Street Journal*, the *Journal of Commerce and Commercial Digest*, *Dun's Statistical Review*, *Dun's Review*, and *Barron's*, and trade publications such as *Textile World*, *Leather Manufacturer*, *Chemical Industry News*, *Chain Store Age*, and many others. Information concerning statistical data from non-governmental sources (as well as governmental sources) may be found in the book by Coman (listed in the Bibliography at the end of this book).

In connection with all these published data, it is important to remember that they are subject to the same possibilities of error and bias as are directly collected data. Thus, "consumers" of published data must always judge very carefully whether any given publication actually provides reliable *disinterested* data.

## STATISTICS, WHAT LIES AHEAD

Earlier in this chapter we indicated how the emphasis in statistics has shifted from its original job of summarizing data by means of charts and tables to its role in inference, that is, making generalizations on the basis of samples. This is not meant to imply, however, that the subject of statistics has now become stable and inflexible, and that it has ceased to grow. Aside from the fact that new statistical techniques are constantly being developed to meet particular needs, the whole philosophy of statistics continues to be in a state of change. Most recently, attempts have been made to treat all problems of statistical inference within the framework of a unified theory called **decision theory**, which, so to speak, covers everything "from cradle to grave." One of the main features of this theory is that we must account for *all of the consequences* which can arise when we base decisions on statistical data. This poses tremendous practical difficulties, and we must say in all fairness that very few of these developments have "filtered through" into elementary texts. To understand some of the difficulties posed by such an all-encompassing theory, we must appreciate the fact that

> No matter how objective one tries to be in the planning and in the performance of an investigation (survey or experiment), it is virtually impossible to eliminate all elements of subjectivity; furthermore, it is generally difficult, if not impossible, to put "cash values" on all possible consequences of one's acts.

So far as the first point is concerned, subjective decisions generally affect the hiring of research personnel, the purchase of equipment, the choice of a particular statistical technique, the number of measurements or observations one decides to make, and the conditions under which they are to be performed. Needless to say, perhaps, it is quite a problem to account for all of these things in one general theory. So far as the second point is concerned, let us merely ask the reader how he would put a "cash value" on the consequences of the decision whether or not to market a new medicine—especially, if the wrong decision may well involve the loss of human lives.

We have mentioned all this primarily to impress upon the reader that statistics, like most other fields of learning, is not static. Indeed, it is difficult to picture what a beginning course in statistics may be like 20 years hence, although it is a pretty fair bet that its contents will differ considerably from the material covered in this book. Certain aspects will probably still be the same, and that includes the role of probability theory in the foundations of statistics as well as certain "bread and butter" techniques which have been very useful in the past and will undoubtedly continue to be widely used also in the future. On the other hand, there may well be much more emphasis on some of the ideas introduced in the final sections of Chapters 3 and 4.

## EXERCISES

1. In three months, John sold 73, 16, and 79 used automobiles, while Sam sold 54, 56, and 70. Which of the following conclusions can be obtained from these figures by means of purely descriptive methods and which require a statistical inference, namely, a generalization? Explain your answers.
   (a) For the given months, John's average sales were 56 used cars per month, while Sam's were 60.
   (b) Sam is a better used-car salesman than John.
   (c) John probably took his annual vacation during the second month.
   (d) If we discard the lowest month's sales of each of these salesmen, John's average sales are higher than Sam's.
   (e) Sam's sales increased from month to month.
   (f) Sam tried harder in each successive month.
   (g) Sam's three monthly sales fluctuate less than John's.
   (h) Next month, Sam's sales will probably be higher than John's.

2. The paid attendance of a hockey team's first four home games was 6,300, 5,030, 7,380, and 6,770 in the year 1971 and 7,270, 6,880, 8,620, and 2,310 in the year 1972. Which of the following conclusions can be obtained from

these figures by means of purely descriptive methods and which require a statistical inference, namely, a generalization? Explain your answers.
   (a) Among these games, the paid attendance for any one game was highest in 1972.
   (b) The average paid attendance for the four 1971 games was 6,370 and the average paid attendance for the four 1972 games was 6,270.
   (c) The difference between the two averages of part (b) is less than 150.
   (d) The difference between the two averages of part (b) is so small that it is impossible to judge whether interest in the hockey team has actually decreased.
   (e) Probably, the fourth 1972 figure was recorded incorrectly and should have been 6,310 or 7,310.
   (f) If the paid attendance at the fourth home game in 1972 had been 7,310, the average for the four 1972 games would have been 7,520.
   (g) The paid attendance at home games in 1971 exceeded 7,000 more often than in 1972.

3. On page 2 we said that the results of costly surveys can be completely useless if questions are asked in the wrong way, of the wrong persons, in the wrong place, or at the wrong time, and we illustrated the first two.
   (a) Give a reasonably realistic example of a survey which is completely useless because it was conducted in the wrong place.
   (b) Give a reasonably realistic example of a survey which is useless because it was conducted at the wrong time.

4. A statistically-minded lawyer has his office on the fourth floor of a very tall office building, and whenever he leaves his office he records whether the first elevator which stops at his floor is going up or coming down. Having done this for some time, he discovers that the vast majority of the time the first elevator which stops is going down. Comment on his conclusion that *fewer elevators are going up than are coming down.*

5. Determine whether the *Federal Reserve Bulletin*, published monthly by the Board of Governors of the Federal Reserve System, is a primary source or a secondary source for the following data:
   (a) U.S. balance of payments;
   (b) gross national product;
   (c) hours and earnings of production workers in manufacturing industries;
   (d) bank debits and deposits turnover;
   (e) margin requirements;
   (f) member bank reserves;
   (g) kinds of U.S. currency outstanding and in circulation.

6. Refer to the *Monthly Labor Review* for the following information:
   (a) the percentage of the civilian labor force unemployed during the years 1967 through 1971;

(b) the Consumer Price Index for all items for the years 1967 through 1971;
(c) the Wholesale Price Index for the same years as in part (b).

**7.** Refer to the *Survey of Current Business* to determine the source of the following data:
(a) U.S. balance of payments summary;
(b) U.S. international transactions by area;
(c) major U.S. government transactions.

**8.** Refer to the *Statistical Abstract of the United States* to determine the sources of the following:
(a) population of the world;
(b) names of the astronauts who manned U.S.S.R. spacecraft from the year 1961 to date;
(c) amount of workmen's compensation payments made in Hawaii.

**9.** Refer to the *U.S. Foreign Agricultural Trade Statistical Report* for fiscal year 1972 to determine
(a) the annual quantity of raw silk imported by the United States from 1968 to 1972;
(b) the country to which the most catsup and chili sauce was exported in 1972 and the amount of this export in thousands of pounds.

# 2

# SUMMARIZING DATA

**INTRODUCTION**

In recent years the collection of statistical data has grown at such a rate that it would be impossible for anyone to keep up even with part of the things which directly affect his life *unless this information were disseminated in "predigested" or summarized form.* The whole matter of putting large masses of data into a usable form has always been important, and it has multiplied greatly in the last few decades. This has been due partly to the development of electronic computers which have made it possible to accomplish in minutes what previously had to be left undone because it would have taken months or even years, and partly to the increasingly quantitative approach of the sciences, especially the social sciences where nearly every aspect of human life is nowadays measured in one way or another.

The most common method of summarizing data is to present them in condensed form in tables or charts, and this used to take up the better part of elementary courses in statistics. Nowadays, the scope of statistics has expanded to such an extent that much less time is devoted to this kind of work—in fact, we shall talk about it only in the next few pages of this chapter. In the remainder of the chapter we shall see how data can be summarized in other ways, namely, by means of a well-chosen statistical description. Since one of the main objectives of this chapter is to impress upon the reader that there are situations in which given data do not require extensive statistical treatment, namely, situations in which the data can "almost" speak for themselves, we shall discuss here only the simplest kinds of statistical descriptions. More complicated kinds of descriptions (the standard deviation, the coefficient of correlation, etc.) will be taken up later, *when needed.*

**FREQUENCY DISTRIBUTIONS**

Grouping, or classifying, measurements and observations is as basic in statistics as it is in every branch of science and, for that matter, in many activities of everyday life. To illustrate its importance, let us consider the problem of an economist who wants to study the earnings of non-supervisory factory workers in the United States. Not even considering the possibility of conducting a survey of his own—the cost would be prohibitive and the data difficult to obtain—he immediately turns to a logical source for this kind of information, the *Statistical Abstract of the United States* (published annually by the Department of Commerce since 1878). It is conceivable that the Bureau of Labor Statistics, the primary source of these data, might make its **raw** (untreated) **data** available to the economist; needless to say, however, this would put him in the unenviable position of having to look at print-outs containing thousands of figures, namely, the average hourly earnings of the many workers included in the government's survey. Thus, he turns to the *Statistical Abstract of the United States* or some other source of statistical data with the hope of finding the desired information in a more "usable" form.

When one deals with large sets of numbers, a good over-all picture and sufficient information can often be conveyed by grouping the data into a number of classes, and the Department of Commerce might publish the figures on the earnings of non-supervisory factory workers in the United States, say, for April, 1970, as in the following table:

| *Average Hourly Earnings* | *Number of Workers* (*thousands*) |
|---|---|
| $1.55–$1.64 | 351 |
| $1.65–$1.74 | 351 |
| $1.75–$1.84 | 488 |
| $1.85–$1.94 | 495 |
| $1.95–$2.09 | 1,005 |
| $2.10–$2.29 | 1,181 |
| $2.30–$2.49 | 1,037 |
| $2.50 or more | 11,058 |

This kind of table is called a **frequency distribution**—it shows how the average hourly earnings of the almost 16 million workers are *distributed* among the chosen classes. Tables of this kind, in which data are grouped according to their numerical size, are called **quantitative distributions**. In contrast, tables like the one given below, in which data are grouped into non-numerical categories, are called **qualitative distributions**:

|  | 1970 Motor Vehicle Registration (millions) |
|---|---|
| United States | 108.4 |
| Other North and Central America | 11.3 |
| South America | 8.1 |
| Europe | 86.8 |
| Africa | 4.5 |
| Asia | 24.8 |
| Oceania | 5.0 |
| Total | 248.9 |

The source for these figures is the *Statistical Abstract of the United States, 1972.*

Frequency distributions present data in a compact form, give a good overall picture, but they do entail a certain loss of information. For instance, we cannot tell from the first of the two tables how many workers averaged exactly $2.15 an hour, we cannot tell the lowest average earnings or the highest, nor can we tell how many of the workers averaged better than $4.00 an hour. Similarly, we cannot tell from the second table how many motor vehicles were registered in California and how many were registered in England or France. *This loss of information is the price we must pay for putting the data into a more usable form, but it is usually a fair exchange.* If a frequency distribution is constructed to meet a specific need, we simply have to make sure that this will be the case; if a frequency distribution is constructed for publication (namely, to be used by others), we have to see to it that it is "compact" yet conveys as much information as is deemed necessary.

The construction of a numerical, or quantitative, distribution consists essentially of the following four steps:

First we choose the classes into which the data are to be grouped; then we sort (tally) the data into the appropriate classes; then we count the number of items in each class; and finally we display the results in a chart or table.

The last step is mainly a matter of artistic ingenuity, although it is, of course, desirable to have a chart or table *tell the whole story*, namely, be usable without making it necessary to check detailed explanations. This means that the labeling of the classes and the column headings must be clear and self-explanatory, and a reference to the source of the data should always be given in the accompanying text or in a footnote to the chart or table.

Since the second and third steps are purely mechanical (and can be handled automatically if the data are on punch-cards or tape), we shall concentrate here on the first step, namely, that of choosing suitable classifications. Essentially, this entails the following two decisions:

> We must decide on the number of classes into which the data are to be grouped, and we must decide "from where to where" each class is to go.

Both of these choices are largely arbitrary, but they depend to some extent on the nature of the data themselves, and also on the ultimate purpose the distribution is to serve. The following are some basic rules which are generally considered "sound practice":

1. We seldom use fewer than six or more than 15 classes. Actually, this choice depends mostly on the number of measurements or observations we have to group. (Clearly, we would lose rather than gain anything if we grouped six observations into 12 classes, and we would probably give away too much information if we grouped 10,000 observations into three or four classes.)

2. We always choose classes which will accommodate all the data. To this end we must make sure that the smallest and largest values fall within the classification, and that none of the values can fall into possible gaps between successive classes.

3. We always make sure that each item can go into only one class. This means that we must avoid successive classes which overlap, namely, classes which have one or more values in common.

4. Whenever feasible, we make the classes cover equal ranges of values. Furthermore, it is generally desirable to make these ranges (or intervals) multiples of 5, 10, 100, ..., or other numbers which facilitate the tally (especially, if the sorting is done by machine), and make the table easy to read.

Note that all but the last of these rules were observed in the construction of the wage distribution on page 10. Actually, the fourth rule was violated in two ways: The first four intervals are *shorter* than the one from $1.95 to $2.09, which in turn is shorter than the next two. Furthermore, the last class is **open;** for all we know, some of the workers may have received $25.00 an hour or more. If a set of data contains a few values that are much larger (or much smaller) than the rest, **open classes** labeled "... or less," "less than ...," "... or more," or "more than ..." will generally help to simplify the overall picture presented by a distribution by *reducing the required number of classes*. Generally speaking, though, open classes should be avoided, for as we shall see on page 39, they can make it impossible (or at least difficult) to give certain further descriptions of the data.

So far as the second rule is concerned, the most important thing to watch besides the smallest and largest values is whether the data are given to the nearest dollar or to the nearest cent, whether they are given to the nearest inch or to the nearest tenth of an inch, whether they are given to the nearest hundredth of an ounce or to the nearest ounce, and so forth. *In general, we must watch the extent to which the numbers are rounded off.* Thus, if we wanted to group data on the sales of gasoline given to the nearest tenth of a gallon, we might use the first of the following three classifications:

| Sales of Gasoline (gallons) | Sales of Gasoline (gallons) | Sales of Gasoline (gallons) |
|---|---|---|
| 5.0– 9.9 | 5– 9 | 5.00– 9.99 |
| 10.0–14.9 | 10–14 | 10.00–14.99 |
| 15.0–19.9 | 15–19 | 15.00–19.99 |
| 20.0–24.9 | 20–24 | 20.00–24.99 |
| 25.0–29.9 | 25–29 | 25.00–29.99 |
| etc. | etc. | etc. |

The second classification would serve the purpose if the sales figures were rounded to the nearest gallon; rounded to the nearest tenth of a gallon, however, a sale of 9.3 gallons would fall between the first two classes, and a sale of 24.2 gallons would fall between the fourth class and the fifth. The third classification, the one on the right, *could* also accommodate sales figures rounded to the nearest tenth of a gallon, but the extra decimal would then give the erroneous impression that the figures are rounded to two decimals.

To give a concrete illustration of the construction of a frequency distribution, suppose that a study made by the manager of a department store yielded

**14** SUMMARIZING DATA                                                                                           CHAP. **2**

the following data on the length of time (in minutes) that was needed to take the orders of 150 customers who called in by telephone:

| 5.0 | 7.5  | 3.9  | 3.5 | 7.7  | 3.7 | 4.2  | 8.7  | 6.3 | 1.8  |
|-----|------|------|-----|------|-----|------|------|-----|------|
| 4.9 | 8.3  | 7.2  | 4.7 | 9.6  | 5.2 | 8.5  | 9.5  | 7.4 | 4.3  |
| 3.2 | 7.1  | 5.5  | 4.8 | 5.7  | 1.4 | 10.7 | 6.1  | 9.8 | 8.2  |
| 5.0 | 4.9  | 3.1  | 6.0 | 7.5  | 6.5 | 7.6  | 8.2  | 5.8 | 4.6  |
| 7.5 | 6.6  | 8.3  | 4.5 | 10.4 | 5.9 | 5.5  | 11.9 | 7.7 | 3.8  |
| 9.3 | 3.9  | 9.2  | 7.2 | 6.2  | 7.0 | 10.3 | 5.0  | 4.6 | 6.0  |
| 6.3 | 8.2  | 3.5  | 4.0 | 4.7  | 5.4 | 3.4  | 1.1  | 6.5 | 8.8  |
| 8.1 | 8.5  | 6.9  | 7.7 | 7.2  | 9.6 | 8.3  | 4.6  | 4.5 | 2.1  |
| 3.9 | 11.6 | 3.6  | 5.4 | 3.7  | 7.7 | 5.2  | 12.4 | 7.5 | 10.1 |
| 2.7 | 3.0  | 8.0  | 4.2 | 6.5  | 4.4 | 6.5  | 8.5  | 5.4 | 6.5  |
| 5.1 | 9.5  | 5.8  | 8.4 | 8.1  | 6.3 | 8.7  | 3.6  | 8.4 | 5.2  |
| 8.7 | 6.8  | 3.8  | 3.7 | 3.9  | 6.7 | 4.1  | 6.3  | 4.5 | 7.4  |
| 8.3 | 7.4  | 12.1 | 8.6 | 8.5  | 5.4 | 6.8  | 6.7  | 1.8 | 4.3  |
| 5.9 | 6.3  | 7.9  | 3.8 | 6.1  | 8.4 | 2.0  | 4.7  | 6.3 | 6.8  |
| 1.8 | 6.3  | 8.7  | 4.4 | 6.5  | 5.4 | 6.9  | 8.2  | 4.0 | 2.8  |

Since the shortest time required was 1.1 minutes and the longest time was 12.4 minutes, we could use the *12* classes 1.0–1.9, 2.0–2.9, 3.0–3.9, . . ., 11.0–11.9, and 12.0–12.9. Another logical choice would be to use the *six* classes 1.0–2.9, 3.0–4.9, 5.0–6.9, 7.0–8.9, 9.0–10.9, and 11.0–12.9. Choosing the latter, performing the actual tally, and counting the number of values in each class, we obtain the results shown in the following table:

| Time Required (minutes) | Tally | Frequency |
|---|---|---|
| 1.0– 2.9 | //// //// | 9 |
| 3.0– 4.9 | //// //// //// //// //// //// //// //// | 39 |
| 5.0– 6.9 | //// //// //// //// //// //// //// //// //// | 45 |
| 7.0– 8.9 | //// //// //// //// //// //// //// //// // | 42 |
| 9.0–10.9 | //// //// / | 11 |
| 11.0–12.9 | //// | 4 |
|  | Total | 150 |

The numbers in the right-hand column are called the **class frequencies,** and they give the number of items in each class. Also, the smallest and largest values that can go into any given class are referred to as its **class limits.** More specifically, 1.0, 3.0, 5.0, 7.0, 9.0, and 11.0 are the **lower limits** of the six classes of the distribution, and 2.9, 4.9, 6.9, 8.9, 10.9, and 12.9 are their **upper limits.**

Since we are dealing here with figures rounded to the nearest tenth of a minute, the first class actually contains all the calls which took from 0.95 minutes to 2.95 minutes (including 0.95 which would be rounded off to 1.0 but *not* 2.95 which would be rounded off to 3.0), the second class contains all the calls which took from 2.95 minutes to 4.95 minutes, ..., and the sixth class contains all those which took from 10.95 minutes to 12.95 minutes.* It is customary to refer to these numbers as the **class boundaries** or as the **"real" class limits.** We also speak of **upper** and **lower class boundaries,** and it should be observed that the upper class boundary of one class is the lower boundary of the next.

It is important to remember that class boundaries should always be "impossible" values, namely, numbers which cannot possibly occur among the values which we want to group. Thus, in the wage distribution on page 10 (where the data were presumably rounded to the nearest cent) the class boundaries are $1.545, $1.645, $1.745, ..., and $2.495. Similarly, if the number of cans of soda pop sold by a vending machine on various days are grouped into the following table:

*Cans of Soda Pop*

0–24
25–49
50–74
75–99
etc.

the lower class limits are 0, 25, 50, 75, ..., the upper class limits are 24, 49, 74, 99, ..., and the class boundaries are the impossible values $-0.5$, 24.5, 49.5, 74.5, 99.5, ....

Two other terms used in connection with frequency distributions are "class mark" and "class interval." **Class marks** are simply the midpoints of the classes, and they are obtained by averaging the respective class limits (or boundaries), namely, by adding the class limits (or boundaries) and dividing by 2. Thus, the class marks of the distribution of the length of time it took for the telephone orders are $\frac{1.0 + 2.9}{2} = 1.95$, $\frac{3.0 + 4.9}{2} = 3.95$, 5.95, 7.95, 9.95, and 11.95. Also, the class marks of the first of the three classifications

---

* When rounding numbers, difficulties may arise when the last digit is a 5. In that case, a businessman may round up or down, for example, depending on whether an amount is owed to him or whether it is something he owes, but a more impartial practice is always *to round so that the last digit remaining on the right is an even number*. This is why we rounded 0.95 to 1.0 and 2.95 to 3.0, but would have rounded $1.545 to $1.54.

16 SUMMARIZING DATA CHAP. 2

given together on page 13 are $\frac{5.0 + 9.9}{2} = 7.45, 12.45, 17.45, \ldots$, and the class marks of the soda-pop table shown above are $\frac{0 + 24}{2} = 12, 37, 62, \ldots$.

A **class interval** is simply the length or width of a class (namely, the range of values that it contains), and it is given by the difference between its class boundaries. Thus, in the wage distribution on page 10 the first four classes have intervals of $0.10, the fifth class has an interval of $0.15, the sixth and seventh classes have intervals of $0.20, and the eighth class is open. In the telephone-order distribution each class has an interval of 2.0 minutes, and we refer to this value as the **interval** of the distribution. *Of course, this term applies only when the class intervals are all equal.* Note that in the soda-pop table the interval is 25 (and not 24, as might have been guessed at a first glance); the class interval is given by the difference between successive class boundaries and not by the difference between the limits of a class.

There are essentially two ways in which frequency distributions can be modified to suit particular needs. One way is to convert the frequencies into percentages, and if we divide each of the class frequencies of the telephone-order distribution by 150 (the total number of calls timed) and then multiply by 100, we obtain the following **percentage distribution**:

| Time Required (minutes) | Frequency | Percentage |
|---|---|---|
| 1.0– 2.9 | 9 | 6.0 |
| 3.0– 4.9 | 39 | 26.0 |
| 5.0– 6.9 | 45 | 30.0 |
| 7.0– 8.9 | 42 | 28.0 |
| 9.0–10.9 | 11 | 7.3 |
| 11.0–12.9 | 4 | 2.7 |

Percentage distributions are often used when it is desired to compare two (or more) distributions. For instance, if we wanted to compare the April, 1970 earnings of non-supervisory factory workers in the United States with corresponding figures for April, 1973, we would convert the distribution on page 10 into a percentage distribution and then compare the percentages associated with the various classes with the corresponding percentages for April, 1973. This would be more meaningful and more informative than comparing the class frequencies themselves, since the total number of non-supervisory factory workers will have changed from 1970 to 1973.

The other way of modifying a frequency distribution is to convert it into an "or more," "more than," "or less," or "less than" **cumulative distribution**.

To this end we simply add the class frequencies, starting either at the top or at the bottom of the distribution; for instance, for the telephone-order distribution on page 14 we obtain the following "less than" cumulative distribution:

| Time Required (minutes) | Cumulative Frequency |
|---|---|
| Less than 1.0 | 0 |
| Less than 3.0 | 9 |
| Less than 5.0 | 48 |
| Less than 7.0 | 93 |
| Less than 9.0 | 135 |
| Less than 11.0 | 146 |
| Less than 13.0 | 150 |

To go one step further, we could convert this distribution into a **cumulative "less than" percentage distribution** by dividing each of the cumulative frequencies by 150 and then multiplying by 100 (or by adding the percentages already given on page 16).

So far we have discussed only quantitative distributions, but the general problem of constructing *qualitative distributions* is very much the same. We must decide how many categories (classes) to use and what kind of items each category is to contain, making sure that there are no ambiguities and that all of the data are accommodated. Since the categories are usually chosen before any data are actually collected, it is a sound practice to include a category labeled "others," or "miscellaneous."

When dealing with categorical distributions we do not have to worry about such mathematical details as class limits or class boundaries, but we can run into serious difficulties in trying to avoid ambiguities. For instance, if we tried to classify items sold at a supermarket into "vegetables," "meats," "baked goods," "frozen foods," etc., it would be difficult to decide where to put frozen vegetables or meat pies. To avoid difficulties like this, it is advisable (when possible) to use standard categories developed by the Bureau of the Census and other government agencies. References to lists of such categories may be found in the book by P. M. Hauser and W. R. Leonard listed in the Bibliography at the end of this book.

## GRAPHS AND CHARTS

When frequency distributions are constructed primarily to condense large sets of data into an "easy to digest" form, it is usually advisable to present them graphically, that is, in a form which has visual appeal. The most common kind of graphical presentation of a frequency distribution is the **histogram,** an example of which is shown in Figure 2.1. Histograms are constructed by representing the quantities which are grouped on a horizontal scale, the

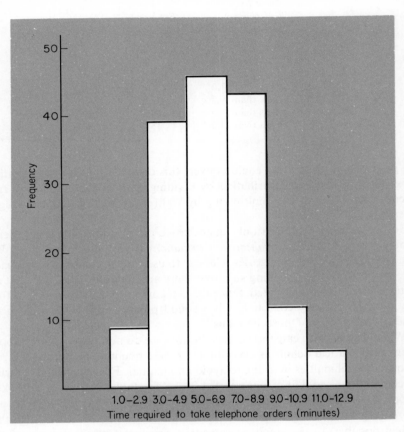

**FIGURE 2.1** Histogram of telephone-order distribution.

class frequencies on a vertical scale, and drawing rectangles whose bases equal the class interval and whose heights are determined by the respective class frequencies. (The markings on the horizontal scale can be the class limits as in Figure 2.1, the class boundaries, or arbitrary key values.) All this assumes that the class intervals are all equal; otherwise, it is best to think of the class frequencies as represented by the *areas* of the rectangles instead of their heights. Unless we do this we can run into all sorts of trouble [see part (h) of Exercise 7 on page 26]; also, the practice of representing class frequencies by means of areas is essential if we want to approximate histograms with smooth curves. For instance, if we approximate the telephone-order distribution with a smooth curve as in Figure 2.2, we can say that the white region under the curve represents the number of telephone orders which required 7.0 minutes or more. Clearly, this area is just about equal to the *sum* of the areas of the corresponding three rectangles.

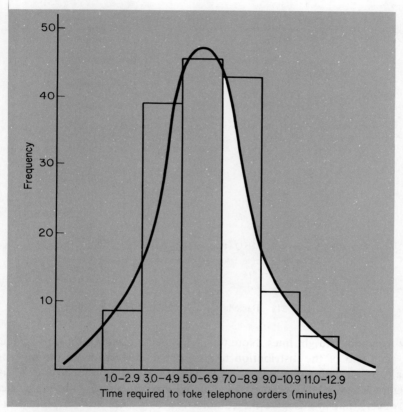

**FIGURE 2.2** Histogram of telephone-order distribution approximated by means of smooth curve.

Similar to histograms are **bar charts**, like the one shown in Figure 2.3; the lengths of the bars are proportional to the class frequencies, but there is no pretence of having a continuous (horizontal) scale.

Another kind of graphical presentation of a frequency distribution is the **frequency polygon,** an example of which is shown in Figure 2.4. Here the class frequencies are plotted at the class marks and the successive points are joined

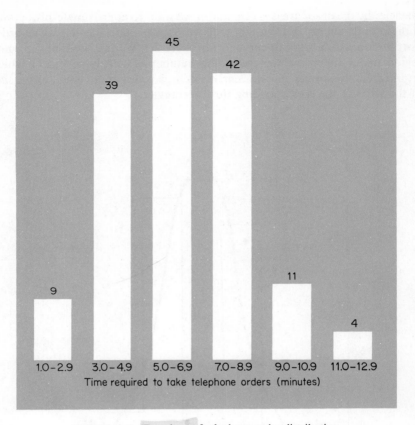

**FIGURE 2.3** Bar chart of telephone-order distribution.

by means of straight lines. Note that we added a class with a zero frequency at each end of the distribution to "tie down" the graph to the horizontal scale. If we apply the same technique to a cumulative distribution, as in Figure 2.5, we obtain what is called an **ogive** (rhymes with "alive"). Observe, however, that in an ogive the cumulative frequencies are *not* plotted at the class marks; it stands to reason that the cumulative frequency corresponding, say, to "less than 5.0 minutes" should be plotted at 5.0, or preferably at the class boundary 4.95, since "less than 5.0 minutes" actually includes here everything up to 4.95 minutes.

Although the visual appeal of histograms, bar charts, frequency polygons, and ogives exceeds that of the corresponding tables, there are various ways in which frequency distributions can be presented more dramatically and, it is hoped, more effectively. We are referring here to the various kinds of pic-

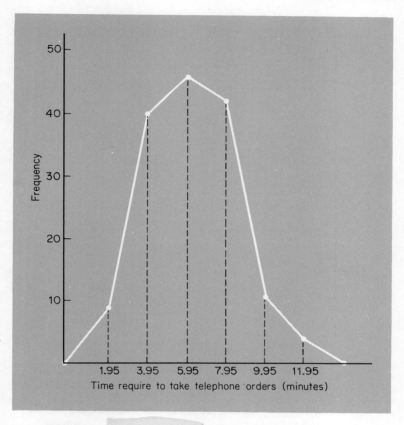

**FIGURE 2.4** Frequency polygon of telephone-order distribution.

torial presentations, like the **pictogram** of Figure 2.6, with which the reader must surely be familiar through newspapers, magazines, and various forms of advertising. *The number of ways in which frequency distributions can be displayed pictorially is practically unlimited, and it depends only on the imagination and the artistic talent of the person who is preparing the presentation.*

Qualitative distributions are often presented graphically as **pie charts** like the one shown in Figure 2.7, where a circle is divided into sectors (pie-shaped pieces) which are proportional in size to the frequencies of the corresponding categories. To construct a pie chart we first convert the distribution into a percentage distribution (by dividing each frequency by the total number of items grouped and multiplying by 100); then we make use of the fact that 1 percent of the data is represented by a sector (pie-shaped piece) with a *central angle* of one hundredth of 360 degrees, namely, an angle of 3.6 degrees. Thus,

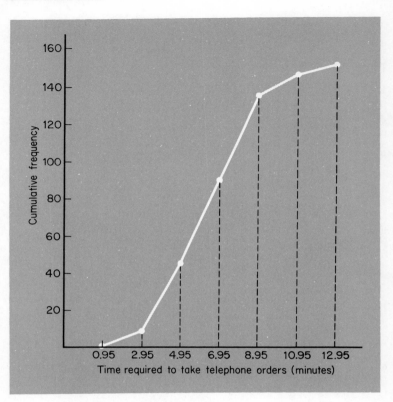

**FIGURE 2.5** Ogive of telephone-order distribution.

for the categorical distribution on page 11 we find that $\frac{108.4}{248.9} \cdot 100 = 43.6$ percent of all the motor vehicles are registered in the United States, $\frac{11.3}{248.9} \cdot 100 = 4.5$ percent are registered in other North or Central American countries, ..., and $\frac{5.0}{248.9} \cdot 100 = 2.0$ percent are registered in Oceania. If we then multiply each of these percentages by 3.6, we get 157, 16, ..., and 7 (rounded to the nearest degree), and these are the central angles of the seven sectors into which we divided the circle in the pie chart of Figure 2.7. Like histograms, pie charts can be improved by using various artistic devices; for instance, by shading the sectors, by giving the whole diagram a three-dimensional effect, or by actually cutting out a piece of the "pie" to draw attention to a particular category.

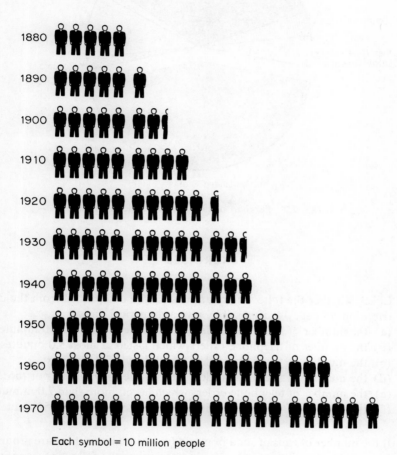

Each symbol = 10 million people

**FIGURE 2.6** Pictogram of the population of the United States.

**24** SUMMARIZING DATA CHAP. 2

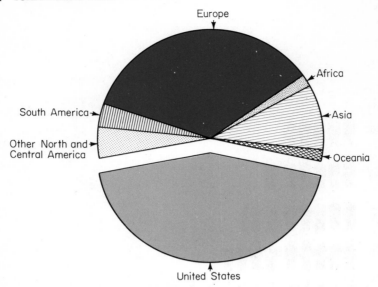

**FIGURE 2.7** Pie chart of 1970 motor vehicle registration.

### EXERCISES

1. Check whether the following quantities can be determined from the distribution on page 14; if possible, give a numerical answer:
   (a) the number of telephone orders which lasted less than 5.0 minutes;
   (b) the number of telephone orders which lasted at most 5.0 minutes;
   (c) the number of telephone orders which lasted 7.3 minutes;
   (d) the number of telephone orders which lasted 7.0 minutes or more;
   (e) the number of telephone orders which lasted more than 7.0 minutes;
   (f) the number of telephone orders which lasted at least 11.5 minutes.

2. If the number of cans of soda pop sold by a vending machine are grouped into a table like that on page 15, check whether the following quantities can be determined from the resulting distribution:
   (a) on how many days at least 75 cans of soda pop were sold;
   (b) on how many days more than 75 cans of soda pop were sold;
   (c) on how many days no cans of soda pop were sold;
   (d) on how many days at least 25 but fewer than 75 cans of soda pop were sold;
   (e) on how many days at most 75 cans of soda pop were sold;
   (f) on how many days fewer than 100 cans of soda pop were sold.

3. If the wages paid to the employees of a certain company in a given month varied from $121.42 to $238.64, indicate the limits of six classes into which these amounts might be grouped.

4. The daily number of hamburgers sold at a drive-in restaurant are grouped into a table having the classes 50–99, 100–149, 150–199, 200–249, 250–299, and 300–349. What are the corresponding class boundaries, the class marks, and the class interval?

5. The class marks of a distribution of the daily number of accident claims reported to an automobile insurance company are 3, 8, 13, 18, 23, and 28. What are the class limits of this distribution and what is its class interval?

6. Class limits and class boundaries have to be interpreted very carefully when we are dealing with ages, for the age group from 5 to 9, for example, includes all those who have passed their fifth birthday but not yet reached their tenth. Taking this into account, what are the boundaries and the class marks of the following age groups: 10–19, 20–29, 30–39, and 40–49?

7. The following is a distribution of the total finance charges which 200 customers paid on their budget accounts at a furniture store:

| Amounts (dollars) | Frequency |
|---|---|
| 0–19 | 28 |
| 20–39 | 52 |
| 40–59 | 67 |
| 60–79 | 34 |
| 80–99 | 19 |

(a) Convert this distribution into a percentage distribution.
(b) Convert this distribution into a cumulative "less than" distribution.
(c) Convert this distribution into a cumulative "more than" distribution.
(d) Convert this distribution into a cumulative "or more" percentage distribution.
(e) Draw a histogram of the original distribution.
(f) Draw a frequency polygon of the original distribution.
(g) Draw an ogive of the cumulative "less than" distribution of part (b).
(h) Draw a histogram of the distribution which is obtained after putting the amounts from $20 to $59, inclusive, into *one* class.

26  SUMMARIZING DATA                                                                CHAP. 2

**8.** The following are the number of business bankruptcies reported in a certain large city in 60 consecutive months:

| 56 | 50 | 58 | 49 | 61 | 52 | 63 | 57 | 54 | 77 | 68 | 55 |
| 57 | 59 | 51 | 65 | 66 | 59 | 48 | 62 | 53 | 71 | 67 | 58 |
| 56 | 73 | 64 | 54 | 75 | 52 | 61 | 48 | 58 | 62 | 56 | 52 |
| 57 | 59 | 60 | 46 | 53 | 55 | 52 | 55 | 54 | 56 | 63 | 57 |
| 82 | 78 | 59 | 51 | 58 | 55 | 68 | 57 | 55 | 53 | 53 | 54 |

(a) Group these figures into a table having the classes 45–49, 50–54, 55–59, ..., and 80–84.
(b) Change the distribution obtained in part (a) into a percentage distribution.
(c) Convert the distribution obtained in part (a) into a cumulative "less than" distribution.
(d) Draw a histogram of the distribution obtained in part (a).
(e) Draw a frequency polygon of the percentage distribution obtained in part (b).
(f) Draw an ogive of the cumulative "less than" percentage distribution obtained in part (c).

**9.** A typing test administered to 80 typists in a large business office yielded the following scores:

| 82 | 42 | 98 | 80 | 50 | 77 | 63 | 72 | 89 | 67 | 45 | 82 | 73 | 95 | 71 | 85 |
| 93 | 82 | 93 | 75 | 88 | 79 | 85 | 62 | 96 | 88 | 59 | 78 | 81 | 95 | 98 | 52 |
| 86 | 49 | 95 | 58 | 94 | 102 | 89 | 67 | 70 | 87 | 78 | 84 | 73 | 77 | 73 | 92 |
| 74 | 87 | 87 | 61 | 87 | 92 | 68 | 102 | 65 | 86 | 96 | 65 | 104 | 79 | 57 | 69 |
| 91 | 85 | 85 | 81 | 78 | 100 | 60 | 83 | 93 | 94 | 82 | 80 | 69 | 86 | 54 | 79 |

(a) Group these scores into a table with a class interval of 10 and draw a histogram of the resulting distribution.
(b) Convert the distribution obtained in part (a) into a percentage distribution and draw its frequency polygon.
(c) Convert the percentage distribution obtained in part (b) into a cumulative "or more" percentage distribution and draw its ogive.

10. A week's records of a bus company show the following values (in dollars) of the gasoline used by each of its 48 buses:

| | | | | | | | |
|---|---|---|---|---|---|---|---|
| 72.17 | 44.63 | 33.21 | 49.30 | 37.80 | 36.45 | 22.16 | 20.45 |
| 28.95 | 43.57 | 18.75 | 42.70 | 33.80 | 71.88 | 33.68 | 56.13 |
| 37.87 | 24.75 | 38.67 | 32.45 | 31.55 | 50.55 | 64.50 | 39.01 |
| 69.49 | 52.83 | 53.41 | 60.75 | 21.45 | 47.82 | 40.58 | 30.56 |
| 37.51 | 34.69 | 41.88 | 30.24 | 15.25 | 27.63 | 24.65 | 45.14 |
| 20.11 | 31.22 | 41.35 | 26.27 | 36.00 | 38.76 | 25.68 | 23.65 |

(a) Group these figures into a table having the classes $15.00–$24.99, $25.00–$34.99, $35.00–$44.99, ..., and $65.00–$74.99.
(b) Convert the distribution obtained in part (a) into a percentage distribution and also into a cumulative "less than" percentage distribution.
(c) Draw a histogram and a bar chart of the distribution obtained in part (a).
(d) Draw an ogive of the cumulative "less than" percentage distribution obtained in part (b).

11. Construct a table showing how many pages of the Sunday edition of a certain large newspaper are devoted primarily to general news, financial news, sports, editorials, comics, women's news, television and other entertainment, classified ads, and so forth.

12. Use a Sunday paper listing prices on the American Stock Exchange to construct a table showing how many of the K, L, M, and N stocks showed a net increase, a net decrease, or no change in price for the preceding week.

13. The following table, based on the *Savings and Loan Fact Book, 1971*, shows the number of private and public housing starts by regions of the United States for the year 1970:

| Region | Number of Units (*thousands*) |
|---|---|
| Northeast | 224 |
| North Central | 300 |
| South | 628 |
| West | 313 |
| Total | 1,465 |

Construct a pie chart of this qualitative distribution.

28  SUMMARIZING DATA  CHAP. 2

14. The following table, based on the *Monthly Review* of the Federal Reserve Bank of New York, August, 1971, shows the total amounts of U.S. securities which it handled during the year 1970:

| Marketable Debt Obligations | Amount (billion dollars) |
|---|---|
| Original issues | 209 |
| Servicing | 783 |
| Redemptions | 170 |
| Total | 1,162 |

where "Servicing" includes denominational exchanges, wire transfers, exchanges of coupons for registered securities, etc. Construct a pie chart which represents this qualitative distribution.

15. The *pictogram* of Figure 2.8 is intended to illustrate the fact that the total

5 billion dollars in 1960

15 billion dollars in 1970

**FIGURE 2.8** Value of corporation stock held by U.S. life insurance companies.

value of corporation stocks held by U.S. life insurance companies has tripled from 1960 to 1970. Does this pictogram convey a "fair" impression of the actual change? If not, how should it be modified?

## STATISTICS

As we are using it here, the word "statistic" is intended to have an entirely different meaning than in the title of this book. In the title of the book we are referring to the subject of statistics, namely, the totality of methods employed in the collection, processing, and analysis of numerical data (or, more generally, to the science of decision making in the face of uncertainty). In the heading of this section, statistics is meant to be the plural of **statistic,** namely, a particular numerical description of a set of data, say, its average. In this sense, a statistic is also referred to as a **statistical measure** or simply as a **statistical description.**

It is customary to classify statistics (that is, statistical measures) according to the particular features of a set of data which they are supposed to describe. Thus, we speak of **measures of location** which may be described crudely as "averages" in the sense that they are generally indicative of the "center," "middle," or the "most typical" of a set of data. Then there are **measures of variation,** which are indicative of the variability, spread, or dispersion of a set of data. Some of these will be introduced in Chapter 7, where we shall also mention **measures of skewness** in an exercise on page 199. In Chapters 9 and 10 we shall study some **measures of association,** which are meant to be indicative of the strength of the relationship (dependence) between two variables, say, a person's income and his intelligence, or the earnings of a corporation's stock and its market price. In addition to these general kinds of descriptions, there are numerous other ways in which statistical data can be summarized. So far as the work of this chapter is concerned, it will be devoted mostly to measures of location.

## THE MEAN

The most popular measure of location is what the layman calls an "average" and what the statistician calls a **mean.** The reason why we gave the word "average" in quotes is that it has all sorts of connotations in everyday language—we speak of a baseball player's batting average, we talk about the average American small town, we describe a criminal suspect's appearance as average, and so forth.

The mean of a set of $n$ numbers is very simply *their sum divided by n*. For instance, if the number of flat tires repaired by a service station on ten consecutive days was 18, 15, 12, 20, 19, 11, 14, 38, 18, and 17, the "average" number of flat tires repaired, namely, the mean, is given by

$$\frac{18 + 15 + 12 + 20 + 19 + 11 + 14 + 38 + 18 + 17}{10} = 18.2$$

To give a formula for the mean which is applicable to any kind of data, it will be necessary to represent the measurements or observations by means of symbols such as $x$, $y$, or $z$. In the above example we might have used the letter $x$, for instance, and denoted the ten values $x_1$ ($x$ sub-one), $x_2$ ($x$ sub-two), $x_3$ ($x$ sub-three), . . ., and $x_{10}$ ($x$ sub-ten). More generally, if we have $n$ measurements or observations which we denote $x_1$, $x_2$, $x_3$, . . ., and $x_n$, their mean can be written as

$$\frac{x_1 + x_2 + x_3 + \cdots + x_n}{n}$$

To simplify our notation, we refer to this expression, namely, the mean of the $x$'s, as $\bar{x}$ ($x$ bar). (Of course, if we refer to the measurements as $y$'s or $z$'s, we would correspondingly write their mean as $\bar{y}$ or $\bar{z}$.) Furthermore, let us introduce the symbol $\Sigma$ (capital *sigma*, the Greek letter for S), which is simply a mathematical shorthand symbol indicating the process of summation, or addition. If we write $\Sigma x$, this means literally "the sum of the $x$'s," and we can thus give the formula for the mean as*

$$\bar{x} = \frac{\Sigma x}{n}$$

The popularity of the mean as a measure of the "center" or "middle" of a set of data is not just accidental. Any time we use a single number to describe a whole set of data, there are certain requirements (namely, certain desirable properties) we must keep in mind. Thus, some of the noteworthy properties of the mean are:

1. It is familiar to most persons, although they may not call it by that name.

2. It always exists, that is, it can be calculated for any kind of data.

3. It is always unique, or in other words, a set of data has one and only one mean.

* See also Exercise 17 on page 38.

4. It takes into account each individual item.

5. It lends itself to further statistical treatment (for instance, the means of several sets of data can always be combined into an over-all mean for all the data; see Exercise 16 on page 37).

6. It is relatively reliable in the sense that for sample data it is generally not as strongly affected by chance as some of the other measures of location.

This question of **reliability** is of fundamental importance when it comes to problems of estimation, hypothesis testing, and making predictions, and we shall study it in quite some detail in Chapter 7.

Whether the fourth property which we have listed is actually desirable is open to some doubt; a single extreme (very small or very large) value can affect the mean to such an extent that it is debatable whether it is really "representative" or "typical" of the data it is supposed to describe. To give an example, suppose that in copying the ten daily figures on the number of flat tires repaired on page 30 we actually made a mistake—the eighth value should have been 8 instead of 38. This means that the mean should have been

$$\frac{18 + 15 + 12 + 20 + 19 + 11 + 14 + 8 + 18 + 17}{10} = 15.2$$

instead of 18.2, and it illustrates how one careless mistake can have a pronounced effect on the mean.

Since the calculation of means is very easy, involving only addition and a division, there is really no need to look for short-cuts or simplifications. However, if the numbers we may want to average have many digits and no adding machine is available, it can be advantageous to group the data first and then calculate the mean on the basis of the resulting distribution. Another reason why we shall devote some time to the calculation of the mean of grouped data is that published data (to which we may have to refer in a study) are generally available only in grouped form.

Actually, results thus obtained may be different, for the *exact* mean of a set of data cannot be determined after they have been grouped; each item, so to speak, loses its identity and we know only how many items there are in each class. Nevertheless, a good approximation can generally be obtained by treating each value in a class as if it equalled the class mark, namely, the midpoint of the class. Although some of the values in a class will usually exceed the class mark, others will be less than the class mark, and all this will more or less "average out." Thus, the nine values in the first class of the telephone-order distribution on page 14 will be treated as if they all equalled 1.95, the

39 values in the second class will be treated as if they all equalled 3.95, the 45 values in the third class will be treated as if they all equalled 5.95, and so on.

To obtain a formula for the mean of a frequency distribution with $k$ classes, let us write the successive class marks as $x_1, x_2, \ldots,$ and $x_k$, and the corresponding frequencies as $f_1, f_2, \ldots,$ and $f_k$. The total that goes into the numerator for the formula for the mean is thus obtained by adding $f_1$ times the class mark $x_1$, $f_2$ times the class mark $x_2, \ldots,$ and $f_k$ times the class mark $x_k$; in other words, it is given by the sum

$$x_1 f_1 + x_2 f_2 + \cdots + x_k f_k$$

Using the $\Sigma$ notation introduced on page 30, we can thus write the formula for the mean of a distribution as*

$$\bar{x} = \frac{\Sigma x \cdot f}{n}$$

where $n$ equals the sum of the class frequencies, namely, $n = \Sigma f$. In words,

To find the mean of a distribution we first add the products obtained by multiplying each class mark by the corresponding class frequency, and then we divide by $n$ (the total number of items grouped).

To illustrate the calculation of the mean of a frequency distribution, let us refer again to the telephone-order distribution on page 14. Writing the class marks in the second column and calculating the necessary products $x \cdot f$, we get

| Time Required (minutes) | Class Mark $x$ | Frequency $f$ | $x \cdot f$ |
|---|---|---|---|
| 1.0– 2.9 | 1.95 | 9 | 17.55 |
| 3.0– 4.9 | 3.95 | 39 | 154.05 |
| 5.0– 6.9 | 5.95 | 45 | 267.75 |
| 7.0– 8.9 | 7.95 | 42 | 333.90 |
| 9.0–10.9 | 9.95 | 11 | 109.45 |
| 11.0–12.9 | 11.95 | 4 | 47.80 |
| Totals | | 150 | 930.50 |

and the mean of the distribution is

$$\bar{x} = \frac{930.50}{150} = 6.20$$

It is of interest to note that the actual mean of the ungrouped data on page

* See also Exercise 17 on page 38.

14 is $\frac{938.7}{150} = 6.26$, and, hence, that the error introduced by calculating the mean on the basis of the distribution is relatively small.

The calculation of the mean of the telephone-order distribution was quite easy, but it could have been made even easier by replacing the class marks with consecutive integers, namely, with numbers which are easier to handle. This process is referred to as **coding,** and we shall illustrate how it is done in Exercise 11 on page 35.

## THE WEIGHTED MEAN

There are many situations in which it would be quite misleading to average quantities without accounting in some way for their relative importance in the over-all picture we are trying to describe. Suppose, for instance, that a student obtained grades of 88, 61, 75, and 84 in four 1-hour examinations in a course in accounting, a 69 in the mid-term examination, and a 46 in the final. The mean of these six grades is $\frac{423}{6} = 70.5$, but this does not mean much because it does not account for the importance which the instructor assigns to the various exams. Similarly, it would be meaningless to calculate the mean of the prices of various food items without accounting in some way for the respective roles which they play in the average family's budget, and we might well get a very misleading picture if we calculated the mean of the average earnings of workers in various industries (construction, manufacturing, mining, etc.) without taking into account how many workers there are in each of the industries.

Returning to the first example, suppose that the instructor decides that each of the four 1-hour examinations should count for 10 percent, the mid-term examination for 20 percent, and the final examination for 40 percent of a student's over-all grade in the course. Thus, the student will get an over-all grade of

$$\frac{10\cdot 88 + 10\cdot 61 + 10\cdot 75 + 10\cdot 84 + 20\cdot 69 + 40\cdot 46}{100} = 63.0$$

and we refer to this kind of average as a **weighted mean**—we "averaged" the six grades giving due weight to their relative importance.

In general, the weighted mean of a set of $n$ numbers $x_1, x_2, \ldots,$ and $x_n$, whose relative importance is measured by a corresponding set of numbers $w_1, w_2, \ldots,$ and $w_n$ called the **weights,** is given by the formula*

$$\bar{x}_w = \frac{\Sigma w \cdot x}{\Sigma w}$$

\* See also Exercise 17 on page 38.

Here $\Sigma w \cdot x$ stands for the sum of the products obtained by multiplying each $x$ by the corresponding weight, while $\Sigma w$ is simply the sum of the weights. Note that when the weights are all equal, the formula for the weighted mean reduces to that of the (ordinary) mean. Also, the mean of grouped data (as given by the formula on page 32) can be looked upon as a weighted mean, with the weights of the class marks being the corresponding class frequencies.

To give another example, suppose that an investor made three purchases of a certain stock. The first purchase was 200 shares at $25.50 per share, the second was 300 shares at $27.25 per share, and the third was 500 shares at $24.75 per share. To find the average price which he paid per share, we substitute $x_1 = 25.50$, $w_1 = 200$, $x_2 = 27.25$, $w_2 = 300$, $x_3 = 24.75$, and $w_3 = 500$ into the formula for the weighted mean, and we get

$$\bar{x}_w = \frac{200(25.50) + 300(27.25) + 500(24.75)}{200 + 300 + 500} = \$25.65$$

Note that we actually divided the *total amount* which the investor spent on the stock by the *total number of shares* which he bought.

The choice of the weights did not pose any problems in either of these examples, but there are situations in which the selection of the weights is far from obvious. For instance, if we wanted to compare figures on the cost of living in different cities (or in different years), it would be quite a job to account for the relative importance of such items as food, rent, entertainment, medical care, ..., in the average person's budget.

## EXERCISES

1. The following are the incomes of ten families residing on a certain street: $12,000, $14,000, $8,000, $15,000, $13,000, $17,000, $6,000, $15,000, $100,000, and $10,000. Comment on the (misleading?) argument that the "average family" on this street has an income of $21,000.

2. Total unemployment in Juneau, Alaska, for the 12 months of the year 1970 was 395, 363, 351, 322, 359, 431, 362, 344, 287, 349, 343, and 327; find the mean monthly unemployment. (Source: Alaska Workforce Estimated, Alaska Department of Labor, July, 1971.)

3. The sales of five food markets in a city were $3,482, $5,694, $12,617, $4,086, and $6,544 on a given day.
    (a) Find the mean, that is, the average sales of these markets on the given day.
    (b) If, altogether, 5,000 customers of these markets made purchases on the given day, how much was the average sale per customer?

4. The records of a city show that the average bid of four contractors on a building construction job was $110,000, and that the bids of three of the contractors were $115,000, $114,000, and $105,000. What must have been the bid of the fourth contractor?

5. During the five consecutive weeks of October 8, October 15, October 22, October 29, and November 5, the average interest rate on prime commercial paper was, respectively, 5.75 percent, 5.63 percent, 5.45 percent, 5.25 percent, and 5.03 percent. If on November 12 the rate was 4.88 percent, by what percentage does the mean rate of the first five weeks exceed that of November 12? (Source: Federal Reserve Bank of St. Louis, November 12, 1971.)

6. A bridge is designed to carry a maximum load of 23,000 pounds. If it is loaded with four trucks having a mean weight of 5,800 pounds, is there any danger that it might be overloaded? Explain your answer.

7. Find the mean of the distribution of finance charges of Exercise 7 on page 25.

8. Find the mean of the monthly business bankruptcies of Exercise 8 on page 26 using (a) the raw (ungrouped) data, and (b) the distribution obtained in the first part of that exercise.

9. Find the mean of the 80 typing-test scores of Exercise 9 on page 26 using (a) the raw (ungrouped) data, and (b) the distribution obtained in the first part of that exercise.

10. Find the mean of the gasoline consumption data of Exercise 10 on page 27 using (a) the raw (ungrouped) data, and (b) the distribution obtained in the first part of that exercise.

11. **CODING** As we indicated on page 33, the calculation of the mean of a distribution can usually be simplified by replacing the class marks with consecutive integers. For the telephone-order distribution on page 14 we might thus get

| Class Marks $x$ | $u$ | $f$ | $u \cdot f$ |
|---|---|---|---|
| 1.95 | −2 | 9 | −18 |
| 3.95 | −1 | 39 | −39 |
| 5.95 | 0 | 45 | 0 |
| 7.95 | 1 | 42 | 42 |
| 9.95 | 2 | 11 | 22 |
| 11.95 | 3 | 4 | 12 |
| Totals | | 150 | 19 |

where we put the 0 of the new scale (which we referred to as the $u$-scale) near the middle of the table *to keep the numbers conveniently small*. Of course, if we use this kind of coding, we must compensate for it in the formula for the mean, which becomes

$$\bar{x} = x_0 + c \cdot \frac{\Sigma u \cdot f}{n}$$

where $x_0$ is the class mark (in the original scale) to which we assign the number 0 in the $u$-scale, $c$ is the class interval, and $n$ is the total number of items grouped. So far as the telephone-order distribution is concerned, we thus get

$$\bar{x} = 5.95 + 2 \cdot \frac{19}{150} = 6.20$$

which is identical (as it should be) with the result obtained on page 32. Use this short-cut technique to rework
(a) Exercise 7;
(b) part (b) of Exercise 8;
(c) part (b) of Exercise 9;
(d) part (b) of Exercise 10.

12. **THE GEOMETRIC MEAN\*** For any set of $n$ positive numbers, the geometric mean is given by the *n*th *root of their product*, and it is used primarily to average ratios or rates of change. To illustrate, the geometric mean of 20, 25, and 16 is $\sqrt[3]{20 \cdot 25 \cdot 16} = \sqrt[3]{8,000} = 20$.
   (a) Find the geometric mean of 6 and 24.
   (b) Find the geometric mean of 4, 25, and 10.
   (c) Find the geometric mean of 1, 2, 8, and 16.
   (d) If the average price of a stock was $1.00 per share in 1971, $2.00 per share in 1972, and $16.00 per share in 1973, we could argue that from 1971 to 1972 its average value multiplied by 2, from 1972 to 1973 its average value multiplied by 8, so that *on the average* its value multiplied by $\frac{2+8}{2} = 5$ from year to year. If we used the geometric mean, on the other hand, we could argue that *on the average* its value multiplied by $\sqrt{2 \cdot 8} = 4$ from year to year. If we apply these two

---

\* To distinguish the "ordinary" mean from the geometric mean of this exercise and the harmonic mean of Exercise 13, the mean is often referred to more explicitly as the **arithmetic mean**.

growth rates to the average 1971 price of the stock, which of the two means would have yielded better (closer) predictions for the actual 1972 and 1973 prices of the stock? What predictions would they yield for the average 1974 price of the stock?

13. **THE HARMONIC MEAN** For any set of $n$ positive numbers, the harmonic mean is given by *n divided by the sum of the reciprocals of the n numbers,* namely, by the expression $n / \left( \Sigma \frac{1}{x} \right)$, and it is used only in very special situations. For instance, if $3.00 is spent on paper clips costing 15 cents a dozen and another $3.00 is spent on paper clips costing 10 cents a dozen, then the average price is *not* $\frac{15 + 10}{2} = 12.5$ cents a dozen; it is 12 cents a dozen, since a total of $6.00 is spent on 50 dozen paper clips. Note that this *is* the harmonic mean of 15 and 10, namely,

$$\frac{2}{\frac{1}{15} + \frac{1}{10}} = \frac{2}{\frac{2}{30} + \frac{3}{30}} = \frac{2}{\frac{5}{30}} = \frac{2 \cdot 30}{5} = 12$$

(a) If an aircraft travels the first 200 miles of a trip at 300 miles per hour and the next 200 miles at 600 miles per hour, how long does it take to travel these 400 miles and what is its average speed? Does the harmonic mean of 300 and 600 give the correct answer for the average speed?

(b) If an investor buys $1,000 worth of capital stock at $50 a share, $1,000 worth at $40 a share, and $1,000 worth at $25 a share, calculate the average price he pays per share, and verify that this is the harmonic mean of $50, $40, and $25.

14. If an instructor counts the final examination in a course four times as much as each hour examination, what is the weighted average grade of a student who received grades of 92, 75, and 73 in three one-hour examinations and a final grade of 87?

15. Use the formula for the weighted mean to determine the average interest rate received by someone who invests $1,000 at 6.5 percent, $1,500 at 7 percent, and $2,500 at 6.25 percent.

16. If $k$ sets of data consisting of $n_1, n_2, \ldots, n_k$ observations have the respective means $\bar{x}_1, \bar{x}_2, \ldots, \bar{x}_k$, the over-all mean of all the data is given by

$$\frac{n_1 \bar{x}_1 + n_2 \bar{x}_2 + \cdots + n_k \bar{x}_k}{n_1 + n_2 + \cdots + n_k} = \frac{\Sigma n \cdot \bar{x}}{\Sigma n}$$

where the numerator represents the actual sum of all the observations, while the denominator represents the total number of observations. *Thus, the over-all mean is the weighted mean of the $\bar{x}$'s with the weights being the number of observations in the corresponding sets of data.*

(a) In a final examination in Business Law, the 38 students in Section 1 averaged 76, the 45 students in Section 2 averaged 81, the 35 students in Section 3 averaged 75, and the 42 students in Section 4 averaged 78. What is the average grade of these 160 students?

(b) In 1970, the five top scoring players in the National Basketball Association scored on the average 31.2, 28.8, 27.5, 26.1, and 25.4 points per game. If they played, respectively, 74, 82, 82, 81, and 80 games, what is the over-all average number of points which these players scored per game?

(c) The average salaries paid in 1972 to the office employees of a company with offices in four cities were, respectively, $121.46, $108.72, $134.16, and $119.43. If the offices employ, respectively, 47, 83, 29, and 74 workers, what is the average salary paid to this company's office employees?

17. The $\Sigma$ notation which we introduced on page 30 is a highly abbreviated notation, and many mathematicians prefer to write the sum $x_1 + x_2 + \cdots + x_n$ more explicitly as $\sum_{i=1}^{n} x_i$. This is meant to indicate that we are adding the $x$'s whose subscript $i$ is 1, 2, ..., and $n$. Similarly, $\sum_{i=1}^{n} x_i^2$ denotes the sum of the squares of the $x$'s with the subscripts 1, 2, ..., and $n$, namely $x_1^2 + x_2^2 + \cdots + x_n^2$, and $\sum_{i=1}^{n} x_i y_i$ denotes the sum of the products $x \cdot y$ with the subscripts of $x$ and $y$ both equal to 1, 2, ..., and $n$, namely, $x_1 y_1 + x_2 y_2 + \cdots + x_n y_n$. Using this more explicit summation notation, rewrite (a) the formula for the mean on page 30, (b) the formula of the mean of a distribution on page 32, and (c) the formula for the weighted mean on page 33. Also write each of the following expressions without the use of summation signs:

(d) $\sum_{i=1}^{5} x_i$;

(f) $\sum_{i=1}^{4} x_i f_i$;

(h) $\sum_{i=2}^{6} x_i$;

(e) $\sum_{i=1}^{6} y_i$;

(g) $\sum_{i=1}^{4} x_i^2$;

(i) $\sum_{i=3}^{5} (y_i - 2)$.

Note that as in parts (h) and (i), sums expressed in this notation need not start with the subscript 1.

## THE MEDIAN

If we are dealing with a distribution with an open class or if we want to avoid the situation described on page 31 where one extreme value gave a very misleading picture, we have no choice but to describe the "middle" or "center" of a set of data by means of a statistical measure other than the mean. One possibility is the **median,** which is obtained by *first arranging the measurements according to size and then choosing the one in the middle, or the mean of the two that are nearest to the middle.* The symbol which we use for the median of a set of $x$'s is $\tilde{x}$ (and, hence, $\tilde{y}$ or $\tilde{z}$ when we refer to the measurements as $y$'s or $z$'s).

When an *odd* number of measurements are arranged according to size, there is always a middle one whose value is the median. For instance, if the dividends paid by five corporations are $1.00, $0.80, $0.00, $1.44, and $0.76 per share, the median is $0.80 (as can easily be verified by first arranging these numbers according to size). Also, if the face amounts of seven life insurance policies sold by a salesman are $20,000, $10,000, $50,000, $1,000, $500, $5,000, and $10,000, the median is $10,000. Note that there are two $10,000's in this example and that we do not refer to either of them as *the* median—the median is a number and not necessarily a particular measurement or observation. Generally speaking, if there are $n$ measurements and $n$ is *odd*, the median is the value of the $\frac{n+1}{2}$ th *largest;* for instance, the median of 35 measurements is given by the value of the $\frac{35+1}{2} = $ 18th largest, the median of 65 measurements is given by the value of the $\frac{65+1}{2} = $ 33th largest, and the median of 113 numbers is given by the value of the $\frac{113+1}{2} = $ 57th largest.

When an *even* number of measurements are arranged according to size, none of them can be exactly in the middle, and the median is defined as the *mean* of the values of the two measurements that are nearest to the middle. For instance, the median of the figures on flat tires repaired on ten consecutive days which we gave on page 30 (namely, 18, 15, 12, 20, 19, 11, 14, 38, 18, and 17) is $\frac{17+18}{2} = $ 17.5. This is halfway between the values of the two figures which are nearest to the middle, and if we interpret it correctly, the formula $\frac{n+1}{2}$ will again give the *position* of the median. For our example we get $\frac{10+1}{2} = $ 5.5, and this means that the median is *halfway between the values of the fifth and sixth largest measurements*. Similarly, the median of 80 measurements is given by the value of the $\frac{80 + 1}{2}$ = 40.5th largest measurement, which means that it is the mean of the 40th largest measurement and the 41st. *It is important to remember that* $\frac{n + 1}{2}$ *is not a formula for the median —it merely tells us where the median is located.*

It should not be surprising that the median of the flat-tire data does not coincide with their mean, which we calculated earlier on page 30. The respective values are 17.5 and 18.2, and their difference is due largely to the one very high value of 38.* If we changed this one value to 28, for example, the mean would become 17.2 whereas the median would remain 17.5, and this illustrates the fact that *the median is generally not as strongly affected by an extreme value as is the mean.* Among the other desirable properties of the median we find that, like the mean, *it can be calculated for any kind of data and it is always unique.* Unlike the mean, on the other hand, *the median of grouped data can generally be found when there are open classes, and the median can even be used to define the middle of a number of objects, properties, or qualities, which do not permit a quantitative description.* It is possible, for example, to rank a number of college courses according to their difficulty and then describe the one in the middle of the list as being of "average difficulty"; also, we might rank samples of ripe tomatoes according to their firmness and then describe the middle one as having "average firmness." Perhaps, the most important distinction between the median and the mean is that in problems of inference (estimation, prediction, etc.) *the median is generally less reliable than the mean.* In other words, the median is generally subject to greater chance fluctuations than the mean, that is, it is apt to *vary more from sample to sample*, as is illustrated in Exercise 7 on page 43. Finally, let us mention that the median is easy enough to find once the data have been arranged according to size, but unless we have automatic equipment *the process of arranging the data according to size can be an extremely tedious job.*

If we want to determine the median of a set of data that has already been grouped, we find ourselves in the same position as on page 31—we can no longer find the *actual* value of the median, although we *can* find the class into which the median will have to fall. The median of a distribution will thus have to be defined in a special way, and Figure 2.9 illustrates how it is done. The median of a frequency distribution is defined as the number, or point on the horizontal scale of Figure 2.9, which is such that *half the total area of the rectangles of the histogram of the distribution lies to its left and the other half*

---

* As we shall see in Exercise 16 on page 199, the very fact that the median and the mean of a set of data *are (or are not) close together* is indicative of a further property of the data namely, whether their distribution is **symmetrical** or **skewed** (that is, lopsided).

*lies to its right*. This means that the sum of the areas of the three rectangles to the left of the dashed line of Figure 2.9 must equal the sum of the areas of the four rectangles which lie to its right. Actually, this definition is equivalent to the assumption that the items are distributed evenly (that is, *spread out evenly*) throughout the class into which the median must fall.

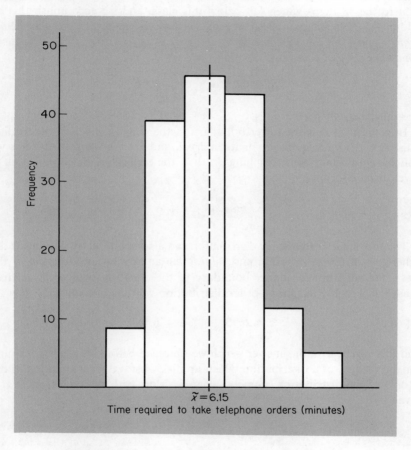

**FIGURE 2.9** The median of the telephone-order distribution.

To find the dividing line between the two halves of the histogram (each of which represents $n/2$ of the items grouped), we must somehow count $n/2$ items starting at either end of the distribution. To illustrate how this may be done, let us refer again to the telephone-order distribution on page 14. Since $n = 150$ in this example, we shall have to count $\frac{150}{2} = 75$ items starting at

**42** SUMMARIZING DATA          CHAP. 2

either end. If we begin at the "bottom" of the distribution, namely, with the smallest values, we find that 48 of the values fall into the first two classes, 93 fall into the first three, so that the median will have to fall into the class whose limits are 5.0–6.9 minutes. Having reached the lower boundary of this class, namely, 4.95, we will thus have to count another $75 - 48 = 27$ items in addition to the 48 items which fall below this class, and we accomplish this by adding $\frac{27}{45}$ of the class interval of 2 to the lower boundary of the class. *Note that we add $\frac{27}{45}$ of the class interval because we want to count 27 of the 45 items which are contained in this class.* Thus, we get

$$\tilde{x} = 4.95 + \tfrac{27}{45} \cdot 2 = 6.15$$

or approximately 6.2 minutes.

In general, if $L$ is the lower boundary of the class into which the median must fall, $f$ is its frequency, $c$ is its interval, and $j$ is the number of items we still have to count after reaching $L$, then the **median of the distribution** is given by the formula

$$\tilde{x} = L + \frac{j}{f} \cdot c$$

The median of a frequency distribution can also be found by starting at the other end of the distribution and *subtracting* an appropriate fraction of the class interval from the upper boundary of the class into which the median must fall. For the telephone-order distribution we thus obtain

$$\tilde{x} = 6.95 - \tfrac{18}{45} \cdot 2 = 6.15$$

and this agrees with the answer which we obtained before. A general formula for the median of a distribution which applies when we start counting at the "top" of the distribution, namely, when we start with the largest values, is given by

$$\tilde{x} = U - \frac{j'}{f} \cdot c$$

where $j'$ is the number of items we still have to count after reaching $U$, the upper boundary of the class into which the median must fall. In our example, the length of 57 of the telephone calls exceeded $U = 6.95$, so that $j' = 75 - 57 = 18$.

The median belongs to a general class of statistical measures called **fractiles,** which are defined as values above or below which certain fractions (or percentages) of the data must fall. Clearly, for the median this fraction is $\frac{1}{2}$ either way. Since the formulas which we gave for the median of a distribution apply also to the calculation of other fractiles, we shall ask for several of them in the exercises which follow.

## EXERCISES

1. In 1970, the 16 counties of the state of Maine had the following populations (in thousands): 91.3, 94.1, 192.5, 22.4, 34.6, 95.2, 29.0, 20.5, 43.5, 125.4, 16.3, 23.5, 40.6, 23.3, 29.9, and 111.6. Find the mean and the median of these population data.

2. Find the median income of the ten families of Exercise 1 on page 34.

3. Find the median size of the sales of Exercise 3 on page 34.

4. The number of games won in 1970 by the 13 teams of the National Football Conference was 12, 10, 6, 6, 10, 9, 8, 6, 3, 10, 9, 4, and 2. Find the median number of games won.

5. During the ten weeks ending November 3, 1971, the certificates of deposit of large commercial banks in the United States totaled, respectively, 30.8, 31.5, 31.9, 32.7, 33.5, 33.0, 33.5, 33.7, 33.1, and 32.8 billion dollars. Find the median and the mean. If the figure for the first week had mistakenly been given as 20.8 (instead of 30.8) billion dollars, how would this have affected the values of the median and the mean?

6. Each of eight magazine subscription salesmen taking part in a sales campaign was assigned a certain quota, and the following are the percentages of their respective quotas which they actually sold: 105, 64, 110, 415, 94, 82, 114, and 107. Calculate the median and the mean of these percentages and indicate which of these measures is a better indication of these magazine salesmen's "average performance."

7. To verify the claim that the mean is generally *more reliable* than the median (namely, that it is subject to smaller chance fluctuations), a gambler conducted an experiment consisting of ten tosses of three dice. The following are his results: 5, 3, and 2; 2, 5, and 5; 1, 6, and 5; 3, 4, and 3; 5, 2, and 2; 6, 1, and 2; 4, 6, and 4; 2, 6, and 5; 4, 1, and 3; 4, 3, and 5.
   (a) Calculate the ten medians and the ten means.
   (b) Group the medians and the means obtained in part (a) into separate distributions having the class boundaries 1.5, 2.5, 3.5, 4.5, and 5.5.
   (c) Draw histograms of the two distributions obtained in part (b) and explain how they illustrate the claim that the mean is generally more reliable than the median.

(d) Repeat the entire "experiment" by repeatedly rolling three dice (or one die three times) and constructing corresponding distributions for the medians and the means. If no dice are available, *simulate* the experiment mentally or by drawing numbered slips of paper out of a hat.

8. Referring to the data of Exercise 8 on page 26, find the median number of bankruptcies reported during the 60 months
   (a) on the basis of the raw (ungrouped) data;
   (b) on the basis of the distribution obtained in part (a) of that exercise.

9. Find the median of the 80 typing-test scores of Exercise 9 on page 26
   (a) on the basis of the raw (ungrouped) data;
   (b) on the basis of the distribution obtained in part (a) of that exercise

10. Find the median of the gasoline expenses of the 48 buses of Exercise 10 on page 27
    (a) on the basis of the raw (ungrouped) data;
    (b) on the basis of the distribution obtained in part (a) of that exercise.

11. **QUARTILES** The three quartiles $Q_1$, $Q_2$, and $Q_3$ are such that 25 percent of the data falls below $Q_1$, 25 percent falls between $Q_1$ and $Q_2$, 25 percent falls between $Q_2$ and $Q_3$, and 25 percent falls above $Q_3$. To find the first and third quartiles of a distribution ($Q_2$ actually *is* the median), we can use either of the two formulas given on page 42. For $Q_1$ we count $\frac{1}{4}$ of the values starting at the bottom of the distribution (or $\frac{3}{4}$ of the values starting at the top), and for $Q_3$ we count $\frac{1}{4}$ of the values starting at the top of the distribution (or $\frac{3}{4}$ of the values starting at the bottom). For instance, for the telephone-order distribution on page 14 we must count $\frac{150}{4} = 37.5$ of the values to reach $Q_1$, and we get $Q_1 = 2.95 + \frac{28.5}{39} \cdot 2 = 4.41$.
    (a) Find $Q_3$ for the telephone-order distribution on page 14.
    (b) Find $Q_1$, $Q_2$, and $Q_3$ for the distribution of finance charges of Exercise 7 on page 25.
    (c) Find $Q_1$ and $Q_3$ for the distribution of bankruptcies obtained in part (a) of Exercise 8 on page 26.
    (d) Find $Q_1$ and $Q_3$ for the distribution of typing-test scores obtained in part (a) of Exercise 9 on page 26.

12. **DECILES** The nine deciles $D_1$, $D_2$, ..., and $D_9$ are such that 10 percent of the data falls below $D_1$, 10 percent falls between $D_1$ and $D_2$, 10 percent falls between $D_2$ and $D_3$, ..., and 10 percent falls above $D_9$. To find the deciles of a distribution we can use either of the formulas on page 42 counting an appropriate fraction of the values instead of the $\frac{1}{2}$ we counted

for the median (which, incidentally, is also the fifth decile, $D_5$). For instance, to determine $D_8$ for the telephone-order distribution, we must count $\frac{2}{10} \cdot 150 = 30$ of the values starting at the top, and we get $8.95 - \frac{15}{42} \cdot 2 = 8.24$.

(a) Find $D_1$ and $D_9$ for the telephone-order distribution on page 14.
(b) Find $D_2$, $D_4$, $D_6$, and $D_8$ for the distribution of finance charges of Exercise 7 on page 25.
(c) Find $D_1$ and $D_9$ for the distribution of bankruptcies obtained in part (a) of Exercise 8 on page 26.
(d) Find $D_3$ and $D_7$ for the distribution of typing-test scores obtained in part (a) of Exercise 9 on page 26.

**13. PERCENTILES** The 99 percentiles of a distribution $P_1, P_2, \ldots,$ and $P_{99}$ are such that 1 percent of the data falls below $P_1$, 1 percent falls between $P_1$ and $P_2$, 1 percent falls between $P_2$ and $P_3, \ldots,$ and 1 percent falls above $P_{99}$. To find the percentiles of a distribution we can use either of the formulas on page 42 counting an appropriate fraction of the values instead of the $\frac{1}{2}$ we counted for the median (which, incidentally, is also the fiftieth percentile, $P_{50}$). For instance, to determine $P_{96}$ for the telephone-order distribution on page 14, we must count 4 percent, or six of the values, starting at the top, and we get $P_{96} = 10.95 - \frac{2}{11} \cdot 2 = 10.59$ minutes.

(a) Find $P_5$ and $P_{15}$ for the telephone-order distribution on page 14.
(b) Find $P_1$ and $P_{99}$ for the distribution of finance charges of Exercise 7 on page 25.
(c) Find $P_2$ and $P_{98}$ for the distribution of bankruptcies obtained in part (a) of Exercise 8 on page 26.
(d) Find $P_1$, $P_5$, $P_{95}$, and $P_{99}$ for the distribution of typing-test scores obtained in part (a) of Exercise 9 on page 26.

[*Hint:* The number of values we have to count in order to reach a fractile (and, hence, $j$ and $j'$ in the formulas on page 42) need not be whole numbers.]

## FURTHER MEASURES OF LOCATION

Besides the mean, the median, and the weighted mean, there are numerous other ways of describing the "middle" or "center" of a set of data. Some of these we already met in Exercises 12 and 13 on pages 36 and 37, and two more worth noting are the **mid-range** and the **mode**. The mid-range of a set of data is simply *the mean of the smallest value and the largest*, and its main advantage is that *it is easy to find*. For the length of the telephone-order data on page 14, the mid-range is $\frac{1.1 + 12.4}{2} = 6.75$, and this is fairly close to the values which we obtained for the median and the mean; however, for the flat-tire data on page 30, the mid-range is $\frac{11 + 38}{2} = 24.5$, and this is much larger than the values which we obtained for the median and the mean.

The mode of a set of data is simply *the value which occurs most often*. Thus, if there are more 19-year-old students attending a certain college than there are students of any other age, we say that 19 is the students' **modal age** (namely, that 19 is the mode); similarly, in the telephone-order distribution on page 14, we refer to the class which has the limits 5.0–6.9 as the **modal class**—its frequency is greater than that of any other class.

An obvious advantage of the mode is that *it requires no calculations*, and its principal value lies in the fact that *it can be used with qualitative data as well as quantitative data*. For instance, if a survey shows that more male students wear blue shirts than shirts of any other color, we can say that blue is their **modal choice**.

A definite disadvantage of the mode is that *it may not exist*, which happens when no two values in a set of data are alike, or that *there may be more than one mode*. For instance, if a set of data consists of the numbers

60, 50, 80, 30, 100, 100, 70, 50, 100, 50, and 70

the numbers 50 and 100 are *both* modes which occur with the maximum frequency of three. Probably, this does not mean very much in this example, but the presence of more than one mode is often indicative of the fact that the data are *not homogeneous*, namely, that they can be looked upon as a *combination of several sets of data*. Thus, if the above numbers are the grades which 11 students received in a test, we might infer that the class is a mixture of two essentially different groups of students—some that are very good and some that are very poor.

So far as modes of grouped data are concerned, we have already pointed out that the class with the highest frequency is referred to as the *modal class*, and if we wanted to be more specific, we could let the class mark of the modal class serve as the mode of the distribution. (Actually, there are *much more refined*, though *very rarely used*, ways of defining the mode of a distribution, and their formulas may be found in some of the older textbooks on statistics.)

The question of what particular "average" should be used in a given situation is not always easily answered, and the fact that there is a good deal of arbitrariness in the selection of statistical descriptions has led some persons to believe that they can take any set of data, apply the magic of statistics, and prove almost anything they want. Indeed, a famous nineteenth century British statesman once said that there are three kinds of lies: *lies, damned lies, and statistics.* To give an example where this kind of criticism might be justified, suppose that the managements of three competing corporations ask their "statisticians" to show on the basis of the following percentage profits earned for five consecutive years that their respective corporations are more profitable than the other two:

    Corporation A; 13.9%  15.8%  15.3%  15.7%  15.2%
    Corporation B; 15.6%  14.3%  15.6%  14.3%  15.5%
    Corporation C; 14.3%  14.5%  14.2%  16.0%  14.8%

The "statistician" of Corporation A will be delighted to find that the respective means for the three corporations are 15.2, 15.1, and 14.8 percent, so that he can claim that his employer's corporation had a higher average (mean) percentage profit than the other two corporations.

Clearly, the "statistician" of Corporation B cannot base the comparison on the means—this would not prove his point—but he does not have to look very far. Trying the medians, he finds that they are 15.3, 15.5, and 14.5 percent, and this provides him with the kind of "proof" he wants. The median is a very "respectable" kind of average, and it enables him to claim that his employer's corporation is preferable to the other two.

Finally, the "statistician" of Corporation C, after trying various measures of location, is lucky enough to find one that does the trick—the mid-range, which we defined on page 45. The mid-ranges of the three sets of data are, respectively, 14.8, 15.0, and 15.1 percent, so that he can claim that his employer's corporation fared best among the three. The moral of this illustration is that

> Methods of describing statistical data should always be decided upon before the data are collected, or at least before they have been inspected.

Of course, there is also the consideration that comparisons based on such small sets of data are generally far from conclusive, and this is a problem to which we shall return in Chapter 8.

## EXERCISES

1. Find the mode (if it exists) of each of the following monthly industrial accident figures reported by three firms:

  Firm A: 25, 29, 26, 27, 29, 27, 28, 28, 30, 28, 28, and 27
  Firm B: 52, 58, 50, 46, 52, 42, 52, 58, 54, 52, 58, and 58
  Firm C: 55, 31, 47, 40, 58, 24, 32, 28, 36, 49, 23, and 22

**2.** Find the mode and the mid-range of the number of football games won by the thirteen teams in Exercise 4 on page 43.

**3.** Find the mid-range of the data on certificates of deposit in Exercise 5 on page 43.

**4.** Find the mode and the mid-range of the family incomes given in Exercise 1 on page 34.

**5.** Find the mode and the mid-range of the ungrouped bankruptcy data of Exercise 8 on page 26.

**6.** Find the mode and the mid-range of the ungrouped typing-test scores of Exercise 9 on page 26.

**7.** A traffic count made at an intersection showed 50 consecutive cars traveling in the following directions: south, west, north, east, west, east, south, west, north, west, east, east, east, south, south, north, west, west, north, south, west, east, south, south, west, south, north, east, south, south, west, west, east, west, east, east, west, south, west, north, east, east, west, north, north, west, west, east, south, and west. What is the modal direction of travel?

**8.** Thirty housewives were asked to test the flavor of a new brand of bread and rate it better, same, or worse than the brand they usually buy. Using the following results, determine their modal choice: worse, better, same, same, worse, same, better, better, same, worse, same, worse, same, better, same, same, same, worse, worse, better, same, same, same, better, better, worse, better, worse, same, and same.

### INDEX NUMBERS

Among the statistical measures that are most useful to businessmen are **index numbers,** which measure the changes that have taken place in the prices, quantities, or values of various commodities. In its simplest form, an index number is a *ratio expressed as a percentage*, and if we are told that in 1960 and in 1971 the average retail price of a pound of butter in the United States was, respectively, 74.9 and 87.6 cents, we find that the 1971 price is $\frac{87.6}{74.9} \cdot 100 = 117.0$ percent of what it was in 1960. In contrast to such a "simple" kind of index referring to a *single* commodity, there are also index numbers expressing changes in such complex phenomena as the cost of living, wholesale prices in general, total industrial production, and so forth. For instance, the widely used and much publicized Consumer Price Index, prepared by the Bureau of Labor Statistics of the U.S. government, has for many years aided businessmen in determining the amount and direction of change in consumer prices; the Wholesale Price Index, also published by the Bureau of Labor Statistics, serves the same purpose for wholesale prices; while the Index of Industrial Production, published by the Federal Reserve Board, provides information concerning changes in volume (or quantity) of industrial production in the United States. During the past few decades, the use of these index numbers, and other *general purpose indexes*, has extended to many new areas of human activity. The Consumer Price Index, for example, is of importance to the members of a certain union, whose hourly wages go up 1 cent for every 0.4 percent increase in the index. It has also found its way into alimony agreements and trust fund payments, which can thus be made to vary with the cost of living.

## UNWEIGHTED INDEX NUMBERS

In the example of the preceding paragraph where we compared the 1960 and 1971 retail prices of butter, we refer to 1960 as the **base year** and 1971 as the **given year.** In general, the year or period which we want to compare is called the *given year* or *given period*, and the year or period relative to which the comparison is made is called the *base year* or *base period*. If the base year is, say, 1967, this is usually expressed by writing 1967 = 100; it means that the 1967 figures represent 100 percent.

Unless the base year is specified, as in our retail-price-of-butter example, it is generally desirable to select a year or period of *economic stability* from the *recent past*. Economic stability is important because years of great instability may be affected by government regulation of prices, by shortages or surpluses of goods, or by other economic irregularities. One reason for selecting a base year or base period in the recent past (or, at least, one that is not in the too distant past) is that constant changes in the availability (or quality) of commercial products may make it difficult, or even impossible, to make meaningful comparisons.

Just as the *average* of a set of data can be described by using the mean, the median, or some other measure of central location, relative changes can be described by means of any one of a large number of formulas, all of which by definition provide index numbers. In this section and the next, we shall study some of these formulas and some of the factors that must be taken into account in choosing one that is appropriate for a given problem. The notation which we shall use consists of denoting all index numbers with the letter $I$, and writing the base-year prices as $p_0$, the given-year prices as $p_n$, the base-year quantities as $q_0$, and the given-year quantities as $q_n$.

## 50 SUMMARIZING DATA

To illustrate one of the simplest methods used in index number construction, let us consider the problem of comparing the 1970 (given year) prices received by farmers for five major crops (in dollars per bushel) with the corresponding prices they received in (the base year) 1969, on the basis of the following data:

|  | 1969 | 1970 |
|---|---|---|
| Wheat | $1.24 | $1.36 |
| Corn | 1.16 | 1.34 |
| Oats | 0.59 | 0.63 |
| Sorghum | 1.07 | 1.13 |
| Soybeans | 2.35 | 2.82 |

A very simple and straightforward index may be constructed by first calculating for each item (commodity) a separate index, called a **price relative,** by means of the formula $\frac{p_n}{p_0} \cdot 100$, and then averaging them, say, by calculating their mean. For our data we would thus obtain

*Price Relatives*

| Wheat | $\frac{1.36}{1.24} \cdot 100 = 109.7$ |
|---|---|
| Corn | $\frac{1.34}{1.16} \cdot 100 = 115.5$ |
| Oats | $\frac{0.63}{0.59} \cdot 100 = 106.8$ |
| Sorghum | $\frac{1.13}{1.07} \cdot 100 = 105.6$ |
| Soybeans | $\frac{2.82}{2.35} \cdot 100 = 120.0$ |

and hence,

$$I = \frac{109.7 + 115.5 + 106.8 + 105.6 + 120.0}{5} = 111.5$$

Symbolically the formula for the *mean of price relatives* is

$$I = \frac{\sum \frac{p_n}{p_0} \cdot 100}{k}$$

where $k$ is the number of items (commodities) whose price relatives are thus combined.

Variations of this kind of index may be obtained by using an "average" other than the (arithmetic) mean. For instance, if we calculate the *median* of the five price relatives we get an index of 109.7, and it should be observed that it is less than the index of 111.5 obtained for the mean, since the latter is more strongly affected by the rather large increase in the price of soybeans (see also discussion on page 31). In general, the arithmetic and geometric means are most commonly used in the construction of this kind of index.

Another way of comparing the given-year prices of a set of commodities with the corresponding base-year prices is to *average (calculate the mean of) the two sets of prices separately, take the ratio of these two means, and then multiply by 100 to express the index as a percentage*. After we cancel the common denominator of $k$ (the number of commodities being compared), the formula for this kind of index, called a **simple aggregative index**, can be written as

$$I = \frac{\sum p_n}{\sum p_0} \cdot 100$$

where $\sum p_n$ is the sum of the given-year prices and $\sum p_0$ is the sum of the base-year prices.

Referring again to the data on page 50, we obtain the following simple aggregative index comparing the 1970 prices of the five crops with those of the year 1969:

$$I = \frac{1.36 + 1.34 + 0.63 + 1.13 + 2.82}{1.24 + 1.16 + 0.59 + 1.07 + 2.35} \cdot 100 = \frac{7.28}{6.41} \cdot 100 = 113.6$$

This tells us that the average price of the five commodities in the given year 1970 was 13.6 percent higher than it was in the base year 1969.

A serious weakness of the simple aggregative index is that it fails the so-called **units test**, namely, that it can produce greatly differing answers if the prices are quoted for different units. To illustrate this, the reader will be asked to show in Exercise 3 on page 52 that the above answer would have been 119.5 instead of 113.6, if the soybean prices had been given in dollars per ton and the other prices in dollars per bushel.

### EXERCISES

1. The following are average prices (per hundred pounds) received by farmers in the United States for selected livestock and poultry items:

|  | 1970 | 1971 |
|---|---|---|
| Hogs | $22.70 | $17.50 |
| Cattle | 34.50 | 36.30 |
| Sheep | 7.69 | 6.58 |
| Chickens | 9.10 | 7.70 |
| Turkeys | 22.60 | 22.10 |

(a) Find the arithmetic mean of the price relatives comparing the 1971 prices of the livestock and poultry items with those of 1970.

(b) Calculate the simple aggregative index which compares the 1971 prices with those of 1970.

**2.** The following are the average annual subscription prices of professional journals in certain fields:

|  | 1960 | 1965 | 1970 | 1971 |
|---|---|---|---|---|
| Business and economics | $ 5.34 | $ 6.39 | $ 9.03 | $ 9.72 |
| Engineering | 5.86 | 7.70 | 12.07 | 13.28 |
| Law | 5.81 | 7.49 | 9.84 | 10.19 |
| Medicine | 10.28 | 14.02 | 23.44 | 27.00 |

(a) Find for 1970 and for 1971 the arithmetic mean of the price relatives of the professional journals with 1960 = 100.

(b) Find for 1970 and for 1971 the median of the price relatives of the professional journals with 1960 = 100.

(c) Calculate for 1970 and for 1971 a simple aggregative index for the prices of the professional journals with 1960 = 100.

(d) Calculate for 1970 and for 1971 a simple aggregative index for the prices of the professional journals with 1965 = 100.

**3.** Show that if in the example on page 51 the price of soybeans were given as $78.33 per ton for 1969 and as $93.99 per ton for 1970 (while the other prices remain in dollars per bushel), the simple aggregative index would be 119.5 instead of 113.6.

**4.** Explain why a mean of price relatives is an index which "passes" the units test.

**5. QUANTITY INDEXES** The index numbers which we discussed in the text were all *price indexes*, but they can easily be converted into *quantity indexes* by substituting $q_n$ for $p_n$ and $q_0$ for $p_0$ wherever these symbols occur in the formulas on pages 50 and 51. Naturally, we would then refer to $\frac{q_n}{q_0} \cdot 100$ as a *quantity relative*.

(a) In September of 1971 and 1972 the total output of selected leather footwear (in thousands of pairs) was

|  | September, 1971 | September, 1972 |
|---|---|---|
| Shoes except athletic | 45,675 | 44,424 |
| Slippers | 10,158 | 9,492 |
| Athletic shoes | 769 | 758 |
| Other footwear | 212 | 196 |

(i) Find the arithmetic mean of the quantity relatives comparing the September, 1972, production of this footwear with that of September, 1971.

(ii) Calculate a simple aggregative index comparing the September, 1972, production of this footwear with that of September, 1971.

(b) The following data on imports to the United States are in millions of pounds:

|  | 1960 | 1970 |
|---|---|---|
| Coffee | 2,943 | 2,609 |
| Tea | 114 | 137 |
| Cocoa and chocolate | 677 | 847 |

(i) Find the median of the quantity relatives comparing the 1970 imports with those of 1960.

(ii) Calculate a simple aggregative index comparing the 1970 imports with those of 1960.

**6. VALUE INDEXES** The index numbers of this section can be converted into *value indexes* by substituting the given-year values for the given-year prices and the base-year values for the base-year prices in the formulas on pages 50 and 51. Use the following data (in millions of dollars) to calculate a simple aggregative index and also a mean of *value relatives* to compare the 1971 sales volume of U.S. life insurance companies with that of 1969:

|  | 1969 | 1971 |
|---|---|---|
| Ordinary | 172,811 | 207,824 |
| Group | 124,124 | 145,199 |
| Industrial | 42,192 | 55,111 |

## WEIGHTED INDEX NUMBERS

As we saw earlier in this chapter, there are many problems where things cannot be averaged without paying attention to their relative importance. For example, if changes in the prices of a bushel of wheat and a gallon of salad dressing were averaged without weighting, this might give a grossly unsatisfactory picture of changes in the general level of prices; clearly, what happens to the price of wheat is many, many times more important to our economy than what happens to the price of salad dressing.

The choice of appropriate weights can be relatively simple, as when we *weight prices by means of the corresponding quantities produced, sold, or consumed* or when we *weight price relatives by means of the total amounts of money spent on the respective items.* It can also be very complicated, as, for example, in the construction of the Consumer Price Index, where the weights, based on periodic surveys, are supposed to reflect the roles which certain goods and services (called a "market basket") play in the average wage earner's or clerical worker's budget.

To illustrate the calculation of a **weighted mean of price relatives,** where the price relatives $\frac{p_n}{p_0} \cdot 100$ are weighted by means of the given-year values $p_n \cdot q_n$, let us refer again to the five crops in the example on page 50, and let us add the information given in the third column of the following table:

|  | Prices 1969 | Prices 1970 | Quantities (millions of bushels) 1970 | $\frac{p_n}{p_0} \cdot 100$ | $p_n \cdot q_n$ |
|---|---|---|---|---|---|
| Wheat | $1.24 | $1.36 | 1,378 | 109.7 | 1,874 |
| Corn | 1.16 | 1.34 | 4,100 | 115.5 | 5,494 |
| Oats | 0.59 | 0.63 | 900 | 106.8 | 567 |
| Sorghum | 1.07 | 1.13 | 697 | 105.6 | 788 |
| Soybeans | 2.35 | 2.82 | 1,136 | 120.0 | 3,204 |
|  |  |  |  | Total | 11,927 |

Here, the price relatives were already calculated on page 50, and the values in the right-hand column are the products of the 1970 prices and the corresponding 1970 quantities.

If we substitute the price relatives and the corresponding given-year values into the formula for the weighted mean on page 33, we get

$$I = \frac{1{,}874(109.7) + 5{,}494(115.5) + 567(106.8) + 788(105.6) + 3{,}204(120.0)}{11{,}927}$$
$$= 114.7$$

and the difference between this value and the (unweighted) mean of 111.5 obtained on page 50 can be attributed mainly to the greater attention paid to the change in the price of soybeans. A general formula for this kind of index can be written as

$$I = \frac{\sum (p_n q_n)\left(\frac{p_n}{p_0} \cdot 100\right)}{\sum p_n q_n}$$

Aggregative index numbers can also be greatly improved by weighting the various prices by means of the corresponding quantities (produced, sold, or consumed) in the given year, the base year, or some other fixed period of time. If we use the given-year quantities as weights we obtain the **weighted aggregative index**

$$I = \frac{\sum p_n q_n}{\sum p_0 q_n} \cdot 100$$

which is sometimes called a **Paasche Index**, and if we use the base-year quantities as weights we obtain the weighted aggregative index

$$I = \frac{\sum p_n q_0}{\sum p_0 q_0} \cdot 100$$

which is sometimes called a **Laspeyres Index.**

To illustrate the use of these two formulas, let us return to the example dealing with the five crops, and let us first compare the 1970 prices with the 1969 prices using the 1970 quantities as weights. Substituting the necessary prices and quantities (from the table on page 54) into the first of the above weighted aggregative index number formulas, we get

$$I = \frac{1.36(1{,}378) + 1.34(4{,}100) + 0.63(900) + 1.13(697) + 2.82(1{,}136)}{1.24(1{,}378) + 1.16(4{,}100) + 0.59(900) + 1.07(697) + 2.35(1{,}136)} \cdot 100$$
$$= 114.6$$

To calculate the Laspeyres Index, we will have to know the 1969 (base year) quantities, and given that these are 1,460, 4,583, 950, 747, and 1,126 million bushels, respectively, for the five crops, we obtain

$$I = \frac{1.36(1{,}460) + 1.34(4{,}583) + 0.63(950) + 1.13(747) + 2.82(1{,}126)}{1.24(1{,}460) + 1.16(4{,}583) + 0.59(950) + 1.07(747) + 2.35(1{,}126)} \cdot 100$$
$$= 114.5$$

## 56 SUMMARIZING DATA

The difference between these two results is very small, but it can be large in problems when there are substantial differences between the two sets of weights. In actual practice, most weighted aggregative indexes used nowadays are **fixed-weight aggregative indexes,** where the weights are quantities referring to some period other than the base year or the given year. This is because most important index numbers are published **in series,** that is, regularly every day, week, month, or year, and it would be very impractical to keep changing the weights, or to be unable to change the base (see Exercise 4 below) without changing the weights.

## EXERCISES

1. The following table gives the prices and production totals of three selected vegetable oils:

|  | Prices (cents per pound) | | | | Quantities (millions of pounds) | |
|---|---|---|---|---|---|---|
|  | 1968 | 1969 | 1970 | 1971 | 1968 | 1971 |
| Cottonseed oil | 16.3 | 14.2 | 17.5 | 19.0 | 1,115 | 1,209 |
| Linseed oil | 12.7 | 12.0 | 11.0 | 8.9 | 307 | 412 |
| Soybean oil | 10.3 | 11.0 | 14.2 | 15.1 | 6,150 | 8,082 |

(a) Using given-year value weights, calculate a weighted mean of price relatives to compare the 1971 prices of these vegetable oils with those of 1968.
(b) Using the 1971 quantities as weights, calculate a weighted aggregative index comparing the 1971 prices of these vegetable oils with those of 1968.
(c) Using the 1968 quantities as weights, calculate weighted aggregative indexes comparing, respectively, the 1969, 1970, and 1971 prices of the three vegetable oils with those of 1968.

2. The following are 1968, 1969, and 1970 prices and production data for three petroleum products:

|  | Prices (cents per gallon) | | | Quantities (hundred-million barrels) | | |
|---|---|---|---|---|---|---|
|  | 1968 | 1969 | 1970 | 1968 | 1969 | 1970 |
| Gasoline | 11.3 | 11.6 | 11.9 | 19.34 | 20.22 | 21.00 |
| Distillate fuel oil | 10.3 | 10.1 | 10.8 | 8.41 | 8.48 | 8.97 |
| Kerosene | 11.3 | 11.1 | 11.8 | 1.02 | 1.03 | 0.96 |

) Using 1970 value weights, calculate a weighted mean of price relatives with 1968 = 100 for the year 1970.

Using given-year quantities as weights and 1968 = 100, calculate weighted aggregative indexes comparing the 1969 and 1970 prices, respectively, with those of 1968.

Using the 1968 quantities as weights and 1968 = 100, calculate weighted aggregative indexes comparing the 1969 and 1970 prices, respectively, with those of 1968.

(d) Using the 1969 quantities as weights and 1968 = 100, calculate *fixed-weight* aggregative indexes comparing the 1969 and 1970 prices, respectively, with those of 1968.

3. Show that if we use base-year values as weights in the formula for a weighted mean of price relatives, we obtain a Laspeyres Index.

4. **SHIFTING THE BASE** It is sometimes necessary (or convenient) to change the base of an index number series from one period to another. To do this, we simply *divide* each value of the series by the value of the index for the new base year, and then *multiply* by 100. For instance, if the 1968 through 1973 values of a price index with 1969 = 100 are 97.1, 100.0, 104.3, 102.5, 107.0, and 111.2, and we want to change the base so that 1972 = 100, we find that the 1968 value of the index becomes $\frac{97.1}{107.0} \cdot 100 = 90.7$, and that the 1969 through 1973 values become, respectively, 93.5, 97.5, 95.8, 100.0, and 103.9.

(a) Shift the base of the price index in the above illustration so that 1970 = 100.
(b) Belgium's index of industrial production with 1965 = 100 was 98.2, 100.0, 101.8, 103.7, 109.2, and 121.1, respectively, for the years 1964 through 1969. Shift the base to 1968.
(c) The 1965 through 1971 values of an index of retail stores' sales in the United States with 1967 = 100 are, respectively, 90.5, 96.9, 100.0, 109.0, 114.1, 119.7, and 130.3. Shift the base so that 1970 = 100.

5. **DEFLATING** If a person earned $100 a week in 1965 and $200 a week in 1970, it is, of course, true that his income has doubled, but it does *not* follow necessarily that he is twice as well off. This is because $200 in 1970 did not buy twice what $100 bought in 1965; in other words, it is due to changes in the *purchasing power of the dollar*. To make such a comparison meaningful, we *divide* each figure by the corresponding value of an appropriate price index and multiply by 100, and if we use the Consumer Price Index, which stood at 94.5 in 1965 and 116.3 in 1970, we obtain $\frac{100}{94.5} \cdot 100 = 105.82$ and $\frac{200}{116.3} \cdot 100 = 171.97$ for the person's wages in

1965 and 1970. Since the base of the Consumer Price Index is 1967, this means that the $100 which the person made in 1965 would have bought the same as $105.82 in 1967, and that the $200 which the person made 1970 would have bought the same as $171.97 in 1967. We are thus expressing the earnings in terms of *constant 1967 dollars*, and this process is referred to as *deflating*. It shows that the person's earnings have not doubled but increased by $\dfrac{171.97}{105.82} \cdot 100 - 100 = 62.5$ percent.

(a) The Gross National Product of the United States was 793.9, 864.2, 930.3, 976.4, and 1,050.4 billion dollars, respectively, in the years 1967 through 1971. Given that the corresponding values of the Consumer Price Index (with 1967 = 100) are, respectively, 100.0, 104.2, 109.8, 116.3, and 121.3, express the 1967 through 1971 figures for the Gross National Product of the United States in terms of constant 1967 dollars. Also find the percentage increase from 1969 to 1971 using (i) the actual dollar amounts, and (ii) the figures in constant 1967 dollars.

(b) The average prices of kitchen chairs produced by a certain furniture manufacturer were $14.87, $15.50, $16.00, $16.95, and $17.75 in the years 1966 through 1970. Use the corresponding values of the Wholesale Price Index with 1967 = 100, which were, respectively, 99.8, 100.0, 102.5, 106.5, and 110.4 for the same years, to judge whether the increases in the prices of these kitchen chairs have fallen behind, kept pace with, or exceeded the general increase in wholesale prices.

# 3

# POSSIBILITIES AND PROBABILITIES

**INTRODUCTION**

We can hardly predict the success of a new steak house unless we know at least with what other restaurants it has to compete, and we cannot very well predict the winner of the Kentucky Derby unless we know at least which horses are going to run. More generally, we cannot make intelligent predictions or decisions unless we know at least what is possible—in other words, *we must know what is possible before we can judge what is probable*. Thus, the next three sections will be devoted to "what is possible" in a given situation, after which the final two sections of the chapter will be devoted to "what is probable."

**THE SAMPLE SPACE**

In statistics, the set of all possible outcomes of an experiment is called the **sample space**, for it usually consists of all the things that can happen when we take a sample. For instance, if a bank auditor must choose three of the 24 banks in a certain city for a surprise audit of their accounts, the sample space consists of all the possible ways in which the three banks can be selected. Also, if the sales manager of a large distributor of hardware has to assign five salesmen to a certain territory, the sample space consists of all the possible ways in which he can select these salesmen from his staff.

To avoid misunderstandings concerning the words "outcome" and "experiment," let us make it clear that statisticians use these terms in a very wide sense. An **experiment** may consist simply of checking whether an employee was absent on a given day; it may consist of determining a day's gasoline production of a refinery or counting the number of food stores in a community; or it may consist of the very complicated process of estimating the effect of a change in the advertising budget on the sales of a certain detergent. Correspondingly, an outcome of an experiment may be a simple "yes or no" answer; it may be the result of a measurement or a count; or it may be an answer obtained from extensive measurements and calculations.

**60** POSSIBILITIES AND PROBABILITIES     CHAP. 3

When we study the outcomes of an experiment, it is usually advantageous to identify the different possibilities with numbers or points, for we can then treat all questions concerning the outcomes *mathematically*, without having to go through long verbal descriptions of what has taken place, is taking place, can take place, or will take place. Actually, this is precisely what we do in sports when we refer to football players by their numbers; it is what the Internal Revenue Service does when it refers to taxpayers by their social security numbers; and it is what the post office does when it uses ZIP codes to identify geographical regions.

The use of points rather than numbers has the added advantage that it makes it easier to *visualize* the various possibilities, and perhaps discover some of the special features which some of the outcomes have in common. As the saying goes, "a picture is worth a thousand words," and it would be difficult to disagree with the statement that "a picture can be worth a thousand numbers." To give an example, suppose that two companies, a company which manufactures clocks and a company which manufactures typewriters, must choose between rail, truck, and air for shipping their products. So far as the clock manufacturer is concerned, we can picture the sample space as in Figure 3.1, where we have identified the three ways of shipping the company's products with three points, to which we have also assigned the (code) numbers

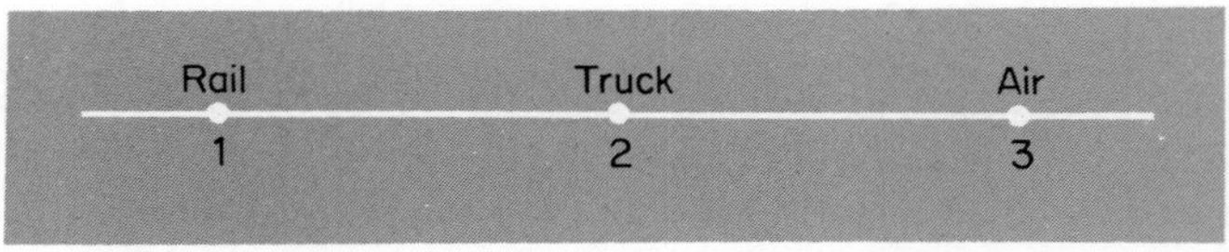

```
    Rail            Truck            Air
     •                •                •
     1                2                3
```

**FIGURE 3.1**  Sample space for one manufacturer.

1, 2, and 3. Actually, these points could have been drawn in any pattern, and we could have assigned to them any arbitrary set of numbers.

If we now consider both companies, the clock manufacturer as well as the typewriter manufacturer, there are nine possibilities, which we have represented by means of the nine points of Figure 3.2. Again, we could have drawn the points in any arbitrary pattern (perhaps, on a circle or in a straight line), but our choice of a two-dimensional system of $x$- and $y$-coordinates has the advantage that each point is easily identified. Using the same coding as before, with 1, 2, and 3 representing shipment by rail, truck, and air, we can say that the point (2, 1) represents the case where the clock manufacturer decides to ship by truck and the typewriter manufacturer decides to ship by rail, (1, 3) represents the case where the clock manufacturer decides to ship by rail and the typewriter manufacturer decides to ship by air, while (2, 2) represents the case where both manufacturers choose to ship by truck. Note that in each case the first number (the *x*-coordinate) represents the clock manufacturer's choice, while the second number (the *y*-coordinate) represents the typewriter manufacturer's choice.

If more than three manufacturers had to choose between shipping by rail, truck, or air, or if they could also ship, say, by barge, the sample space would be larger and, perhaps, even impossible to visualize. For example, if four man-

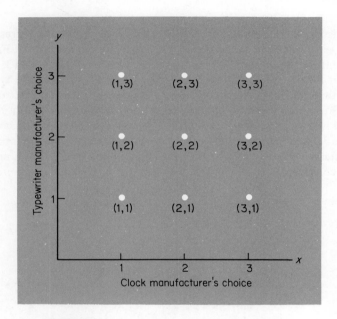

**FIGURE 3.2** Sample space for two manufacturers.

ufacturers had to choose between shipping by rail, truck, or air, there would be more than 80 possibilities (see Exercise 11 on page 77), among which (1, 2, 2, 3) would represent the case where the first manufacturer decides to ship by rail, the second and third decide to ship by truck, and the fourth decides to ship by air. (Since each point is now represented by *four* numbers, we cannot picture the sample space as easily as we pictured those of Figures 3.1 and 3.2; in fact, this would require a space of *four dimensions*.)

Generally, we classify sample spaces according to the number of points which they contain. The ones we studied in Figures 3.1 and 3.2 contained three points and nine points, respectively, and they are referred to as **finite**,

**62** POSSIBILITIES AND PROBABILITIES  CHAP. 3

meaning that they contain a limited, or fixed, number of points. (An interesting example of a much larger, though still finite, sample space is the one for all possible 13-card bridge hands which can be dealt with an ordinary deck of 52 playing cards—there are more than 635 *billion* possibilities.) In this chapter and in Chapter 4 we shall consider only sample spaces that are finite; those with infinitely many points will be taken up later on.

## OUTCOMES AND EVENTS

In statistics, we refer to the outcomes of experiments and the corresponding points of a sample space as **events**. For instance, the point (1, 1) of Figure 3.2 is the event that both manufacturers decide to ship by rail, while (3, 2) is the event that the clock manufacturer decides to ship by air while the typewriter manufacturer decides to ship by truck. Events can also consist of several points of a sample space; *in fact, the term "event" is applied to any subset, or part, of a sample space.* For instance, the three points (1, 1), (1, 2), and (1, 3) of Figure 3.3 *together* constitute the event which we shall call *T*, that *the clock manufacturer decides to ship by rail;* the three points (1, 1), (2, 1), and (3, 1) *together* constitute the event which we shall call *U*, that *the typewriter manu-*

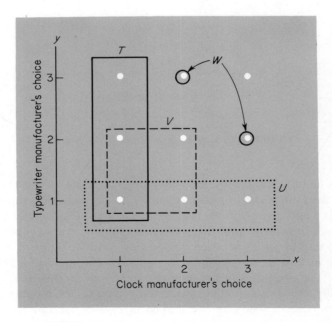

**FIGURE 3.3** Sample space for two manufacturers.

*facturer decides to ship by rail;* the four points (1, 1), (1, 2), (2, 1), and (2, 2) *together* constitute the event which we shall call $V$, that *neither manufacturer decides to ship by air;* and the two points (2, 3) and (3, 2) *together* constitute the event which we shall call $W$, that *one of the two manufacturers decides to ship by truck while the other decides to ship by air.*\* In Figure 3.3 we indicated the points which constitute $T$ by means of a solid line, those which constitute $U$ by means of a dotted line, and those which constitute $V$ by means of a dashed line; the two points which constitute event $W$ are circled. Observe that events $V$ and $W$ have no points in common; they are referred to as **mutually exclusive events,** which means that they cannot both occur at the same time. (Obviously, if neither manufacturer decides to ship by air, it is impossible for one to ship by truck and the other to ship by air.)

There are several reasons why we introduced capital letters to stand for the various events. Not only does this make it easier to refer to the respective events, but it also enables us to demonstrate how two or more events can be combined to form other events. Suppose, for instance, that we are interested in the event that *at least one of the two manufacturers will decide to ship by rail,* namely, the event which is given by the points (1, 1), (2, 1), (3, 1), (1, 2), and (1, 3). We shall denote this set $T \cup U$ and refer to it as the **union** of $T$ and $U$; it reads "$T$ union $U$" or simply "$T$ or $U$." Generally speaking,

> The union of two events is the event which consists of all the outcomes (points) contained in either or both of the two events.

Thus, if $A$ is the event that a certain person will take his vacation in July and $B$ is the event that he will take his vacation in August, then $A \cup B$ is the event that this person will take his vacation in July, in August, or in both of these months.

Another event which can be formed by combining events $T$ and $U$ is the event that *both* of the manufacturers will decide to ship by rail, namely, the event which consists only of the point (1, 1). We shall denote this event $T \cap U$ and refer to it as the **intersection** of $T$ and $U$; it reads "$T$ intersection $U$" or simply "$T$ and $U$." Generally speaking,

> The intersection of two events is the event which consists of all the outcomes (points) which the two sets have in common.

Thus, if $C$ is the event that a driver will be cited for speeding and $D$ is the event that he will be cited for going through a red light, then $C \cap D$ is the event that the driver will be cited for both, speeding and going through a red

---

\* An alternate way of describing event $V$ is that each manufacturer decides to ship either by rail or by truck.

light. With regard to our earlier examples, $A \cap B$ is the event that the person will take his vacation in July and August, and $V \cap W$ is an event which cannot possibly occur since $V$ and $W$ are mutually exclusive. We refer to sets like $V \cap W$ as **empty sets** and denote them with the symbol $\emptyset$.

Still referring to Figure 3.3, we might also consider the event that the clock manufacturer will decide *not* to ship by rail, namely, the event which consists of the points (2, 1), (2, 2), (2, 3), (3, 1), (3, 2), and (3, 3). We shall denote this event $T'$ and refer to it as the **complement** of $T$; it reads "$T$ prime" or "non-$T$." Generally speaking,

> **The complement of an event consists of all the outcomes (points) of the sample space which are not included in the event.**

Sample spaces and events, particularly relationships among events, are often pictured by means of **Venn diagrams** like those of Figures 3.4 and 3.5.

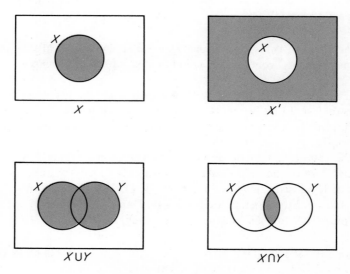

**FIGURE 3.4** Venn diagrams.

In each case the sample space is represented by a rectangle, while events are represented by regions within the rectangle, usually by circles or parts of circles. Thus, the shaded regions of the four Venn diagrams of Figure 3.4 represent, respectively, event $X$, the complement of $X$, the union of two events $X$ and $Y$, and the intersection of $X$ and $Y$. For instance, if $X$ is the event that a new play is a commercial success and $Y$ is the event that it is an artistic success, then the region shaded in the first diagram represents the event that it is a commercial success, the region shaded in the second diagram represents the event that it is not a commercial success, the region shaded in the third diagram represents the event that it is a commercial success and/or an artistic success, and the region shaded in the fourth diagram represents the event that it is a commercial success as well as an artistic success. Incidentally, the region outside both circles, which we can denote $X' \cap Y'$, represents the event that the play is neither a commercial success nor an artistic success.

When we deal with three events, it is customary to draw the circles as in Figure 3.5, where $K$, $L$, and $M$ are the respective events that a motorist interviewed at a gasoline station stops there regularly, is satisfied with the service,

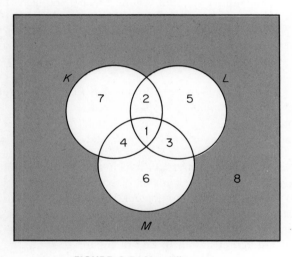

**FIGURE 3.5** Venn diagram.

and likes the mileage he gets with the station's brand of gasoline. As can be seen from this diagram, the circles divide the sample space into eight regions (which we numbered 1 through 8) and it is fairly easy to identify the corresponding events. Region 2, for example, is inside $K$, inside $L$, and outside $M$, so that it represents the event that the motorist interviewed stops there regularly, is satisfied with the service, but does not like the mileage he gets with their gasoline. Similarly, Region 6 represents the event that the motorist does not stop there regularly, is not satisfied with the service, but likes the mileage he gets with their gasoline. It will be left to the reader to identify some of the other regions of Figure 3.5 (and also some of their combinations) in Exercise 6 below.

**EXERCISES**

1. Referring to the sample space of Figure 3.2, describe *in words* the event which is represented by each of the following sets of points:
   (a) (1, 1), (2, 2), and (3, 3);
   (b) (1, 2), (2, 2), and (3, 2);
   (c) (1, 3) and (3, 1);
   (d) (1, 2), (1, 3), (2, 1), (2, 3), (3, 1), and (3, 2);
   (e) (1, 2), (1, 3), (3, 2), and (3, 3);
   (f) (2, 1), (2, 2), (3, 1), and (3, 2).

2. Referring to the sample space of Figure 3.3, describe each of the following events *in words* and list the points which it contains:
   (a) $T'$;             (d) $U \cap V$;      (g) $T \cap V$;
   (b) $V'$;             (e) $T \cap W$;      (h) $T' \cap U'$;
   (c) $U \cup V$;       (f) $T \cup V$;      (i) $T' \cup U'$.

3. Referring to the sample space of Figure 3.3, which of the following pairs of subsets represent *mutually exclusive events:*
   (a) $T$ and $U$;                  (c) $T$ and $V'$;
   (b) $T$ and $W$;                  (d) $V'$ and $T \cap U$?

4. In a survey of gasoline service stations in a certain town, the stations are classified according to the number of gasoline pumps which they have and also according to the number of air pumps. It is known that each service station has one, two, or three gasoline pumps, at least one air pump, and at least as many gasoline pumps as air pumps.
   (a) Draw a diagram similar to that of Figure 3.3 (with the $x$-coordinate denoting the number of gasoline pumps and the $y$-coordinate the number of air pumps) which shows the six different ways in which a gasoline service station can thus be classified. For instance (3, 1) represents the case where a gasoline service station has three gasoline pumps and one air pump, and (2, 2) represents the case where a gasoline service station has two gasoline pumps and two air pumps.
   (b) Describe *in words* the event which is represented by each of the following sets of points: event $E$ which consists of the points (2, 2) and (3, 1); event $F$ which consists of the points (2, 1), (3, 1), and (3, 2); and event $G$ which consists of the points (1, 1), (2, 1), and (2, 2). Also indicate these three sets on the diagram of part (a) by enclosing the corresponding points by means of solid, dotted, or dashed lines.

(c) Referring to part (b), describe each of the following events *in words* and list the points which it contains:
  (i) $F'$;  (iii) $E \cap F$;  (v) $E \cap G$;
  (ii) $G'$;  (iv) $F \cup G$;  (vi) $E \cup G'$.
(d) Referring to part (b), which of the following pairs of subsets represent *mutually exclusive events:*
  (i) $E$ and $G$;  (iii) $F'$ and $G$;
  (ii) $F$ and $G$;  (iv) $E$ and $F \cap G$?

5. A certain market is fully staffed when there are three check-out clerks, each stationed at a cash register, and three stock boys placing merchandise on the shelves. Of course, the market is not always fully staffed, but there is always at least one check-out clerk.
  (a) Draw a diagram similar to that of Figure 3.3 (with the *x*-coordinate denoting the number of check-out clerks and the *y*-coordinate the number of stock boys) which shows the 12 different ways in which this market can be fully or partially staffed. For instance, the point (2, 0) represents the event that there are two check-out clerks and no stock boys, while (1, 2) represents the event that there is one check-out clerk and two stock boys.
  (b) Describe *in words* the event which is represented by each of the following sets of points: event $A$ which consists of the points (1, 1), (2, 2), and (3, 3); event $B$ which consists of the points (1, 2), (2, 2), and (3, 2); event $C$ which consists of the points (1, 0), (2, 0), and (1, 1); and event $D$ which consists of the points (1, 0), (2, 0), (3, 0), (2, 1), (3, 1), and (3, 2). Also indicate these four sets on the diagram of part (a) by enclosing the corresponding points by means of solid, dotted, or dashed lines.
  (c) Referring to part (b), describe each of the following events *in words* and list the points which it contains:
    (i) $B'$;  (iv) $A \cap B$;  (vii) $B \cap C$;
    (ii) $C'$;  (v) $A \cup D$;  (viii) $B \cap D$;
    (iii) $D'$;  (vi) $A \cap D$;  (ix) $A \cap C'$.
  (d) Referring to part (b), which of the following pairs of subsets represent *mutually exclusive events:*
    (i) $B$ and $C$;  (iii) $B$ and $D$;
    (ii) $A$ and $C$;  (iv) $D$ and $A \cap C$?

6. With reference to Figure 3.5 explain *in words* what events are represented by the following regions of the Venn Diagram:
  (a) Region 1;  (d) Regions 1 and 4 together;
  (b) Region 3;  (e) Regions 2 and 5 together;
  (c) Region 5;  (f) Regions 2, 5, 7, and 8 together.

**68** POSSIBILITIES AND PROBABILITIES  CHAP. 3

7. With reference to the Venn diagram of Figure 3.5, list the regions or combinations of regions which represent the following events:
   (a) the event that the motorist stops there regularly, likes the mileage he gets with their gasoline, but is not satisfied with the service;
   (b) the event that the motorist stops there regularly, but likes neither the service nor the mileage he gets with their gasoline;
   (c) the event that the motorist stops there regularly and likes the service;
   (d) the event that the motorist likes the mileage he gets with their gasoline, but is not satisfied with the service;
   (e) the event that the motorist does not stop there regularly.

8. Suppose that the owner of a defective washing machine is waiting for the repair man and that $H$ is the event that he will come today, $I$ is the event that he can fix the washing machine, and $J$ is the event that his bill will be fair. With reference to the Venn diagram of Figure 3.6 list the regions or combinations of regions which represent the following events:

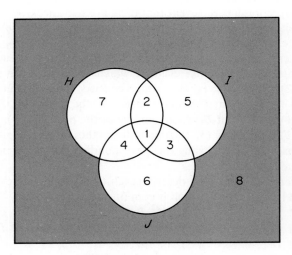

**FIGURE 3.6** Venn diagram.

   (a) the event that the repair man will come today, that he can fix the washing machine, but that his bill will not be fair;
   (b) the event that the repair man will come today, that he will not be able to fix the washing machine, and that his bill will not be fair;
   (c) the event that the repair man will come today and that his bill will be fair;

CHAP. 3　　　　　　LISTING OUTCOMES AND COUNTING EVENTS　　　69

(d) the event that the repair man will not come today, but that his bill will be fair;
(e) the event that the repair man will not come today, that he will not be able to fix the washing machine, and that his bill will not be fair.

**9.** With reference to Exercise 8 and the Venn diagram of Figure 3.6, explain *in words* what events are represented by the following regions:
(a) Region 1;
(b) Region 4;
(c) Region 7;
(d) Regions 1 and 3 together;
(e) Regions 4 and 7 together;
(f) Regions 5 and 8 together.

**10.** Which of the following events are mutually exclusive? Explain your answers:
(a) being a licensed physician and not having attended college;
(b) wearing a blue shirt and having brown eyes;
(c) being under 18 years of age and being a United States Senator;
(d) being a woman and being president of a large corporation;
(e) scoring a touchdown and hitting a home run in the same game.

## LISTING OUTCOMES AND COUNTING EVENTS

In this section we shall discuss two kinds of problems related to sample spaces, outcomes, and events. *First, there is the problem of listing everything that can happen in a given experiment,* and although this may sound simple, it is often easier said than done. Sometimes there are so many outcomes that the job of listing them all becomes an Herculean task, and even when the number of outcomes is fairly small, it can be quite difficult to make sure that none of them have been left out. *The second kind of problem is that of determining the total number of outcomes (or the total number of events of a given kind) without actually constructing a complete list.* This is important because there are many problems in which we really do not need a complete list and, hence, can save a lot of work. For instance, on page 62 we mentioned that more than 635 *billion* different 13-card bridge hands can be dealt with an ordinary deck of 52 playing cards, and it would hardly seem necessary to point out that this result was obtained without actually listing all of these hands.

To give an example in which the listing of all possible outcomes is not straightforward, suppose that two college students are looking for part-time jobs and that they are planning to devote three days to finding suitable employment. The problem is to list what can happen on these three days, assuming that we are interested only in how many of the students find a job each day. Clearly, there are many possibilities: both of the students might find jobs on the first day; they might both fail to get jobs on all three days; one of them might get a job on the second day while the other fails to get a job each day; one of them might get a job on the first day and the other on the third day; and so on. Being careful, we may be able to complete this list, but the chances are that we will leave out at least one of the possibilities.

To handle this kind of problem systematically, it is helpful to refer to a **tree diagram** like that of Figure 3.7. This diagram shows that first there are

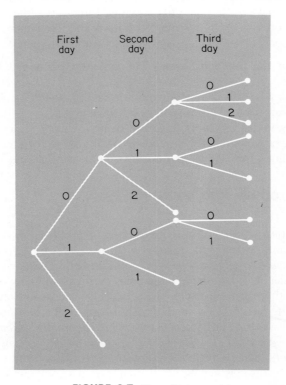

**FIGURE 3.7** Tree diagram.

three possibilities (three branches) corresponding to 0, 1, or 2 of the students getting a job on the first day. For the second day there are three branches growing out of the top branch, two out of the middle branch, and none out of the bottom branch. Clearly, there are again three possibilities (0, 1, or 2) when none of the students gets a job on the first day, two possibilities (0 or 1) when one of the students gets a job on the first day, and there is no need to go any further when both students get jobs on the first day. The same sort of reasoning applies also to the third day, and we find that (going from left to right) there are altogether ten different paths along the "branches" of the tree. In other words, *the "experiment" has ten possible outcomes.*

To consider another example in which a tree diagram can be of some aid, suppose that a student plans to either walk, drive his car, or take a bus to *one* of the following places: the movie theater, coffee house, record shop, bowling alley, bookstore, or library. What we would like to know is *the number of different things he can do.* Looking at the tree diagram of Figure 3.8, it is

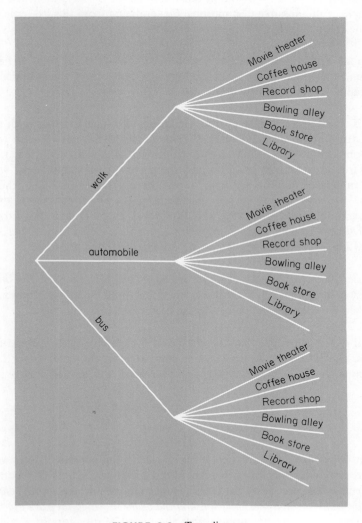

**FIGURE 3.8** Tree diagram.

apparent that the answer is 18, corresponding to the 18 distinct paths along the branches of the "tree." Starting at the top, the first path represents walking to the movie theater, . . ., the seventh path represents the student driving his car to the movie theater, . . ., and the bottom path represents a ride to the library by bus.

Note that the answer which we obtained in this example is the *product* of 3 and 6, namely, the product of the number of ways in which the student can choose a means of transportation and the number of ways in which he can select one of the possible destinations. Generalizing from this example, let us state the following rule:

> **If a selection consists of two separate steps, of which the first can be made in $m$ ways and the second in $n$ ways, then the whole selection can be made in $m \cdot n$ ways.**

Thus, if a restaurant offers 12 different desserts, which it serves with coffee, Sanka, tea, milk, or hot chocolate, then there are $12 \cdot 5 = 60$ different ways in which one can order a dessert and a drink. Also, if a chain of drug stores has four warehouses and 22 retail outlets, there are $4 \cdot 22 = 88$ different ways in which an item can be shipped from one of the warehouses to one of the retail outlets.

Using suitable tree diagrams, it is fairly easy to generalize the above rule so that it applies to selections involving more than two steps. If there are $k$ steps, where $k$ is a positive integer, we obtain the following rule:

> **If a selection consists of $k$ separate steps, of which the first can be made in $n_1$ ways, the second in $n_2$ ways, . . ., and the $k$th in $n_k$ ways, then the whole selection can be made in $n_1 \cdot n_2 \cdot \ldots \cdot n_k$ ways.**

Thus, if a new-car buyer has the choice of four body styles, three different engines, and 12 colors, we find that he can make his choice in $4 \cdot 3 \cdot 12 = 144$ different ways. Also, if a true-false test consists of 25 questions, there are

$$\underbrace{2 \cdot 2 \cdot 2 \cdot 2 \cdot 2 \cdot \ldots \cdot 2}_{25 \text{ factors}} = 2^{25} = 33{,}554{,}432$$

different ways in which one can mark the test—unfortunately, only one of these corresponds to the case where all of the questions are answered correctly.

The rule of the preceding paragraph is often applied when *several selections are made from one and the same set, and the order in which they are made is of importance*. To illustrate, suppose that the board of directors of a company has to select a president and a vice-president from among their five leading executives: Mr. Smith, Miss Jones, Mr. Brown, Mrs. Henry, and Mr. Forbes.

To determine the number of ways in which this can be done, suppose that the board of directors first chooses the president and then the vice-president. Clearly, the president can be chosen in five ways, and since this leaves only four of the executives for the choice of the vice-president, the whole selection can be made in $5 \cdot 4 = 20$ different ways. If the board of directors had to pick a president, a first vice-president, a second vice-president, a third vice-president, and a fourth vice-president, from among the five leading executives, they could have done so in $5 \cdot 4 \cdot 3 \cdot 2 \cdot 1 = 120$ different ways; clearly, *after each choice there is one less to choose from for the next selection.* To consider another example, if a first prize, a second prize, and a third prize are to be awarded to three of the 35 dogs entered in a dog show, these prizes could be distributed in $35 \cdot 34 \cdot 33 = 39{,}270$ different ways; again, there is one less to choose from after each selection.

Generally speaking, if $r$ different objects are selected from a set of $n$ objects, any particular arrangement of these $r$ objects is referred to as a **permutation**. For instance, 2 3 1 5 4 is a permutation of the first five positive integers; Idaho, New Mexico, and Utah is a permutation (a particular ordered arrangement) of three of the eight Mountain States; D A H E may be looked upon as a permutation of four letters selected from among the first eight letters of the alphabet; and if we are asked to list all possible permutations of two of the five letters used to spell the word "count," the answer would be

co cu cn ct ou on ot un ut nt
oc uc nc tc uo no to nu tu tn

So far as the counting of possible permutations is concerned, direct application of the second rule on page 72 leads to the following result:*

> The number of permutations of $r$ objects selected from a set of $n$ objects is
>
> $$n(n - 1)(n - 2) \cdot \ldots \cdot (n - r + 1)$$

To prove this formula we have only to observe that the first selection is made from the whole set of $n$ objects, the second selection is made from the $n - 1$ objects which remain after the first selection has been made, the third selection is made from the $n - 2$ objects which remain after the first two selections have been made, ..., and the $r$th and final selection is made from the

$$n - (r - 1) = n - r + 1$$

---

* The following are some of the symbols used in other books to denote the number of permutations of $r$ objects selected from a set of $n$ objects: $_nP_r$, $P(n, r)$, $P_r^n$, and $(n)_r$.

objects which remain after the first $r - 1$ selections have been made. Thus, the number of ways in which four parties can be assigned to four of the seven booths of a restaurant is $7 \cdot 6 \cdot 5 \cdot 4 = 840$, and the number of ways in which a child can make a first, second, and third choice from among 25 kinds of breakfast cereal is $25 \cdot 24 \cdot 23 = 13,800$.

Since products of consecutive integers arise in many problems dealing with permutations and other kinds of special arrangements, we write $5 \cdot 4 \cdot 3 \cdot 2 \cdot 1$, for instance, as 5! and refer to it as "5 factorial." More generally, $n!$, which we call "**$n$ factorial**," denotes the product

$$n(n - 1)(n - 2) \cdot \ldots \cdot 3 \cdot 2 \cdot 1$$

for any positive integer $n$.* Thus, $1! = 1$, $2! = 2 \cdot 1 = 2$, $3! = 3 \cdot 2 \cdot 1 = 6$, $4! = 4 \cdot 3 \cdot 2 \cdot 1 = 24$, $5! = 5 \cdot 4 \cdot 3 \cdot 2 \cdot 1 = 120$, $6! = 6 \cdot 5 \cdot 4 \cdot 3 \cdot 2 \cdot 1 = 720$, and so on. Using this notation, we can now say that

**The number of permutations of $n$ objects taken all together is $n!$.**

This is simply a special case of the formula on page 73 with $r = n$, so that the last factor becomes $n - r + 1 = n - n + 1 = 1$. Thus, the number of ways in which seven bus drivers can be assigned to seven buses is $7! = 5,040$, and the number of ways in which the 10 football teams in a conference can finish the season is $10! = 3,628,800$ (not counting possible ties in the final standings).

To illustrate the concept of a sample space in the beginning of this chapter, we referred to the one which consists of all the possible ways in which three of the 24 banks in a certain town can be selected for a surprise audit. Well, the number of ways in which they can be selected is 2,024, but this is *not* the number of permutations of three objects selected from a set of 24 objects. If we were interested also in the *order* in which the three banks are selected, the answer would be $24 \cdot 23 \cdot 22 = 12,144$, but each possible set of three banks would then be counted *six times*. For instance, if we let $A$, $U$, and $V$ denote the three banks chosen for a special audit, then

$$AUV \quad AVU \quad UAV \quad UVA \quad VAU \quad VUA$$

constitute six *different permutations*, but they all represent the sample of three banks chosen from among the 24 banks in the given town. To go from the number of permutations to the number of different samples, we shall thus have to divide by six, and we get $\dfrac{12,144}{6} = 2,024$, which is the answer we gave above.

* Also, it is customary to let $0! = 1$ *by definition*.

The number of different samples (or subsets) of $r$ objects which can be chosen from a set of $n$ objects is generally referred to as the number of **combinations** of $r$ objects selected from a set of $n$ objects—thus, *combinations differ from permutations insofar as we disregard the order in which the selection is made*. To obtain a formula for the number of combinations of $r$ objects taken from a set of $n$ objects, we have only to observe that any $r$ objects can be arranged in $r!$ different ways; in other words, *the $r!$ permutations of any $r$ objects count only as one combination*. Hence, the $n(n-1)(n-2) \cdot \ldots \cdot (n-r+1)$ permutations of $r$ objects selected from a set of $n$ objects contain each combination $r!$ times, and it follows that we obtain the number of combinations by dividing the number of permutations by $r!$. Thus, we arrive at the following rule:

> The number of ways in which a sample of $r$ objects can be selected from a set of $n$ objects, namely, the **number of combinations of $r$ objects selected from a set of $n$ objects, is given by**
>
> $$\frac{n(n-1)(n-2) \cdot \ldots \cdot (n-r+1)}{r!}$$

Symbolically, we shall let $\binom{n}{r}$ denote the number of combinations of $r$ objects selected from a set of $n$ objects, and an easy way of determining these quantities is indicated in Exercise 22 on page 79.* Otherwise, the values of $\binom{n}{r}$ for $n = 2$ through $n = 20$ can be obtained from Table VIII at the end of the book. As is explained in Exercise 23 on page 80 the quantities $\binom{n}{r}$ are also referred to as **binomial coefficients**.

The following are some straightforward applications of the formula for the number of combinations. If a department store manager wants to select four of his 16 sales clerks to staff a new department, he can do so in

$$\binom{16}{4} = \frac{16 \cdot 15 \cdot 14 \cdot 13}{4!} = 1{,}820$$

different ways; if an airline wants to choose five of its 20 aircraft for an assignment, it can do so in

$$\binom{20}{5} = \frac{20 \cdot 19 \cdot 18 \cdot 17 \cdot 16}{5!} = 15{,}504$$

---

* The following are some alternate symbols used to denote the number of combinations of $r$ objects selected from a set of $n$ objects: $_nC_r$, $C(n, r)$, and $C_r^n$.

different ways; and if an official wants to check the smokestacks of 6 of 24 factories for excessive emission of certain chemicals, he can select the six factories in

$$\binom{24}{6} = \frac{24 \cdot 23 \cdot 22 \cdot 21 \cdot 20 \cdot 19}{6!} = 134{,}596$$

different ways.* Of course, if the order *does* matter, the official can choose the six factories in $24 \cdot 23 \cdot 22 \cdot 21 \cdot 20 \cdot 19 = 96{,}909{,}120$ different ways. Note also that in the first two of these examples we could have obtained the answer directly from Table VIII at the end of the book.

## EXERCISES

1. With reference to the illustration on page 73, draw a tree diagram which shows the 20 ways in which the board of directors can choose the president and the (first) vice-president from among the five leading executives.

2. If a professional association has applications from New York, Chicago, New Orleans, Phoenix, and Atlanta for hosting its annual convention in 1975 and 1976, draw a tree diagram showing the 20 ways in which these annual conventions can be held. (Assume that each city can host the convention in either year but not in both years.)

3. A game manufacturer decides to put $100,000 into *one* new type of game each year, so long as the number of commercially unsuccessful games in which he invests does not exceed the number of commercial successes. (It will be assumed here that each game can be classified as being either a success or a failure.)
   (a) Draw a tree diagram which shows the eight possible situations that can arise in the first four years the plan is in operation. (Note that if the first year's game is a commercial failure, that branch of the tree diagram ends right then and there.)
   (b) In how many of the situations described in part (a) will the game manufacturer continue to invest in a new game in the fifth year?

* In problems like these it helps to cancel as many factors as possible before performing any multiplications. For instance, in the last example we could have written

$$\binom{24}{6} = \frac{\overset{}{24} \cdot 23 \cdot 22 \cdot \overset{7}{\cancel{21}} \cdot \overset{2}{\cancel{20}} \cdot 19}{\cancel{6} \cdot \cancel{5} \cdot \cancel{4} \cdot \cancel{3} \cdot \cancel{2} \cdot 1} = 23 \cdot 22 \cdot 7 \cdot 2 \cdot 19 = 134{,}596$$

(c) If the game manufacturer loses his total investment in a commercial failure and doubles his money in a commercial success, in how many of the situations described in part (a) will he be exactly even after four years?

4. A furrier stocks two fur coats of the same kind, reordering at the end of each day (for delivery the next morning) if and only if both coats have been sold. Construct a tree diagram which shows that if he starts on a Monday with two of the fur coats in stock, there are altogether eight different ways in which he can make sales on Monday and Tuesday of that week.

5. If a college cafeteria offers five different main courses and six different desserts, in how many different ways can a student choose one of each?

6. The proprietor of a shop has three different lamp shades and five different lamp bases. In how many different ways can he combine these lamp shades and lamp bases to make up a lamp?

7. A jeweler sells four different $1.00 charms and two different $2.00 charms. In how many different ways can a customer buy one of each kind
   (a) if we do not care about the order in which they are selected;
   (b) if we do care about the order in which they are selected?

8. If a college offers 2-, 3-, and 4-person dormitory rooms, which may be had on five different floors and with or without telephone, in how many different ways can a student make his choice?

9. A landlord classifies his tenants according to whether or not they are married, according to whether they have 0, 1, 2, 3, or more than 3 children, and according to whether they own 0, 1, or 2 automobiles. In how many different ways can a tenant thus be classified?

10. In an ice cream parlor a customer can order a banana split with one, two, or three flavors of ice cream, with or without syrup, with or without whipped cream, and with or without nuts. In how many different ways can a person order a banana split at this ice cream parlor?

11. In the example on page 61, each manufacturer had to choose between shipping by rail, truck, or air. How many points would there have been to the sample space if there had been (a) three manufacturers, (b) four manufacturers, and (c) five manufacturers?

12. If an electrical stage-lighting panel has 20 switches, in how many different ways can the theater electrician throw the switches; that is, in how many different ways can he turn the switches either on or off?

**13. THE NUMBER OF SUBSETS OF A SET** When we speak of the number of subsets of a set, we always include the *empty set* which has no elements at all, and also the whole set, itself. To select a subset from any given set we must decide for each element whether or not it is to be included, and for a set with $n$ elements there are, therefore,

$$\underbrace{2 \cdot 2 \cdot 2 \cdot 2 \cdot \ldots \cdot 2 \cdot 2}_{n \text{ factors}} = 2^n$$

possibilities, namely, $2^n$ different subsets.

(a) List the $2^3 = 8$ different subsets of the set which consists of the three major automobile manufacturers: General Motors, Ford, and Chrysler.

(b) List the $2^4 = 16$ different subsets of the set which consists of the following four sectors of the economy: manufacturing, mining, public utilities, and transportation.

(c) In how many different ways can one select a subset from the set which consists of the New England States: Massachusetts, Rhode Island, Connecticut, Maine, New Hampshire, and Vermont?

(d) In how many different ways can one select a subset from the set which consists of the ten stocks which were most actively traded during the year 1973 on the New York Stock Exchange?

**14.** A student is required to write reports on three of eight books on a reading list. In how many different ways can he select these three books?

**15.** If the drama club of a college has to choose four of ten half-hour skits to present on one evening from 8:00 to 10:00, in how many different ways can they arrange their schedule?

**16.** In how many different ways can the 18 paintings displayed at an art exhibition be awarded first prize, second prize, and third prize?

**17.** In how many different ways can a television director schedule eight half-hour programs during the eight half-hour intervals from 8 A.M. to noon?

**18.** Four fathers and their only sons bought eight seats in a row for a football game.
(a) In how many different ways can they be seated?
(b) In how many different ways can they be seated if each father and son are to sit together with the father to the left of the son?
(c) In how many different ways can they be seated if all the fathers are to sit together and all the sons are to sit together.

**19.** In Exercise 12 we asked for the total number of ways in which a theater electrician can throw the switches of a stage-lighting panel which has 20

switches. In how many different ways can the electrician throw the switches of this panel so that
(a) two switches are off and 18 are on;
(b) 15 of the switches are off and five are on;
(c) eight of the switches are off and 12 are on?
Check your answers in Table VIII.

20. If a carton of 12 flashlight batteries contains one that is defective, in how many ways can one choose three of these batteries so that
    (a) the defective battery is not included;
    (b) the defective battery is included?

21. In hiring his office staff, an office manager has to choose three typists from among six applicants, and two file clerks from among four applicants. What is the total number of ways in which he can select these five employees?

22. **PASCAL'S TRIANGLE** The number of combinations of $r$ objects selected from a set of $n$ objects, namely, the quantities $\binom{n}{r}$, can be determined by means of the following arrangement called *Pascal's triangle*;

$$
\begin{array}{ccccccccc}
 & & & & 1 & & & & \\
 & & & 1 & & 1 & & & \\
 & & 1 & & 2 & & 1 & & \\
 & 1 & & 3 & & 3 & & 1 & \\
1 & & 4 & & 6 & & 4 & & 1 \\
\end{array}
$$
..........................

where each row begins with a 1, ends with a 1, and each other entry is the *sum* of the nearest two entries in the row immediately above.
(a) Use Table VIII to verify that the *fourth* row of the triangle contains the values of $\binom{3}{r}$ for $r = 0, 1, 2,$ and 3.
(b) Use Table VIII to verify that the *fifth* row of the triangle contains the values of $\binom{4}{r}$ for $r = 0, 1, 2, 3,$ and 4.
(c) Construct the *sixth* row of the triangle and verify in Table VIII that it gives the values of $\binom{5}{r}$ for $r = 0, 1, 2, 3, 4,$ and 5.
(d) Construct the *seventh* row of the triangle and verify in Table VIII that it gives the values of $\binom{6}{r}$ for $r = 0, 1, 2, 3, 4, 5,$ and 6.

**23.** The quantities $\binom{n}{r}$ are referred to as *binomial coefficients* because they are, in fact, the coefficients in the binomial expansion of $(a + b)^n$. Verify that this is true for $n = 2$, 3, and 4 by expanding $(a + b)^2$, $(a + b)^3$, and $(a + b)^4$, and comparing the coefficients with the corresponding values of $\binom{n}{r}$ in Table VIII.

**24.** It is convenient to let $0! = 1$ *by definition*, for we can then write the formula for the number of permutations on page 73 as $\dfrac{n!}{(n - r)!}$ and the formula for the number of combinations on page 75 as $\dfrac{n!}{(n - r)!r!}$.
Use these formulas to rework
(a) Exercise 14;
(b) Exercise 15;
(c) the second part of Exercise 19;
(d) both parts of Exercise 20.

## PROBABILITIES AND ODDS

So far we have studied only *what is possible* in a given situation. In some instances we listed all possibilities and in others we merely determined how many different possibilities there are. *Now we shall go one step further and judge also what is probable and what is improbable.* The most common way of measuring the uncertainties connected with events (say, the commercial success of a new breakfast food, the outcome of a presidential election, the effectiveness of a new drug, or the total points we may roll with a pair of dice) is to assign them **probabilities,** or to specify the **odds** at which it would be fair to bet that the events will occur. [As we shall see on page 82, these two concepts (probabilities and odds) are very closely related—in fact, if we know one we can always calculate the other.]

Among the different theories of probability—and there are many—most widely held is the **frequency concept of probability,** according to which the probability of an event is interpreted as the *proportion of the time that such kind of event will occur in the long run.*[*] If we say that there is a probability of 0.82 that a jet from New York to Chicago will arrive on time, this means that such flights arrive on time about 82 percent of the time. More generally, we say that an event has a probability of, say, 0.90, in the same sense in which we might say that our car will start in cold weather about 90 percent of the time. *We cannot guarantee what will happen on any one try, the car may start*

---
[*] Another popular concept of probability, based on "equally likely" events, will be discussed later in Chapter 4.

and then it may not, but it would be reasonable to bet $9.00 against $1.00 or 90 cents against a dime (*namely, at odds of 9 to 1*) that the car will start at any given try. This would be "reasonable" or "fair," because we would win $1.00 (or a dime) about 90 percent of the time, lose $9.00 (or 90 cents) about 10 percent of the time, and, hence, can expect to break even in the long run.

In accordance with the frequency concept of probability, we *estimate* the probability of an event by observing how often (what part of the time) similar events have occurred in the past. For instance, if airline records show that over a certain period of time 492 of 600 jets from New York to Chicago arrived on time, we *estimate* the probability that any such flight from New York to Chicago (perhaps, the next one) will arrive on time as $\frac{492}{600} = 0.82$. Similarly, if (over a number of years) 210 of 1,050 restaurants in a certain city failed during their first year of operation, we *estimate* the probability that any restaurant in this city will fail during its first year of operation as $\frac{210}{1,050} = 0.20$.

Since we defined probabilities in terms of what happens to similar events in the long run, let us examine briefly whether it is at all meaningful to talk about the probability of a particular event which can happen only once, say, the probability that a certain major-league baseball player's broken arm will heal within a month, or the probability that a certain major-party candidate will win an upcoming presidential election.

If we put ourselves in the position of the baseball player's doctor, we could check medical records, observe that 39 percent of the time such fractures have healed within a month, and apply this figure to the baseball player's arm. This may not be of much comfort to the baseball player, but it does give a meaning to a probability statement about his broken arm—the probability that it will heal within a month is 0.39. *Thus, when we make a probability statement about a specific (non-repeatable) event, the frequency concept of probability leaves us no choice but to refer to a set of similar events.* This can lead to complications, for the choice of "similar" events is often neither obvious nor easy. With reference to the baseball player's arm, for example, we might count as "similar" only those cases where the fracture was in the same (left or right) arm, we might only count those cases in which the patients were of the same age as the baseball player, we might count only those cases in which the patients were also professional athletes, or we might count only those cases in which the patients were also of the same height and weight. Ultimately, this is a matter of individual judgment, but it should be observed that *the more we narrow things down, the less information we will have for estimating the corresponding probabilities.*

So far as the second example is concerned, the one concerning the presidential election, suppose that we ask a person who has conducted a poll "how sure" he is that the given candidate will actually win. If he replies that he is "99 percent sure," namely, that he assigns the candidate's election a probability of 0.99, this is not meant to imply that the candidate would win 99 percent of the time if he ran for office a great many times. No, *he means that his conclusions are based on methods which (in the long run) will "work" 99 percent of the time.* In this sense, many of the probabilities which we use in statistics to express our faith in predictions or decisions are simply "success ratios" that apply to the methods we have employed.

An alternate point of view, which is currently gaining favor, is to interpret probabilities as **personal** or **subjective** evaluations. For instance, a businessman may "feel" that the odds for the success of a new venture, say, a new oil well being drilled, are 3 to 2. This means that he would be willing to bet (or consider it fair to bet) $300 against $200, or perhaps $3,000 against $2,000, that the venture will succeed (strike oil). In this way he expresses the *strength of his belief* regarding the uncertainties connected with the success of the new oil well. This method of dealing with uncertainties works nicely (and is certainly justifiable) in situations where there is very little *direct evidence*, and there may be no choice but to consider collateral (indirect) information, "educated" guesses, and perhaps intuition and other subjective factors. Thus, the businessman's odds concerning the success of the new oil well may be based on his analysis of geological conditions, on the success of other oil wells drilled nearby, the opinions of experts, and perhaps a good dose of optimism.

Regardless of how we interpret probabilities and odds, subjectively or objectively (namely, as personal judgments or in terms of frequencies or proportions), the mathematical relationship between the two is always the same. It is given by the following rule:

If somebody considers it fair or equitable to bet $a$ dollars against $b$ dollars that a given event will occur, he is, in fact, assigning the event the probability $\dfrac{a}{a+b}$.

Thus, the businessman who is willing to give odds of 3 to 2 that the drilling of the new well will be a success is actually assigning its success a probability of $\dfrac{3}{3+2} = 0.60$. Also, if the odds are 7 to 2 that a student *will not* get an A in a certain accounting course, then the probability that he will not get an A is $\dfrac{7}{7+2} = \dfrac{7}{9}$; correspondingly, the odds that he *will* get an A are 2 to 7, the probability that he will get an A is $\dfrac{2}{2+7} = \dfrac{2}{9}$, and it should be observed that the *sum* of these two probabilities is $\tfrac{7}{9} + \tfrac{2}{9} = 1$.

To illustrate how probabilities are converted into odds, let us refer back to the example on page 81, where we dealt with the question of whether or not we could start our car. As we pointed out at the time, the probability of 0.90 implies that we should win (namely, get the car started in cold weather) about 90 percent of the time, and lose (fail to get it started) about 10 percent of the time. In other words, *we should win about nine times as often as we should lose*, so that the proper odds are 9 to 1. In general,

> If the probability of an event is $p$, then the odds for its occurrence are $p$ to $1 - p$ and the odds against its occurrence are $1 - p$ to $p$.

(This result can also be obtained by showing *algebraically* that $p = \dfrac{a}{a+b}$ leads to $\dfrac{a}{b} = \dfrac{p}{1-p}$.) For instance, if the probability that an item lost in a department store will never be claimed is 0.15, then the odds are 0.15 to 0.85, or 3 to 17, that such a lost item will never be claimed (and they are 17 to 3 that it *will* be claimed). Also, if the probability that we will not find a parking space in a certain block is 0.25, then the odds are 0.25 to 0.75, or 1 to 3 that we *will not* find a parking space in the block (and they are 3 to 1 that we *will* find a parking space there). Note that in both of these examples we followed the common practice of quoting odds as *ratios of positive integers* (having no common factors).

## EXERCISES

1. A study made by a downtown department store showed that 2,014 of 5,300 customers turned left at a certain aisle and that 3,286 turned right. Estimate the probability that any customer who approaches this aisle will turn left.

2. Statistics compiled by the manager of a pharmacy show that 378 of 600 customers who entered his store during a special sale purchased at least one item that was not "on sale." Estimate the probability that a customer who enters this store during a sale will buy at least one item which is not "on sale."

3. If 779 of 1,025 voters interviewed in a poll say that they are for a certain candidate, what is the probability that any one voter will favor this candidate?

4. Statistics compiled by the weather bureau of a certain Eastern community show that 56 times it did not rain on the Fourth of July during the last 70 years.
   (a) Estimate the probability that it will not rain in this community on the next Fourth of July.
   (b) What are the odds that it will rain there on the next Fourth of July?
   (c) What are the odds that it will not rain there on the next Fourth of July?

5. Among the 288 times that a camp counselor has gone on a picnic, he has come back without suffering a single mosquito bite only 96 times.
   (a) Estimate the probability that he will come back from the next picnic without a single mosquito bite.
   (b) What are the odds that he will get at least one mosquito bite?
   (c) If someone offered the camp counselor *even money* (that is, odds of 1 to 1) that he will *not* get bitten by a mosquito, who would be favored by this bet?

6. In a sample of 110 lakes, ponds, and streams in a certain state, 77 were found to contain excessive amounts of pollutants.
   (a) Estimate the probability that any one of the lakes, ponds, or streams in this state will contain excessive pollutants.
   (b) What are the odds that any one of the lakes, ponds, or streams in this state will contain excessive pollutants?
   (c) What are the odds that any one of the lakes, ponds, or streams in this state will not contain excessive pollutants?
   (d) If we offered to bet a fisherman $7.00 against his $2.00 that the next lake, pond, or stream tested in this state will be found to contain excessive pollutants, would this be a smart thing to do?

7. If a college senior feels that 2 to 1 are fair odds that he will be admitted to a certain graduate school, what is his personal probability that he will be admitted to this graduate school?

8. If the odds are 13 to 12 that a newly-hired unmarried airline stewardess will get married before she has been with the company for two years, what is the probability that this will be the case.

9. If somebody claims that the odds are 5 to 11 that he will get a speeding ticket during any one year, what probability does he assign to his *not* getting a speeding ticket during any one year?

10. If somebody claims that the odds are 5 to 3 that a certain executive will get to his office on time, what probability does he assign to the executive's being late?

11. A friend applying for a new job is anxious to bet us $13.50 against our $1.50 that he will get the job. What does this tell us about the personal probability which he assigns to his getting the job? (*Hint:* The answer should read "at least . . . .")

12. A student refuses to bet $1.00 against $4.00 that he will *not* get stuck in freeway traffic while driving home from class. What does this tell us about the personal probability he assigns to his getting stuck? (*Hint:* The answer should read "greater than . . . .")

## EXPECTATIONS AND DECISIONS

If an insurance agent tells us that a person aged 20 can expect to live 53 more years, this does not mean that he really "expects" the person to live until his 73rd birthday and then drop dead on the next day. Similarly, if we read that a person living in the United States can expect to eat 17.9 pounds of ice cream and 98.5 pounds of fresh vegetables per year, that an embezzler confined in a federal prison can expect to serve 11.4 months of his sentence, or that a school child can expect to be absent 5.4 days per year because of illness or injury, it must be apparent that the word "expect" is not being used in its colloquial sense. A child cannot very well be absent 5.4 days, and it would be surprising, indeed, if we found somebody who actually eats exactly 17.9 pounds of ice cream a year. So far as the first statement is concerned, some persons aged 20 will live another 14 years, some will live another 29 years, some will live another 67 years, . . ., and the life expectancy of "53 more years" can only be interpreted as an average, namely, as a **mathematical expectation.**

Originally, the concept of a mathematical expectation arose in connection with games of chance, and in its simplest form it is *the product of the probability that a player will win and the amount he stands to win.* Thus, if we stand to receive $4.00 if a balanced coin comes up *heads*, our mathematical expectation is $4(\frac{1}{2}) = \$2.00$. Similarly, if we buy one of 1,000 raffle tickets for a color television set worth $650.00, the probability for each ticket is $\frac{1}{1,000}$ and our mathematical expectation is $650 \left(\frac{1}{1,000}\right) = \$0.65$, or at least the equivalent of this amount in merchandise. Thus, it would be foolish to pay more than 65 cents for the ticket, unless, of course, the proceeds of the raffle went to a worthy cause. Note that in this last example 999 of the tickets will not pay anything at all, one of the tickets will pay $650.00, so that altogether the 1,000 tickets pay $650.00, or *on the average* 65 cents per ticket—*this is the mathematical expectation for each ticket.*

To demonstrate how the concept of a mathematical expectation can be made more general, let us change the preceding example so that there is also a second prize of a record player worth $150.00 and a third prize of a radio worth $50.00. We can then argue that 997 of the tickets will not pay anything at all, one ticket will pay $650.00 (in merchandise), another will pay the equivalent of $150.00, and a third will pay the equivalent of $50.00; altogether, the 1,000 tickets will thus pay the equivalent of 650 + 150 + 50 = $850.00, or *on the average* 85 cents per ticket. *As before, this is the mathematical expectation for each ticket.* Looking at the problem in a different way, we could argue that if the raffle were repeated many times, we would not win anything 99.7 percent of the time, and each of the three prizes about 0.1 percent of the time. On the average, we should thus win about

$$0(0.997) + 650(0.001) + 150(0.001) + 50(0.001) = \$0.85$$

which is the sum of the products obtained by multiplying each amount by the corresponding proportion or probability. Generalizing from this example, let us now make the following definition:

**If the probabilities of obtaining the amounts $a_1, a_2, \ldots$, or $a_k$, are, respectively, $p_1, p_2, \ldots$, and $p_k$, then the mathematical expectation is**

$$a_1 p_1 + a_2 p_2 + \cdots + a_k p_k$$

Each amount is multiplied by the corresponding probability, and the mathematical expectation is the sum of the products thus obtained; symbolically, we could write the formula for a mathematical expectation as $\Sigma\, a \cdot p$. So far as the $a$'s are concerned, it is important to keep in mind that they are *positive* when they represent profits, winnings, or gains (namely, amounts we stand to receive), and that they are *negative* when they represent losses, penalties, or deficits (namely, amounts we have to pay). For instance, if we bet $10.00 on the flip of a coin (that is, we either win $10.00 or lose $10.00 depending on the outcome), the amounts $a_1$ and $a_2$ are $+10$ and $-10$, the corresponding probabilities are $p_1 = 0.50$ and $p_2 = 0.50$, and the mathematical expectation is

$$10(0.50) + (-10)(0.50) = 0$$

This is what it should be in an **equitable game**, namely, in a game which does not favor either player.

It is also important to remember that although we referred to the $a$'s as "amounts," they need not be *cash* winnings, losses, penalties, or rewards. For instance, when we say that a household in the United States can be expected to have 3.19 members, the $a$'s refer to the possible numbers of members, and 3.19 is the sum of the products obtained by multiplying 0, 1, 2, 3, 4, ..., by the respective probabilities that a household in the United States has that many members. Similarly, if we say that a person living in the United States can be expected to make 0.81 flights a year on commercial airlines, the $a$'s refer to the number of flights, and 0.81 is the sum of the products obtained by multiplying 0, 1, 2, 3, 4, ..., by the respective probabilities that a person living in the United States will make that many flights.

To consider another numerical example, suppose that a businessman is offered $1,000,000 for his business, and that he must make up his mind whether to accept this offer or turn it down with the hope of getting a better price. To help him make up his mind, he consults an investment banker, who tells him that if he turns down the offer, the probabilities that he will ultimately get $1,100,000, $1,050,000, $1,000,000, or only $800,000 are, respectively, 0.10, 0.20, 0.40, and 0.30. Accepting these figures as correct, the businessman finds that his mathematical expectation is

$$1,100,000(0.10) + 1,050,000(0.20) + 1,000,000(0.40) + 800,000(0.30)$$
$$= \$960,000$$

and unless he is a *confirmed optimist* (or a somewhat *reckless gambler*), it stands to reason that he should accept the $1,000,000 he is offered for his business.

The argument we have used here is based on the assumption that it is "rational" to choose whichever alternative has the highest mathematical expectation (or, at least, it was assumed that "$1,000,000 in one's pocket is preferable to a mathematical expectation of $960,000"). On the surface, the idea of maximizing one's mathematical expectation looks like a very reasonable criterion for making decisions, but we should point out that it involves quite a few difficulties. To illustrate, let us consider the following situation: A furniture manufacturer has to decide whether to spend a considerable sum of money to build a new factory. He knows that if the new factory is built and the furniture business has a good sales year, there will be a $328,000 profit; if the factory is built and the furniture business has a poor sales year, there will be a deficit of $80,000. On the other hand, if the old factory is used and the furniture business has a good sales year, there will be a $160,000 profit; and if the old factory is used and the furniture business has a poor sales year, there will be a profit of $16,000 (mostly because of lower overhead costs). Schematically, all this information can be presented as in the following table:

**88** POSSIBILITIES AND PROBABILITIES  CHAP. 3

|  | Good Sales Year | Poor Sales Year |
|---|---|---|
| Build New Factory | $328,000 | −$80,000 |
| Use Old Factory | $160,000 | $16,000 |

Evidently, it will be advantageous to build the new factory only if the furniture business is going to have a good sales year, and the decision whether or not to build the new factory will, therefore, have to depend largely on the chances of having a good year. Suppose, for instance, that the furniture manufacturer judges (on the basis of years of experience) that the odds are 2 to 1 that the furniture business will have a poor sales year, or that in accordance with the rule on page 82 the probability for a good sales year is $\frac{1}{1+2} = \frac{1}{3}$ and the probability for a poor sales year is $\frac{2}{2+1} = \frac{2}{3}$. He could then argue that if the new factory is built the *expected profit* is

$$328,000 \cdot \frac{1}{3} + (-80,000) \cdot \frac{2}{3} = \$56,000$$

and if the old factory is used the *expected profit* is

$$160,000 \cdot \frac{1}{3} + 16,000 \cdot \frac{2}{3} = \$64,000$$

Now, since an expected profit of $64,000 is obviously preferable to an expected profit of only $56,000, it stands to reason that the manufacturer should decide to continue using the old factory. Or should he? What if he were hasty in assessing the odds as 2 to 1 against there being a good sales year? What if the odds should have been 3 to 2 instead of 2 to 1? In that case the *expected profit* would be

$$328,000 \cdot \frac{2}{2+3} + (-80,000) \cdot \frac{3}{3+2} = \$83,200$$

if he builds the new factory, or

$$160,000 \cdot \tfrac{2}{5} + 16,000 \cdot \tfrac{3}{5} = \$73,600$$

if he uses the old factory, and it can be seen that his decision would be reversed. This demonstrates that if mathematical expectations are to serve as criteria for making decisions, *it is essential to know the correct values of all relevant probabilities*. Not only that, but *we must also know the correct values of the "payoffs" which are associated with the various possibilities*. For instance, what should the manufacturer do if the $328,000 profit (corresponding to a good sales year and the new factory) had to be changed to $320,000 or $400,000; or if the deficit of $80,000 (corresponding to a poor sales year and the new factory) had to be changed to $60,000 or $120,000? We shall leave it to the reader to answer some of these questions in Exercise 12 on page 91. The method which we used in this last example constitutes the rudiment of what is called a **Bayesian analysis**; in this kind of analysis probabilities are assigned to the conditions about which uncertainties exist, the so-called "states of nature" (which in our example were a good sales year and a bad sales year), and *the alternative which is ultimately decided upon is the one which has the greatest expected profit or gain*.

## EXERCISES

1. A prize of a television set valued at $450 is offered to a person who will deposit his contest entry at a downtown department store. What is each entrant's mathematical expectation if 2,000 persons filed their names and one of the names is randomly drawn? Would this make it worthwhile for a contestant to spend 25 cents on gasoline to drive to the department store?

2. Repeat Exercise 1 with the modification that there is also a second prize of a stereo set valued at $300.

3. Four catering companies are attempting to obtain the concession to operate a college cafeteria, which promises a profit of $24,000. If one of the caterers figures that the odds are 4 to 1 against his getting the concession, what is his mathematical expectation? Would he be better off if he made a secret agreement with the other caterers to divide the profit evenly regardless of who is awarded the concession?

4. An importer has arranged for a shipment of seasonal merchandise which promises a profit of $120,000 with a probability of 0.60 or a loss of $45,000 (due to late delivery of the shipment) with a probability of 0.40. What is the importer's mathematical expectation?

5. If it is extremely cold in New York City, a hotel in Miami, Florida, will have 480 guests during a mid-winter week-end; if it is cold (but not

extremely cold) in New York City, the hotel will have 414 guests, and if the weather is moderate in New York City, the hotel will have only 264 guests. How many guests can they expect if the probabilities for extremely cold, cold, or moderate weather in New York City at that time are, respectively, 0.34, 0.58, and 0.08?

**6.** The parents of a college student promise him $1,000 towards his expenses if he passes all of his subjects, $500 if he fails only one subject, and otherwise no money at all. What is the student's mathematical expectation if the probability of his passing all of his subjects is 0.65, while the probability of his failing exactly one subject is 0.20?

**7.** An insurance company agrees to pay the promoter of a state fair $60,000 in the event of rain on the date of the fair. If the company's actuary figures that their expected loss for this risk is $7,680, what probability does he thus assign to the possibility that it will rain on the day of the fair?

**8.** The following table gives the probabilities that a person browsing in a book store will buy 0, 1, 2, 3, or 4 books:

| Books | 0 | 1 | 2 | 3 | 4 |
|---|---|---|---|---|---|
| Probability | 0.40 | 0.30 | 0.15 | 0.10 | 0.05 |

How many books can a person browsing in this book store be expected to buy?

**9.** A land speculator is offered the choice of either taking an immediate cash payment of $10,000 for his land, or gambling that a certain manufacturing corporation will buy the land for a factory building. If the manufacturing corporation buys the land, the speculator will receive $40,000, but if the manufacturing corporation does not buy it, the speculator will have to sell the land as a residential lot for only $8,000. What can we say about the probability which the speculator assigns to the sale of the land to the manufacturing corporation if he figures that taking the gamble has the higher mathematical expectation? (*Hint:* Let the probability of the sale of the land to the manufacturing corporation be $p$, and set up an inequality which compares the mathematical expectation associated with the gamble with the immediate cash payment of $10,000.)

**10.** A young man plans to drive his car to the campus where he is supposed to meet his date. He remembers that he is supposed to meet her either at her dormitory or at the library, but he cannot remember which. If he drives to the dormitory, his gasoline expense will be 42 cents; if he drives to the library, the gasoline expense will be 36 cents. Furthermore, if he drives

to the wrong place it will cost him an extra 10 cents in gasoline to correct his error.
(a) Present all this information in a table like that on page 88.
(b) Where should he drive if he wants to minimize his expected expenses and feels that the odds are 5 to 1 that he was supposed to meet his date at her dormitory?
(c) Repeat part (b) when the odds are 2 to 1 instead of 5 to 1.
(d) Repeat part (b) when the odds are 4 to 1 instead of 5 to 1.

11. The management of a mining company must decide whether to continue an operation at a certain location. If they continue and are successful, this will be worth $1,000,000; if they continue and are not successful, this will entail a loss of $600,000; if they do not continue but would have been successful, this will entail a loss of $400,000 (for competitive reasons); and if they do not continue and would not have been successful anyhow, this will be worth $100,000 to the company (because funds allocated to the operation remain unspent).
(a) Present all this information in a table like that on page 88.
(b) What decision would maximize the company's expected gain if the odds against success are 3 to 2?
(c) Repeat part (b) when the odds against success are 4 to 1.
(d) Repeat part (b) when the odds against success are 2 to 1.

12. With reference to the illustration on page 87, what decision should the furniture manufacturer make if
(a) the $328,000 profit is replaced by $400,000 and the probability for a good sales year is $\frac{1}{3}$;
(b) the $80,000 deficit is replaced by a deficit of $120,000 and the probability for a good sales year is $\frac{2}{5}$?

13. **OPPORTUNITY LOSSES** Suppose that the furniture manufacturer of the example in the text is the kind of person who always worries about *missing out on a good deal*. Looking at the table on page 88, he would argue that if he decides to use the old factory and there is a good sales year, he would have been better off by $328,000 − $160,000 = $168,000 if he had decided to build the new factory, and we refer to this quantity as the **opportunity loss** (or **regret**) associated with this case. Similarly, the opportunity loss would be $16,000 − (−$80,000) = $96,000 if he decides to build the new factory and the sales year is poor. In the other two cases the opportunity loss is zero, since in each case he makes the more profitable decision.
(a) What action should the furniture manufacturer take so as to hold the greatest possible opportunity loss to a minimum?

(b) With reference to Exercise 10, find the opportunity losses that are associated with the four possibilities; also, determine whether the young man should drive to the dormitory or to the library if he wants to hold the greatest possible opportunity losses to a minimum.

(c) With reference to Exercise 11, find the opportunity losses that are associated with the four possibilities; also, determine what the management of the mining company should do so as to hold the greatest possible opportunity loss to a minimum.

**14. THE EXPECTED PROFIT WITH PERFECT INFORMATION** With reference to the example in the text, suppose that the furniture manufacturer has the option of hiring an "infallible forecaster" for $40,000, so that he will know for certain whether the sales year will be good or poor. Of course, this raises the question whether it is worthwhile to spend the $40,000. To answer it, let us take the furniture manufacturer's original odds of 2 to 1 that there will be a poor sales year. If he knew *for sure* whether the sales year will be good or poor, the right decision would yield a profit of either $328,000 or $16,000 (the corresponding entries in the table on page 88), and since the corresponding probabilities are $\frac{1}{3}$ and $\frac{2}{3}$, his *expected profit* is

$$328{,}000 \cdot \tfrac{1}{3} + 16{,}000 \cdot \tfrac{2}{3} = \$120{,}000$$

This is what is called the **expected profit with perfect information**; in general, it is *the amount one can expect to gain, profit, or win in a given situation if one always makes the right decision.* Since the value obtained here *exceeds* by $120{,}000 - 56{,}000 = \$64{,}000$ and $120{,}000 - 64{,}000 = \$56{,}000$ the expected profits determined on page 88, it follows that the $40,000 fee of the "infallible forecaster" would be well spent.

(a) With reference to Exercise 10 and the odds of part (b), what is the expected profit with perfect information? Would it be worthwhile to spend 10 cents on a telephone call to find out whether he was supposed to meet his date at her dormitory or at the library?

(b) With reference to Exercise 11 and the odds of part (b), what is the expected profit with perfect information? If it were possible to continue the operation for three months at a cost of $300,000, after which it would be known *for certain* whether or not the operation will be a success, would it be worthwhile to spend these additional funds?

# 4

# SOME RULES OF PROBABILITY

## INTRODUCTION

Familiarity with the rules of probability can be of great value to such diverse groups as businessmen, military strategists, doctors, and many others, including, of course, the gamblers for whose "benefit" the theory was originally developed. Businessmen must invest capital in raw materials and inventories months before they can be used or sold, military strategists must commit men and material to the hazards of battle, doctors must diagnose ailments to determine the form of treatment, gamblers must decide whether to bet on a particular hand of cards, etc., and in each case the rules of probability make it possible, or at least easier, to "live with the corresponding uncertainties."

In the study of probability there are three basic questions. First, there is the question of *what we mean* when we say that the probability of an event is, say, 0.90, 0.74, or 0.05; then there is the question of *how the values of probabilities are determined* in actual practice; and finally there is the question of *how the probabilities of simple events can be used to calculate those of more complicated kinds of events.* Having discussed the first two questions briefly in Chapter 3, we shall devote most of this chapter to investigating the third. [So far as the first question is concerned, we stated that if an event is assigned a probability of, say, 0.90, many people interpret this as a *proportion*, namely, the proportion of the time that such events will occur in the long run—to others it means that they feel *subjectively* that 9 to 1 would be fair odds for betting that the event will occur. Correspondingly, if probabilities are interpreted as "long run proportions" they are *estimated* in terms of observed proportions, and if they are interpreted subjectively their values are calculated on the basis of the odds at which a person would be willing to bet (or consider it fair to bet) on the corresponding events.]

## SOME BASIC RULES

To tackle the third question which we listed in the preceding section, we shall first have to investigate some of the simpler rules according to which probabilities "behave." Since probabilities are *measures of uncertainty* and since most of the things which we measure (distances, areas, weights, electric currents, barometric pressures, etc.) are given by non-negative numbers, let us begin with the following rule:

> **The probability of any event must be a positive real number or zero.**

It is important to note that this rule is in complete agreement with the frequency interpretation as well as the subjective concept of probability. Clearly, proportions are always positive or zero, and so long as the amounts $a$ and $b$ which we bet for and against an event are positive, the probability of the event, $\dfrac{a}{a+b}$, cannot be negative.

If we think of probabilities in terms of proportions, we are immediately led to several other basic rules. For instance:

> **Probabilities can never be greater than 1; a probability of 1 implies that the event is certain to occur; and a probability of 0 implies that the event is certain not to occur.**

Evidently, an event cannot possibly happen more than 100 percent of the time; an event is certain to occur if it happens 100 percent of the time, and an event is certain *not* to occur if it happens 0 percent of the time. In actual practice, we also assign a probability of 1 to an event for which we are "practically certain" that it will occur, and a probability of 0 to an event for which we are "practically certain" that it will not occur. Thus, we would assign a probability of 1 to the event that at least one person will make a mistake in calculating his income tax for the year 1975, and we should assign a probability of 0 to the event that there will be no fatalities caused by automobile accidents in the United States during the next Labor Day weekend. So far as subjective probabilities are concerned, we would be willing to give "better and better" odds when we become more and more certain that an event will occur—say, 100 to 1, 1,000 to 1, or perhaps even 1,000,000 to 1. The corresponding probabilities are $\dfrac{100}{100+1}$, $\dfrac{1,000}{1,000+1}$, and $\dfrac{1,000,000}{1,000,000+1}$ (or approximately 0.99, 0.999, and 0.999999), and it can be seen that *the more certain we are that an event will occur, the closer its probability will be to 1*. A corresponding argument leads to the result that *the more certain we are that an event will not occur, the closer its probability will be to 0*.

The next rule is especially important, and it applies only to events which cannot both occur at the same time, namely, events which are *mutually exclusive*:

**If two events are mutually exclusive, the probability that one or the other will occur equals the sum of their probabilities.**

For instance, if the respective probabilities that there will be exactly two or three persons in an automobile at a drive-in movie theater are 0.47 and 0.15, then the probability that there will be either two or three persons in an automobile is 0.47 + 0.15 = 0.62. This definitely agrees with the frequency interpretation of probability: If one kind of event occurs 47 percent of the time, another kind of event occurs 15 percent of the time, and they cannot both occur at the same time, then one or the other will occur 47 + 15 = 62 percent of the time. So far as subjective probabilities are concerned, this last rule does not follow from the definition which we gave in Chapter 3, but it is generally imposed as the so-called **consistency criterion.** In other words, if a person's subjective probabilities "behave" in accordance with the rule, he is said to be *consistent;* otherwise, he is said to be *inconsistent* and his probability judgments must be taken with a grain of salt (see Exercises 4 and 5 on page 101).

Following the practice of denoting events with capital letters as in Chapter 3, and writing the probability of event $A$ as $P(A)$, the probability of event $B$ as $P(B)$, . . . , we can express the last rule symbolically as

$$P(A \cup B) = P(A) + P(B)$$

keeping in mind, of course, that it applies only when $A$ and $B$ are mutually exclusive events. This formula is sometimes referred to as the **Special Addition Rule** for probabilities; several other addition rules, namely, more general formulas, will be discussed on pages 96 and 100.

Another basic rule of probability concerns the probability of the *complement $A'$* of an event $A$ (which we defined on page 64). Leaving the proof of the rule to the reader in Exercise 19 on page 103, let us merely state it as follows:

**The sum of the probabilities of an event $A$ and its complement $A'$ is always equal to 1; symbolically,**

$$P(A) + P(A') = 1$$

for any event $A$, and hence $P(A') = 1 - P(A)$.

This certainly agrees with the frequency interpretation of probability, for if 12 percent of a certain kind of washing machine require repairs under the guarantee, then 88 percent will *not* require repairs under the guarantee, and if 61 percent of all business executives over 45 play golf, then 39 percent do not play golf. So far as subjective probabilities are concerned, if the odds *for* the occurrence of an event are $a$ to $b$, then the odds *against* its occurrence are $b$ to $a$, the corresponding probabilities are $\frac{a}{a+b}$ and $\frac{b}{b+a}$, according to the rule on page 82, and their sum is always equal to 1.

## FURTHER ADDITION RULES

The addition rule of the preceding section was limited to two mutually exclusive events, but it can easily be generalized so that it applies also when there are more than two mutually exclusive events. In that case we have:

If $k$ events are all mutually exclusive, the probability that one of them will occur equals the sum of their respective probabilities; symbolically, if the events are denoted $A_1, A_2, \ldots,$ and $A_k$, we have

$$P(A_1 \cup A_2 \cup \ldots \cup A_k) = P(A_1) + P(A_2) + \ldots + P(A_k)$$

For instance, if Mrs. Smith is planning to read a book on a flight to Dallas, and the probabilities that she will read a love story, a murder mystery, or a biography are, respectively, 0.31, 0.25, and 0.18, then the probability that she will read one of these types of books is $0.31 + 0.25 + 0.18 = 0.74$; the probability that she will read another type of book or forget the idea altogether is $1 - 0.74 = 0.26$. Similarly, if the probabilities that the Standard and Poor's Corporation will rate a given company's stock $A+$, $A$, $A-$, or $B+$ are, respectively, 0.09, 0.15, 0.10, and 0.07, then the probability that the company will get one of these ratings is $0.09 + 0.15 + 0.10 + 0.07 = 0.41$, and the probability that it will get a lower rating is $1 - 0.41 = 0.59$.

The job of assigning probabilities to *all* the events that are possible in a given situation can be very tedious, to say the least. Even if a sample space has as few as five points corresponding, say, to a trainee getting an $A$, $B$, $C$, $D$, or $F$ in a management training course, there are already $2^5 = 32$ possible events in accordance with the rule of Exercise 13 on page 78: the event that he will get an $A$, the event that he will get a $B$ or a $D$, the event that he will get a $C$, $D$, or $F$, and so on. Things get worse very rapidly when a sample space has more than five points—for 10 points, for example, there are 1,024 possible events, as the reader was asked to show in part (d) of

Exercise 13 on page 78, and if we follow the rule of that exercise we could also show that for a sample space with 25 points there are 33,554,432 possible events. Fortunately, it is seldom necessary to assign probabilities to all possible events, and the following rule makes it easy to find the probability of any particular event on the basis of the probabilities which are assigned to the points (individual outcomes) of a sample space:

**The probability of any event $A$ is given by the sum of the probabilities of all the individual outcomes which are included in $A$.**

This rule is illustrated in Figure 4.1, where the dots represent the individual (mutually exclusive) outcomes; the fact that the probability of $A$ is given by the sum of the probabilities of the individual points in $A$ follows immediately from the addition rule on page 96.

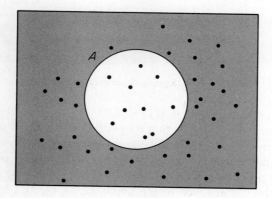

**FIGURE 4.1** A sample space.

To illustrate this rule, suppose that the probabilities of the trainee mentioned in the preceding paragraph getting an $A$, $B$, $C$, $D$, or $F$ are, respectively, 0.10, 0.18, 0.47, 0.19, and 0.06. Then, the probability that he will get a $B$ or a $D$ is $0.18 + 0.19 = 0.37$, and the probability that he will get a $C$, $D$, or $F$ is $0.47 + 0.19 + 0.06 = 0.72$. To consider another illustration let us refer again to the clock and typewriter manufacturers who had to decide whether to ship their products by rail, truck, or air, and let us suppose that the nine possibilities have the probabilities shown in Figure 4.2 (which is otherwise identical with Figure 3.2 on page 61, with 1 denoting rail, 2 denoting truck, and 3 denoting air). If we are interested in the probability that at least one of the two manufacturers will decide to ship by truck, we have only to add the probabilities associated with the points (1, 2), (2, 1), (2, 2), (2, 3), and

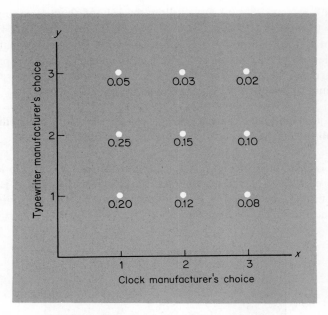

**FIGURE 4.2** Sample space with probabilities.

(3, 2), and we get $0.25 + 0.12 + 0.15 + 0.03 + 0.10 = 0.65$. Similarly, if we are interested in the probability that they will both decide to ship their products the same way, we simply add the probabilities assigned to the points (1, 1), (2, 2), and (3, 3), getting $0.20 + 0.15 + 0.02 = 0.37$, and if we are interested in the probability that neither will decide to ship by rail, we simply add the probabilities assigned to the points (2, 2), (3, 2), (2, 3), and (3, 3), getting $0.15 + 0.10 + 0.03 + 0.02 = 0.30$.

The situation is even simpler when the outcomes are all **equiprobable,** which is often the case in games of chance. In that case, it follows from the rule on page 96 that:

> If there are $n$ possible outcomes which are all equiprobable and $s$ of these outcomes are included in event $A$, then the probability of $A$ is $s/n$.

This ratio of the number of "successes" to the total number of outcomes is sometimes used as a *definition* of probability, but aside from the logical difficulty of defining "probability" in terms of "equiprobable events," it has the shortcoming that it applies only when the individual outcomes all have the same probability. Nevertheless, this is often the case in games of chance, where we can assume that each card in the deck has the same chance of being drawn, that each face of a coin has the same probability of coming up, and likewise for each of the six sides of a balanced die. Thus, the probability of drawing a *spade* from an ordinary deck of playing cards is $\frac{13}{52}$ (there are 13 spades among the 52 cards), the probability of getting *tails* with a balanced coin is $\frac{1}{2}$, and the probability of rolling a 1 *or* a 2 with a balanced die is $\frac{2}{6}$. Note that the special rule for equiprobable events applies also, say, when each bank in a chain has the same chance of being selected for an audit, when each salesman has the same chance of being selected for promotion to sales manager, when each electric light bulb coming off a production line has the same chance of being chosen for inspection, or when each taxicab in a fleet of taxicabs has the same chance of being involved in an accident.

To develop an addition rule which applies also to non-mutually exclusive events, let us take a brief look at a survey which shows that 39 percent of all shoppers in a certain department store visit the furniture department and 18 percent visit the sporting goods department. We cannot conclude from this that $39 + 18 = 57$ percent visit either department—the two events are not mutually exclusive—and to find the correct answer let us refer to the Venn diagram of Figure 4.3, where we indicated also that 10 percent of all shoppers in the department store visit both departments. We can now argue that in addition to the 10 percent who visit both departments, $39 - 10 = 29$ percent visit the furniture department but not the sporting goods department, $18 - 10 = 8$ percent visit the sporting goods department but not the furniture department, so that $10 + 29 + 8 = 47$ percent visit either department.

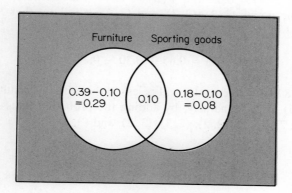

**FIGURE 4.3** Venn diagram.

This result could also have been obtained by *subtracting* from the original total of $39 + 18 = 57$ percent the 10 percent which we *inadvertently included twice*—once among the shoppers who visit the furniture department and once among the shoppers who visit the sporting goods department. In fact, if we translate the percentages into proportions and, hence, probabilities, and if we let $F$ and $G$ denote the respective events that a shopper in the given department store will visit the furniture department and the sporting goods department, we can write

$$P(F \cup G) = P(F) + P(G) - P(F \cap G)$$
$$= 0.39 + 0.18 - 0.10$$
$$= 0.47$$

Since this argument holds for any two events $A$ and $B$, we can now state the following addition rule, called the **General Addition Rule** for probabilities, which applies regardless of whether $A$ and $B$ are mutually exclusive:

$$P(A \cup B) = P(A) + P(B) - P(A \cap B)$$

Note that when $A$ and $B$ *are* mutually exclusive, then $P(A \cap B) = 0$ since $A$ and $B$ have no outcomes in common, and the General Addition Rule reduces to the Special Addition Rule on page 95. To illustrate the new formula, let us refer again to the sample space of Figure 4.2 on page 98. If $A$ is the event that at least one of the two manufacturers will decide to ship by truck and $B$ is the event that neither will decide to ship by rail, we already know from page 98 that $P(A) = 0.65$ and $P(B) = 0.30$. Furthermore, since $A \cap B$ consists of the points $(2, 2)$, $(2, 3)$, and $(3, 2)$, it follows that $P(A \cap B) = 0.15 + 0.03 + 0.10 = 0.28$, and, hence, that

$$P(A \cup B) = 0.65 + 0.30 - 0.28 = 0.67$$

In Exercise 18 below, the reader will be asked to verify this result by calculating $P(A \cup B)$ directly, namely, by adding the probabilities assigned to the six points $(1, 2)$, $(2, 1)$, $(2, 2)$, $(2, 3)$, $(3, 2)$, and $(3, 3)$.

## EXERCISES

1. Explain why there must be a mistake in each of the following statements:
    (a) The probability that a certain stock will go up in price is 0.42 and the probability that it will go down in price is 0.68.
    (b) The probability that a fish caught in Canyon Lake will be below the legal limit is 0.38 and the probability that it will be at or over the legal limit is 0.52.

(c) The probability that an absent-minded professor will forget his umbrella is 0.15 and the probability that he will forget his umbrella and also his hat is 0.25.
(d) The probability that exactly five contractors will bid for a certain job is 0.26 and the probability that at least five contractors will bid for the job is 0.22.

2. Given the *mutually exclusive* events $A$ and $B$ for which $P(A) = 0.18$ and $P(B) = 0.67$, find (a) $P(A')$, (b) $P(B')$, (c) $P(A \cap B)$, (d) $P(A \cup B)$, and (e) $P(A' \cap B')$. (*Hint:* Draw a Venn diagram and fill in the probabilities associated with the various regions.)

3. Given the *mutually exclusive* events $C$ and $D$ for which $P(C) = 0.27$ and $P(D) = 0.48$, find (a) $P(C')$, (b) $P(D')$, (c) $P(C \cap D)$, (d) $P(C \cup D)$, and (e) $P(C' \cap D')$. (*Hint:* Draw a Venn diagram and fill in the probabilities associated with the various regions.)

4. Asked about his chances of earning commissions of $10,000 for the year or commissions of $20,000, a salesman replies that the odds are 5 to 1 *against* his earning commissions of $10,000 and 2 to 1 *against* his earning commissions of $20,000; furthermore, he feels that it is an even bet (the odds are 1 to 1) that he will earn commissions of either $10,000 or $20,000. Discuss the *consistency* of the corresponding probabilities.

5. Asked about the chances that business conditions will improve or remain the same, an economist replies that the odds are 2 to 1 that business conditions will improve and 3 to 1 that they will *not* remain the same; furthermore, he feels that the odds are 5 to 1 that either business conditions will improve or remain the same. Discuss the *consistency* of the corresponding probabilities.

6. Sometimes the cook in a certain restaurant puts too much salt in the soup, sometimes she puts in too little, sometimes she forgets to put any salt in the soup at all, and sometimes she puts in exactly the right amount. If the corresponding probabilities are 0.22, 0.08, 0.15, and 0.55, find the probabilities that
    (a) the cook will put too much salt or too little salt in the soup;
    (b) the cook will put too little salt, no salt at all, or exactly the right amount of salt in the soup;
    (c) the cook will put exactly the right amount or no salt at all in the soup.

7. The probabilities that a student will get an $A$, $B$, $C$, $D$, or $F$ in a course in accounting are, respectively, 0.06, 0.22, 0.44, 0.22, and 0.06.
    (a) What is the probability that he will receive at least a $B$?
    (b) What is the probability that he will receive at most a $C$?

(c) What is the probability that he will receive a *B* or a *C*?
(d) What is the probability that he will receive neither an *A* nor an *F*?

8. Referring to the probabilities of Figure 4.2 on page 98 and the events *T*, *U*, *V*, and *W* described in Figure 3.3 on page 62, evaluate each of the following probabilities and express *in words* what they represent:
  (a) $P(T)$;  (c) $P(V)$;  (e) $P(T')$;  (g) $P(T \cup U)$;
  (b) $P(U)$;  (d) $P(W)$;  (f) $P(V')$;  (h) $P(T' \cup U')$.

9. Suppose that in Exercise 4 on page 66 the points $(1, 1)$, $(2, 1)$, $(3, 1)$, $(2, 2)$, $(3, 2)$, and $(3, 3)$ of the sample space are assigned the probabilities 0.08, 0.21, 0.13, 0.14, 0.38, and 0.06.
  (a) What are the respective probabilities that one of the gasoline service stations will have 1, 2, or 3 gasoline pumps?
  (b) Referring to part (b) of that exercise, what are the probabilities of events *E*, *F*, and *G*? Also state *in words* what each of these probabilities represents.

10. Referring to Exercise 5 on page 67, suppose that each of the 12 points of the sample space has the same probability of $\frac{1}{12}$.
  (a) What are the respective probabilities of finding 0, 1, 2 or 3 stock boys staffing the market?
  (b) What are the respective probabilities of finding 1, 2, or 3 check-out clerks staffing the market?
  (c) Referring to part (b) of that exercise, what are the probabilities of events *A*, *B*, *C*, and *D*? Also state *in words* what each of these probabilities represents.

11. If each of the 52 cards of an ordinary deck of playing cards has a probability of $\frac{1}{52}$ of being drawn, what is the probability of getting (a) a black ace, (b) a queen, king, or ace, (c) a card which is neither a queen nor a king, (d) a spade or a club, and (e) an ace, 2, or 3 of any suit?

12. When we roll a balanced die, what is the probability of getting (a) a 3 or a 4, (b) an odd number, and (c) less than 5?

13. If H stands for *head* and T for *tail*, the eight possible outcomes for three flips of a coin are HHH, HHT, HTH, HTT, THH, THT, TTH, and TTT. Assuming that these eight possibilities are all equally likely, what are the respective probabilities of getting 0, 1, 2, or 3 heads in three flips of a (balanced) coin?

14. Given two events *A* and *B* for which $P(A) = 0.58$, $P(B) = 0.32$, and $P(A \cap B) = 0.12$, find (a) $P(A')$, (b) $P(B')$, (c) $P(A \cup B)$, (d) $P(A' \cap B)$, (e) $P(A' \cup B)$, and (f) $P(A' \cap B')$. (*Hint:* Draw a Venn diagram and fill in the probabilities associated with the various regions.)

15. The probability that a person patronizing a given restaurant will order a hamburger is 0.25, the probability that he will order a cup of coffee is 0.40, and the probability that he will order a hamburger as well as a cup of coffee is 0.20. What is the probability that a person patronizing this restaurant will order a hamburger, a cup of coffee, or both?

16. For a typist and a file clerk working in the same office, the probability that the typist will get to work late is 0.09, the probability that the file clerk will get to work late is 0.16, and the probability that they will both get to work late is 0.05. What is the probability that at least one of them will get to work late?

17. Among 60 corporation presidents in a certain industry, 39 hold college degrees, 33 of the whole group devote at least part of their time to sales, and 27 of the 39 with college degrees devote at least part of their time to sales. If one of these corporation presidents is chosen by lot to serve on an administrative committee of their trade association (that is, each of the presidents has a probability of $\frac{1}{60}$ of being selected), what is the probability that the one who is chosen does *not* hold a college degree and does *not* devote any of his time to sales?

18. As suggested on page 100, verify the result $P(A \cup B) = 0.67$ by directly adding the probabilities assigned to the six points (1, 2), (2, 1), (2, 2), (2, 3), (3, 2), and (3, 3) in Figure 4.2 on page 98.

19. Prove the formula $P(A') = 1 - P(A)$ for any event $A$ by making use of the fact that $A$ and $A'$ are *by definition* mutually exclusive and $A \cup A'$ is certain to occur.

## CONDITIONAL PROBABILITIES

It can be quite confusing to speak of the probability of an event without specifying the sample space with which we happen to be concerned. For instance, if we ask for the probability that a lawyer makes more than $25,000 a year, we may get many different answers and they could all be correct. One of the answers might apply to all lawyers who live in a certain large city, another might apply to all lawyers who are employed by business corporations, a third might apply to all lawyers who are engaged in private practice, and so forth. Since the choice of the sample space (namely, the set of all possibilities under consideration) is rarely self-evident, it helps to use the symbol $P(A|S)$ which denotes the **conditional probability** of event $A$ relative to the sample space $S$. In other words, the symbol $P(A|S)$ indicates not only that we are concerned with event $A$, but also that we are referring to the specific sample space $S$. This makes it preferable to the abbreviated symbol $P(A)$ unless the tacit choice of $S$ is clearly understood; the more explicit notation $P(A|S)$ is also of special value *when we are dealing with more than one sample space in one and the same problem.* (For instance, we may be interested in all lawyers who are employed by business corporations and *also* in all lawyers who live in California in one and the same statistical investigation.)

To illustrate some of the ideas related to conditional probabilities, let us consider the following situation faced by the proprietor of an accounting firm: There are 20 applicants for an accounting position who all meet required standards, but since the income-tax season is about to start there is no time to screen them any further. Thus, we shall assume that each of the applicants has the probability $\frac{1}{20}$ of getting the job. Among the applicants, some are Certified Public Accountants, some have at least five years of accounting experience, with the actual breakdown being as follows:

|  | C.P.A. | Not a C.P.A. |
|---|---|---|
| With Less Than 5 Years Accounting Experience | 2 | 8 |
| With At Least 5 Years Accounting Experience | 4 | 6 |

Well then, if we let $E$ denote the selection of an applicant with at least five years of accounting experience and $C$ the selection of an applicant who is a Certified Public Accountant, it follows from the above table (and the rule for equiprobable outcomes on page 98) that

$$P(E) = \frac{4+6}{20} = \frac{1}{2} \quad \text{and} \quad P(C) = \frac{2+4}{20} = \frac{3}{10}$$

Also, the probability that the applicant who is selected will have at least five years of accounting experience *and* be a Certified Public Accountant is

$$P(E \cap C) = \tfrac{4}{20} = \tfrac{1}{5}$$

Suppose now that the proprietor of the accounting firm decides to consider only those applicants who are Certified Public Accountants. This means that the number of applicants (the number of possibilities, or the *size* of the sample space) is reduced to six, and if we assume that each of these applicants has an equal chance, we now have

$$P(E|C) = \tfrac{4}{6} = \tfrac{2}{3}$$

This is the conditional probability that the applicant who is selected will have had at least five years of accounting experience *given that he is a Certified Public Accountant*. Note that $P(E|C)$ is greater than $P(E)$, which simply expresses the fact that there is a greater proportion of applicants with at least five years of accounting experience among those who are Certified Public Accountants.

The result which we obtained for $P(E|C)$ is actually the ratio of the number of applicants who have at least five years of accounting experience and are Certified Public Accountants to the number of applicants who are Certified Public Accountants. Replacing these numbers by the corresponding proportions, that is, dividing both figures by 20, we could actually have written

$$P(E|C) = \frac{\frac{4}{20}}{\frac{6}{20}} = \frac{P(E \cap C)}{P(C)}$$

and, generalizing from this example, we can make the following definition of a conditional probability which applies to any two events $A$ and $B$:

If $P(B)$ is not equal to zero, then the conditional probability of $A$ relative to $B$ is given by

$$P(A|B) = \frac{P(A \cap B)}{P(B)}$$

Although we used equiprobable outcomes in our example, this definition is not limited to situations involving equally likely possibilities—the formula for $P(A|B)$ is quite general and it applies regardless of what probabilities we assign to the individual outcomes that are included in $A$ and $B$. The only restriction is that $P(B)$ must not equal zero, for division by zero is never permissible.

Returning to our numerical example, we could also argue that if the selection were limited to applicants with at least five years of accounting experience, the probability that the applicant who is selected will be a Certified Public Accountant *given that he has had at least five years of accounting experience* would be

$$P(C|E) = \frac{P(C \cap E)}{P(E)} = \frac{\frac{4}{20}}{\frac{10}{20}} = \frac{2}{5}$$

Of course, this result could have been obtained directly (and more easily) by observing that four of the ten applicants with at least five years of accounting experience are Certified Public Accountants.

To consider an example in which the outcomes are *not all equiprobable*, let us refer again to the sample space of Figure 4.2 on page 98. Proceeding as before and letting $Y$ and $Z$ denote the respective events that both the clock manufacturer and the typewriter manufacturer decide to ship by rail, we find that

$$P(Y) = 0.20 + 0.25 + 0.05 = 0.50$$

$$P(Z) = 0.20 + 0.12 + 0.08 = 0.40$$

and

$$P(Y \cap Z) = 0.20$$

Thus, we get

$$P(Y|Z) = \frac{P(Y \cap Z)}{P(Z)} = \frac{0.20}{0.40} = 0.50$$

for the probability that the clock manufacturer will decide to ship by rail *given that the typewriter manufacturer decides to ship by rail*. This serves to illustrate the use of the formula for conditional probabilities when the outcomes are not all equiprobable.

Note also that $P(Y|Z) = 0.50 = P(Y)$ in the preceding example, which means that the probability of the clock manufacturer deciding to ship by rail is the same regardless of whether or not the typewriter manufacturer will ship by rail, and we say that event $Y$ is **independent** of event $Z$. *Intuitively speaking, we say that one event is independent of another if the occurrence of the first is in no way affected by the occurrence or non-occurrence of the other.*

Had we tried to calculate $P(Z|Y)$ in our example, namely, the probability that the typewriter manufacturer will decide to ship by rail *given that the clock manufacturer is going to ship by rail*, we would have obtained

$$P(Z|Y) = \frac{P(Z \cap Y)}{P(Y)} = \frac{0.20}{0.50} = 0.40$$

which shows that event $Z$ is also independent of event $Y$ since $P(Z|Y) = P(Z)$. As it can be shown in general that *if one event is independent of a second then the second is also independent of the first*, we simply say that **the two events are independent.** Note that in the illustration dealing with the 20 applicants for the accounting position the two events $E$ and $C$ are *not* independent; we had $P(E) = \frac{1}{2}$ and $P(E|C) = \frac{2}{3}$, so that the two events are **dependent.**

So far we have used the formula

$$P(A|B) = \frac{P(A \cap B)}{P(B)}$$

only to calculate conditional probabilities, and this was, of course, the purpose for which it was introduced. However, if we multiply the expressions on both sides of the equation by $P(B)$, we get

$$P(A \cap B) = P(B) \cdot P(A|B)$$

and we now have a formula, called the **General Multiplication Rule**, for determining the probability that two events will *both occur*. In words, the formula states that

> The probability that two events will both occur is the product of the probability that one of the events will occur and the conditional probability that the other event will occur given that the first event has occurred (occurs, or will occur).

As it does not matter which of the two events we refer to as $A$ and which we refer to as $B$, the above multiplication rule can also be written as

$$P(A \cap B) = P(A) \cdot P(B|A)$$

To illustrate the use of these formulas, suppose that we randomly select two ball point pens, one after the other, from a carton containing 12 ball point pens of which three are defective. *What is the probability that both of the ball point pens we pick are defective?* If we assume equal probabilities for each choice (which is, in fact, what we mean by "randomly selecting the pens"), the probability that the first pen we pick is defective is $\frac{3}{12}$, and the probability that the second pen we pick is defective *given that the first one was defective* is $\frac{2}{11}$. Clearly, there are only two defective pens among the 11 pens which remain after one defective pen has been removed. Hence, the probability of picking two defective pens is

$$\frac{3}{12} \cdot \frac{2}{11} = \frac{1}{22}$$

**108** SOME RULES OF PROBABILITY  CHAP. 4

Using the same kind of argument we find that the probability of getting two *good* pens is

$$\tfrac{9}{12} \cdot \tfrac{8}{11} = \tfrac{6}{11}$$

and it follows, by subtraction, that the probability of getting one good pen and one defective pen is $1 - \tfrac{1}{22} - \tfrac{6}{11} = \tfrac{9}{22}$ (see also Exercise 11 on page 110).

If two events $A$ and $B$ are *independent*, we can substitute $P(A)$ for $P(A|B)$ in the first of the multiplication rules on page 107, or $P(B)$ for $P(B|A)$ in the second, and we obtain the **Special Multiplication Rule**

$$P(A \cap B) = P(A) \cdot P(B)$$

This formula can be used, for example, to find the probability of getting two heads in a row with a balanced coin, or the probability of drawing two aces in a row from an ordinary deck of 52 playing cards *provided the first card is replaced before the second is drawn*. For the two flips of the coin we get $\tfrac{1}{2} \cdot \tfrac{1}{2} = \tfrac{1}{4}$ and for the two aces we get $\tfrac{4}{52} \cdot \tfrac{4}{52} = \tfrac{1}{169}$. (Note, however, that *if the first card is not replaced*, the probability for the two aces would be $\tfrac{4}{52} \cdot \tfrac{3}{51} = \tfrac{1}{221}$, for there are only three aces among the 51 cards which remain after one ace has been removed from the deck.) Similarly, if the probability that a typical college student will receive money from his parents in today's mail is 0.03, then the probability that two (totally unrelated) college students in a dormitory will both receive money from their parents in today's mail is $(0.03)(0.03) = 0.0009$; and the probability that neither of the students will receive money from his parents in today's mail is $(1 - 0.03)(1 - 0.03) = 0.9409$.

The special multiplication can easily be generalized so that it applies to the occurrence of more than two independent events—*we simply multiply all of the respective probabilities*. For instance, the probability of getting three *heads* in a row with a balanced coin is $\tfrac{1}{2} \cdot \tfrac{1}{2} \cdot \tfrac{1}{2} = \tfrac{1}{8}$, and the probability of first rolling four *ones* and then another number in five rolls of a balanced die is $\tfrac{1}{6} \cdot \tfrac{1}{6} \cdot \tfrac{1}{6} \cdot \tfrac{1}{6} \cdot \tfrac{5}{6} = \tfrac{5}{7,776}$. For dependent events, the formula becomes somewhat more complicated, as is illustrated in Exercise 18 on page 112.

## EXERCISES

**1.** If $M$ is the event that an assembly worker received a low grade in a manual dexterity test and $A$ is the event that he does not meet his pro-

duction quota, express *in words* what probability is represented by
(a) $P(A|M)$, (b) $P(M|A)$, (c) $P(A|M')$, (d) $P(M|A')$, and (e) $P(M'|A')$.

2. If $Y$ is the event that a bond has a low yield (effective interest rate) and $R$ is the event that it has a high rating (high rating means low risk), express each of the following probabilities in symbolic form:
   (a) the probability that a bond with a low yield will have a high rating;
   (b) the probability that a bond with a high rating will have a low yield;
   (c) the probability that a bond which does not have a low yield will have a high rating;
   (d) the probability that a bond which does not have a high rating will not have a low yield.

3. If $E$ is the event that an applicant for a sales position has had prior sales experience, $A$ is the event that he owns an automobile, and $C$ is the event that he has a college degree, state *in words* what probability is represented by (a) $P(C|E)$, (b) $P(A|C)$, (c) $P(C|E')$, (d) $P(E|A')$, (e) $P(A'|C')$, (f) $P(A \cap C|E)$, and (g) $P(C|E \cap A)$.

4. Referring to Exercise 3, express each of the following probabilities in symbolic form:
   (a) the probability that one of the applicants who owns an automobile will also have a college degree;
   (b) the probability that one of the applicants without a college degree will have prior sales experience;
   (c) the probability that one of the applicants who has neither a college degree nor an automobile will have prior sales experience;
   (d) the probability that one of the applicants without prior sales experience will have neither a college degree nor an automobile.

5. A consumer research organization studied the level of customer service provided by gasoline stations carrying advertised brands of gasoline and those carrying only unadvertised brands. In its report it summarized its findings about 46 gasoline stations carrying only advertised brands and 34 gasoline stations carrying only unadvertised brands by means of the following table:

|  | Low Level of Service | Moderate-to-High Level of Service |
|---|---|---|
| Advertised Brands | 18 | 28 |
| Unadvertised Brands | 22 | 12 |

If a motorist drives into one of these gasoline stations at random (that is, each of these stations has a probability of $\frac{1}{80}$ of being selected), $L$ and $L'$ denote the selection of a station with a low level of service or with a moderate-to-high level of service, while $A$ and $A'$ denote the selection of a station which carries only advertised brands or only unadvertised brands, determine the following probabilities *directly* from the table: (a) $P(A)$, (b) $P(A')$, (c) $P(L)$, (d) $P(L')$, (e) $P(A \cap L)$, (f) $P(A' \cap L')$, (g) $P(A|L)$, (h) $P(L|A)$, (i) $P(A'|L')$, and (j) $P(L'|A')$.

6. Use the results of Exercise 5 to verify that

(a) $P(A|L) = \dfrac{P(A \cap L)}{P(L)}$;  (c) $P(L|A) = \dfrac{P(A \cap L)}{P(A)}$;

(b) $P(A'|L') = \dfrac{P(A' \cap L')}{P(L')}$;  (d) $P(L'|A') = \dfrac{P(A' \cap L')}{P(A')}$.

7. Use the probabilities given in Exercise 14 on page 102 and some of the results obtained in that exercise to determine $P(A'|B)$ and $P(A'|B')$.

8. Referring to Exercise 15 on page 103, find (a) the probability that a person patronizing the restaurant who orders a hamburger will also order a cup of coffee, and (b) the probability that a person patronizing the restaurant who orders a cup of coffee will also order a hamburger.

9. For corpulent couples living in a certain suburb, the probability that the husband will go on a weight-reducing diet is 0.13, the probability that his wife will go on a weight-reducing diet is 0.17, and the probability that they will both go on such a diet is 0.10. Find (a) the probability that the wife will go on a diet given that her husband is going on a diet, and (b) the probability that the husband will go on a diet given that his wife is going on a diet.

10. The proprietor of a shoe store in a certain town has the following information: the probability that it will snow prior to New Year's Day is 0.70, and the probability that the store will sell a large number of overshoes if it snows before New Year's Day is 0.80. What is the probability that it will snow prior to New Year's Day *and* that the store will sell a large number of overshoes?

11. On page 108 we showed that the probability of getting one good ball point pen and one defective ball point pen is $\frac{9}{22}$ by subtracting from 1 the respective probabilities of getting two good pens or two defective pens. Verify this result by *adding* the probabilities of the mutually exclusive alternatives of drawing the good pen first or drawing the defective pen first.

CHAP. 4    CONDITIONAL PROBABILITIES    111

12. The foreman of a group of ten construction workers wants to talk with two members randomly selected from among this group about their new safety helmets. If six of these workers like the new helmets while the other four dislike them, what are the probabilities that the foreman will get the opinions of (a) two construction workers who like the new helmets, (b) two construction workers who dislike the new helmets, and (c) one construction worker of each kind? [*Hint:* For part (c) follow the method of Exercise 11.]

13. Among 20 invoices prepared by a billing department, 12 contain errors while the others do not. If we randomly check two of these invoices, what are the probabilities that (a) both invoices will contain errors, (b) neither invoice will contain errors, and (c) one will contain errors while the other one does not?

14. Which of the following pairs of events are independent and which are dependent?
    (a) forgetting to study and failing the examination;
    (b) getting sevens in two successive rolls of a pair of dice;
    (c) being an accountant and having small feet;
    (d) running out of gasoline and being late for work;
    (e) running out of gasoline and wearing a blue sweater;
    (f) having a large income and filing an income tax return;
    (g) living in Arizona and being a stamp collector;
    (h) any two mutually exclusive events.

15. If the probability that a professor will be in his office between 7:45 A.M. and 8:00 A.M. is 0.20 and the probability that a student will visit his office during this time is 0.05, what is the probability that on some morning the professor will be in his office *and* also be visited by a student?

16. Referring to Figure 4.2, we showed on page 98 that the clock manufacturer's decision to ship by rail was independent of the typewriter manufacturer's decision to ship by rail. Check also whether
    (a) the clock manufacturer's decision to ship by truck is independent of the typewriter manufacturer's decision to ship by rail;
    (b) the typewriter manufacturer's decision to ship by rail is independent of the clock manufacturer's decision to ship by air;
    (c) the clock manufacturer's decision to ship by air is independent of the typewriter manufacturer's decision to ship by truck.

17. As we indicated on page 108, the probability that any number of independent events will occur is given by the product of their respective probabilities. Use this rule to find
    (a) the probability of getting three *tails* in a row with a balanced coin;

**112** SOME RULES OF PROBABILITY  CHAP. 4

(b) the probability that a salesman will sell a pair of shoes to three customers in a row, given that the probability of his selling a pair of shoes to any one customer is 0.15;

(c) the probability of drawing (with replacement) four spades in a row from an ordinary deck of 52 playing cards;

(d) the probability that five shipments of raw material will arrive on time, given that the probability of any one shipment arriving on time if 0.90.

**18.** The problem of determining the probability that any number of events will occur becomes more complicated when the events are *not independent*. For three events, for example, which we shall arbitrarily refer to as the first, second, and third, the probability that they will all occur is obtained by multiplying the probability that the first event will occur by the probability that the second event will occur *given that the first has occurred*, and then multiplying by the probability that the third event will occur *given that the first and second have occurred*. For instance, the probability of drawing (without replacement) three aces in a row from an ordinary deck of 52 playing cards is $\frac{4}{52} \cdot \frac{3}{51} \cdot \frac{2}{50} = \frac{1}{5,525}$; clearly, there are only three aces among the 51 cards which remain after the first ace has been drawn, and only two aces among the 50 cards which remain after the first two aces have been drawn.

(a) Referring to the illustration on page 107, find the probability that if three ball point pens are randomly selected, they are all defective.

(b) Referring to Exercise 12, find the probability that if the foreman randomly selects four of the construction workers, they all dislike the new kind of helmet.

(c) Referring to Exercise 13, find the probabilities that if four invoices are randomly selected, they will (i) all contain errors, and (ii) all be without errors.

(d) In a certain Alaskan city during the month of May, the probability that a rainy day will be followed by another rainy day is 0.70 and the probability that a sunny day will be followed by a rainy day is 0.60. Assuming that each day is classified as being either rainy or sunny and that the weather on any given day depends only on the weather the day before, find the probability that in the given city a rainy day in May is followed by two more rainy days, then a sunny day, and finally another rainy day.

**19. MARKOV CHAINS** Part (d) of the preceding exercise dealt with a situation called a *Markov chain*, namely, with a sequence of experiments (interpreted loosely as on page 59) in which the outcome of each experi-

ment depends only on what happened in the preceding experiment. The following are some business applications of Markov chains:
(a) The only supermarket in a small town offers two brands of frozen orange juice, Brand A and Brand B. Among its customers who buy Brand A one week, 80 percent will buy Brand A and 20 percent will buy Brand B the next week, and among its customers who buy Brand B one week, 40 percent will buy Brand B and 60 percent will buy Brand A the next week. To simplify matters, it will be assumed that each customer buys frozen orange juice once a week.
  (i) What is the probability that a customer who buys Brand A one week will buy Brand B the next week, Brand B the week after that, and Brand A the week after that?
  (ii) What is the probability that a customer who buys Brand B one week will buy Brand B the next two weeks, and Brand A the week after that?
  (iii) What is the probability that a customer who buys Brand A in the first week of a month will also buy Brand A in the third week of that month? (*Hint:* Add the probabilities associated with the two mutually exclusive possibilities corresponding to his buying Brand A or Brand B in the second week.)
(b) A department store which bills its charge-account customers once a month has found that if a customer pays promptly one month, the probability is 0.90 that he will also pay promptly the next month; however, if a customer does not pay promptly one month, the probability that he will pay promptly the next month is only 0.50.
  (i) What is the probability that a customer who pays promptly one month will also pay promptly the next three months?
  (ii) What is the probability that a customer who does not pay promptly one month will also not pay promptly the next three months and then make a prompt payment the month after that?
  (iii) What is the probability that a customer who pays promptly one month will also pay promptly the third month after that? (*Hint:* Add the probabilities associated with the *four* mutually exclusive possibilities corresponding to what he did in the first and second months after he made the prompt payment.)
(c) Suppose that if the value of a given stock went up on a given business day, the probabilities that its value will go up, remain the same, or go down on the next business day are, respectively, 0.50, 0.30, and 0.20. Also, if the value of the stock remained the same on a given business day, the probabilities that its value will go up, remain the same, or go down on the next business day are, respectively, 0.20, 0.60, and 0.20; and if the value of the stock went down on a given business day, the probabilities that its value will go up, remain the

same, or go down on the next business day are, respectively, 0.10, 0.40, and 0.50.
  (i) What is the probability that if the value of the stock went up on a given business day, it will remain the same on the next two business days?
  (ii) What is the probability that if the value of the stock went down on a given business day, it will go down the next two business days and then remain the same the next two business days after that?
  (iii) What is the probability that if the value of the stock remained the same on a given business day, it will also remain the same on the second business day after that? (*Hint:* Add the probabilities associated with the *three* mutually exclusive possibilities corresponding to what the stock did on the business day in between.)

**20. MARKOV CHAINS** (*continued*) Suppose that in part (d) of Exercise 18 we had wanted to know the probability that it will rain in the given city on any day in May *regardless of what happened the day before*. Since it was either rainy or sunny on the day before, there are the two mutually exclusive possibilities $RR$ and $SR$, where $RR$ represents the case where the rainy day was preceded by a rainy day and $SR$ represents the case where the rainy day was preceded by a sunny day. If we let $p$ denote the probability that it will rain on any one day regardless of what happened the day before, we find that the probability for $RR$ is $p$ times 0.70 while that for $SR$ is $1 - p$ times 0.60. Since the *sum* of these two probabilities must equal $p$, the probability that it will rain on any given day, we get

$$0.70p + 0.60(1 - p) = p$$

which yields $0.70p + 0.60 - 0.60p = p$, $0.60 = p - 0.70p + 0.60p$, $0.60 = 0.90p$, and, finally, $p = \dfrac{0.60}{0.90} = \dfrac{2}{3}$.

  (a) Using this technique in connection with part (a) of Exercise 19, show that if a customer has been shopping at the supermarket for some time, the probability is 0.75 that he will buy Brand A any given week. In other words, it can be expected that Brand A's *share of the market* will be 75 percent.
  (b) Using this technique in connection with part (b) of Exercise 19, show that the probability is $\frac{5}{6}$ that a person who has been charging for some time at the department store will pay promptly. In other words, the department store can expect $\frac{5}{6}$ of the bills to be paid promptly.

## BAYES' RULE

Although the symbols $P(A|B)$ and $P(B|A)$ may seem very much alike, the same cannot be said about the probabilities which they represent. On page 105 we saw that $P(C|E)$ was the probability that the proprietor of the accounting firm will select an applicant who is a Certified Public Accountant *given that he must have had at least five years accounting experience*, while $P(E|C)$ was the probability that the proprietor will select an applicant with at least five years of accounting experience *given that he must be a Certified Public Accountant*. Similarly, if $R$ represents the event that a person committed a certain robbery and $G$ is the event that he is judged guilty, then $P(G|R)$ is the probability that the person will be judged guilty *given that he actually committed the robbery*, and $P(R|G)$ is the probability that the person actually committed the robbery *given that he is judged guilty*.

Since there are many situations in which we must consider such pairs of probabilities, let us derive a formula which expresses $P(A|B)$ in terms of $P(B|A)$, and vice versa. All we really have to do is combine the formula for $P(A|B)$ on page 105 with the second form of the General Multiplication Rule on page 107, namely, $P(A|B) = \dfrac{P(A \cap B)}{P(B)}$ and $P(A \cap B) = P(A) \cdot P(B|A)$. Substituting $P(A) \cdot P(B|A)$ for $P(A \cap B)$ into the formula for $P(A|B)$, we thus get

$$P(A|B) = \frac{P(A) \cdot P(B|A)}{P(B)}$$

which is a very simple (special) form of a rule called the **Rule of Bayes**. To illustrate its use, suppose that the manager of a business firm is advised by a credit investigator that one of his customers is a bad risk. Knowing that credit checks are not infallible, the manager of the business firm will have to judge whether the customer actually is a bad risk—in other words, he will have to determine the probability $P(B|R)$, where $B$ denotes the event that the customer is a bad risk, while $R$ denotes the event that the customer is rated a bad risk. In order to calculate this probability, the manager of the business firm will have to know the values of the probabilities $P(R|B)$, $P(B)$, and $P(R)$, and, for the sake of argument, let us suppose that the values of these probabilities are

$$P(R|B) = 0.96, \quad P(B) = 0.05, \quad P(R) = 0.08$$

This means that 96 percent of the customers who are bad risks would be rated bad risks by the credit investigator; 5 percent of the customers whose credit is checked by the investigator actually are bad risks; and 8 percent of the customers whose credit is thus checked are rated bad risks. Now, if we substitute all these probabilities into the formula, we get

$$P(B|R) = \frac{P(B) \cdot P(R|B)}{P(R)} = \frac{(0.05)(0.96)}{0.08} = 0.60$$

for the probability that a customer who is rated a bad risk actually *is* a bad risk. Since this value is fairly low, the manager of the business firm would probably be well advised to conduct further inquiries before refusing to grant credit to the customer. [In Exercise 6 on page 119 the reader will be asked to continue this problem and show that almost 3.4 percent of the customers who are *not* bad risks are rated bad risks by the investigator, namely, that $P(R|B') = 0.034$.]

Even though the rule we used here was easy to derive and there is no question about its *validity*, extensive criticism has always surrounded its application. This is due largely to the fact that the rule involves an "inverse" sort of reasoning, namely, reasoning *from effect to cause*. In our numerical example we used the formula to calculate the probability that a customer who is rated a bad risk actually is a bad risk, and earlier we suggested that we might be interested in the probability $P(R|G)$ that a person who is judged guilty of a robbery actually did commit the crime. All we can really say is that *the formula must be used with care* and that *the significance of the probability which we calculate as well as the probabilities whose values we substitute must be clearly understood*.

In most practical applications there are more than two possible "causes" and the formula for Bayes' Rule must be written in a slightly different (that is, more general) form. If $A_1$, $A_2$, ..., and $A_k$ denote the $k$ mutually exclusive "causes" which could have led to the event $B$, then the probability that it was $A_i$ (for $i = 1, 2, \ldots,$ or $k$) is given by

$$P(A_i|B) = \frac{P(A_i) \cdot P(B|A_i)}{P(A_1) \cdot P(B|A_1) + P(A_2) \cdot P(B|A_2) + \cdots + P(A_k) \cdot P(B|A_k)}$$

Referring to Figure 4.4, we might say that $P(A_i|B)$ is the probability that event $B$ is reached along the $i$th branch of the "tree" *given that the event must be reached along one of the $k$ branches of the tree*. Note, furthermore, that the value of $P(A_i|B)$ is given by the *ratio* of the probability associated with the $i$th branch to the *sum* of the probabilities associated with all of the $k$ branches of the "tree."

To illustrate this more general form of Bayes' rule, suppose that an automobile manufacturing company assembles automobiles in three different

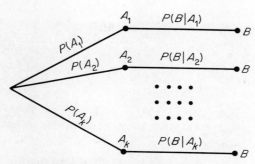

**FIGURE 4.4** Tree diagram for Bayes' rule.

plants, with 30 percent of its output coming from its eastern plant, $A_1$, 49 percent from its Detroit plant, $A_2$, and 21 percent from its western plant, $A_3$. If the probabilities that the three plants assemble an automobile defectively are, respectively, 0.005, 0.030, and 0.020, we would like to know the respective odds that a defectively-assembled car which was returned by a customer came from the eastern plant, from Detroit, or from the western plant. If $B$ is the event that one of the automobiles turns out to be defective, we thus have $P(B|A_1) = 0.005$, $P(B|A_2) = 0.030$, $P(B|A_3) = 0.020$, and if we substitute these values together with $P(A_1) = 0.30$, $P(A_2) = 0.49$, and $P(A_3) = 0.21$ into the formula for the rule of Bayes, we get

$$P(A_1|B) = \frac{(0.30)(0.005)}{(0.30)(0.005) + (0.49)(0.030) + (0.21)(0.020)}$$
$$= 0.074$$

for the probability that the defective car came from the eastern plant. Correspondingly the odds are about 13 to 1 (0.926 to 0.074 to be more exact) that the defective automobile did *not* come from the eastern plant. Picturing this problem as in Figure 4.5, it can be seen that the probabilities associated with the three branches are $(0.30)(0.005) = 0.0015$, $(0.49)(0.030) = 0.0147$, and $(0.21)(0.020) = 0.0042$. The sum of these three probabilities is 0.0204, so that we get

$$P(A_2|B) = \frac{0.0147}{0.0204} = 0.72 \quad \text{and} \quad P(A_3|B) = \frac{0.0042}{0.0204} = 0.206$$

besides $P(A_1|B) = 0.074$. Thus, the odds are about 5 to 2 that the defectively-assembled automobile came from the Detroit plant, and about 4 to 1 that it did *not* come from the western plant.

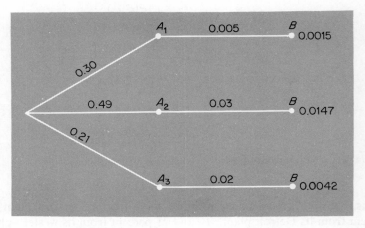

**FIGURE 4.5** Tree diagram for Bayes' rule.

## EXERCISES

1. The supervisor of an electronics plant knows from past experience that a new worker who has attended the company's training program has a probability of 0.80 of meeting his production quota, while a new worker who has not attended the training program has a probability of only 0.20. If 60 percent of all new workers attend the training program, what is the probability that a new worker who meets the production quota attended the training program?

2. There is a fifty-fifty chance that Harry will get the loan for which he applied to save his business. If he gets the loan, the odds are 2 to 1 that he will be able to save his business, but if he does not get the loan the odds are 4 to 1 that he will *not* be able to save his business. If we hear later on that Harry was able to save his business, what is the probability that he got the loan?

3. A study made by a management consultant shows that among new businesses having difficulties, 75 percent have difficulties mainly because they are undercapitalized and 25 percent have difficulties mainly because of poor management. If the study also shows that in the first case the odds are 2 to 1 that such a business will fail, while in the second case the odds are 4 to 1, what is the probability that a new business with difficulties that failed had its difficulties due mainly to undercapitalization?

4. The probabilities that a producer of a new kind of fruit juice will decide to market the product in frozen, canned, or bottled form are, respectively, 0.40, 0.40, and 0.20. If the producer decides to sell the fruit juice in frozen form, the probability of high sales is 0.25; if he decides to sell the fruit juice in cans, the probability of high sales is 0.30; and if he decides to sell the fruit juice in bottles, the probability of high sales is 0.10. If an executive of the company producing this new fruit juice leaves for a long vacation before a decision is reached and hears indirectly that the new product is enjoying high sales, what odds should he be willing to give that the firm did not choose to bottle the juice?

5. Four attendants employed at a gasoline service station are supposed to wash the windshield of every customer's car. Jack, who services 30 percent of all cars, fails to wash the windshield 1 time in 50; James, who also services 30 percent of all cars, fails to wash the windshield 1 time in 40; Jerry, who services 20 percent of all cars, fails to wash the windshield 1 time in 10; and George, who also services 20 percent of the cars, fails to wash the windshield 1 time in 20. If a customer complains later on that his windshield was not washed, what are the probabilities that
   (a) his car was serviced by Jack;
   (b) his car was serviced by Jerry or George?

6. Use the probabilities given in the illustration on page 115, as well as the result, to show that almost 3.4 percent of the customers who are *not* bad risks are rated bad risks by the investigator, namely, that $P(R|B')$ is about 0.034.

# 5

# CHANCE VARIATION: PROBABILITY FUNCTIONS

**INTRODUCTION**

We might well have called this chapter "How to Live with Uncertainties," for this is undoubtedly the most important lesson to be learned from modern statistics. Uncertainties face us wherever we turn—we cannot be certain that an experiment will succeed, we cannot be certain that the weather will stay nice, we cannot be certain that the milk we buy in a sealed carton is fresh, we cannot be certain that a rocket will get off the ground, we cannot be certain that a marriage will last, we cannot be certain that an investment will pay off, and so on. Actually, life without uncertainties would probably be quite dull and most uncertainties cannot be eliminated, but we shall see that **chance fluctuations** (namely, changes, variations, or differences "caused" by chance) are quite often *predictable*, that they can be *measured*, and that sometimes they can even be *controlled* to some extent.

If the values which a quantity takes on are somehow dependent on chance, we refer to it as a **random variable**, and this definition makes the term applicable to a great variety of phenomena: It applies to the price of a share of A.T. & T.; it applies to the moisture content of the air, the wind velocity at Kennedy Airport, and the annual production of corn; it applies to the number of eggs a chicken will lay each week or the number of mistakes a secretary will make in typing a report; and it also applies to the size of an audience at a football game or the number of leaves that are left on a tree after a mild frost. Needless to say, perhaps, the fluctuations of such random variables are *not* predictable in the same way in which a fortune teller predicts events—all we can do is determine the probabilities that random variables will take on specific values, the probabilities that their values will fall on given intervals, or the probabilities that their values will be less than (or greater than) given numbers. For instance, we *cannot* say for sure whether we will get two *heads* and two *tails* in four flips of a balanced coin, but we *can* determine the probabilities of getting zero *heads* and four *tails*, one *head* and three *tails*, two *heads* and two *tails*, and so forth. Similarly, we *cannot* say for sure whether a new tire we buy for our car will last 15,000 miles, but we *can* quote probabilities that without unduly rough treatment it will last at most 15,000 miles, anywhere from 12,000 to 18,000 miles, at least 10,000 miles, and so on. Finally, we *cannot* predict for sure whether a newly-married couple will eventually have two children, but we can determine the probabilities that they will have 0, 1, 2, 3, 4, or more.

The problem of determining the probabilities which are associated with the different values of a random variable will be the subject matter of the next four sections. Then we shall turn to the problem of constructing a "yardstick" which will enable us to *measure* the size of chance fluctuations, and in the last section we shall see how this "yardstick" can actually be used to *predict* and to some extent *control* chance variations to which we are exposed.

## PROBABILITY FUNCTIONS

The table which follows pertains to one of the many random variables mentioned in the preceding section, namely, the number of *heads* we might get in four flips of a balanced coin. Its purpose is to illustrate what we mean by a **probability function**, namely, *a correspondence which assigns probabilities to the values of a random variable.* We calculated the five probabilities shown in the right-hand column of the table by considering as equally likely the sixteen possibilities HHHH, HHHT, HHTH, HHTT, HTHH, HTHT, HTTH, HTTT, THHH, THHT, THTH, THTT, TTHH, TTHT, TTTH, and TTTT, where H stands for *head* and T for *tail*. Then we counted the number of H's in each case, and obtained

| Number of Heads | Probability |
|---|---|
| 0 | $\frac{1}{16}$ |
| 1 | $\frac{4}{16}$ |
| 2 | $\frac{6}{16}$ |
| 3 | $\frac{4}{16}$ |
| 4 | $\frac{1}{16}$ |

Although this example may seem trivial and of interest only to someone addicted to flipping coins, *it can provide answers to questions relating to entirely different matters.* Consider, for example, the owner of an antique shop who has just been notified that a valuable shipment of matched antique cups and saucers has been damaged and that only four pieces remain unbroken. The unbroken cups are valued at $40.00 each, the unbroken saucers are valued at $30.00 each; an unbroken matched set is valued at $90.00. The owner of the shop is unable to obtain further information as to whether the unbroken pieces are cups or saucers, but assuming that each piece is as likely to be a cup as a saucer, he wants to determine the expected value of the undamaged portion of the shipment. Thus, he might modify the above table, with "Number of Cups" substituted for "Number of Heads," getting

| Number of Cups | Probability | Value of Unbroken Items |
|---|---|---|
| 0 | $\frac{1}{16}$ | $120.00 |
| 1 | $\frac{4}{16}$ | $150.00 |
| 2 | $\frac{6}{16}$ | $180.00 |
| 3 | $\frac{4}{16}$ | $170.00 |
| 4 | $\frac{1}{16}$ | $160.00 |

Note that in the right-hand column $4(30.00) = \$120.00$ is the value of 0 cups and 4 saucers; $\$90.00 + 2(30.00) = \$150.00$ is the value of 1 cup and 3 saucers (namely, one matched set and two saucers); $2(90.00) = \$180.00$ is the value of 2 cups and 2 saucers (namely, two matched sets); and so on. Using the formula for a *mathematical expectation* on page 86, he would thus find that the expected value of the unbroken items is

$$120.00(\tfrac{1}{16}) + 150.00(\tfrac{4}{16}) + 180.00(\tfrac{6}{16}) + 170.00(\tfrac{4}{16}) + 160.00(\tfrac{1}{16}) = \$165.00$$

Whenever possible, we try to express probability functions by means of mathematical formulas (equations) which enable us to calculate the probabilities associated with the various values of a random variable. For the number of *heads* obtained in four flips of a balanced coin, we could have used the formula

$$f(x) = \frac{\tfrac{3}{2}}{x!(4-x)!} \quad \text{for } x = 0, 1, 2, 3, \text{ or } 4$$

where the *factorials* $x!$ and $(4-x)!$ are as defined on page 74, and $0!$ is taken to equal 1 as in Exercise 24 on page 80. For instance, for $x = 1$ and $x = 2$ we get

$$f(1) = \frac{\tfrac{3}{2}}{1!3!} = \frac{\tfrac{3}{2}}{1 \cdot 6} = \frac{1}{4} \quad \text{and} \quad f(2) = \frac{\tfrac{3}{2}}{2!2!} = \frac{\tfrac{3}{2}}{2 \cdot 2} = \frac{3}{8}$$

which agrees with the corresponding values of $\tfrac{4}{16}$ and $\tfrac{6}{16}$ shown in the table on page 121. In case the reader is curious to know *how we got the formula for this* probability function, let us merely point out at this time that it is of the special kind which we shall study in the next section.

To conclude this preliminary discussion of probability functions, let us point out the following general rules which the values of any probability function must obey:

> Since the values of probability functions are probabilities, they must always be numbers between 0 and 1, inclusive; also, since a random variable has to take on one of its values (for instance, we have to get either 0, 1, 2, 3, or 4 heads in four flips of a coin), the sum of all the values of a probability function must always be equal to 1.

Let us also point out that although some statisticians make a rather fine distinction between **probability functions** and **probability distributions,** we shall use these terms interchangeably in this book; indeed, a probability function tells us how the total probability of 1 is *distributed* among the possible values a random variable can take on.

## THE BINOMIAL DISTRIBUTION

There are many problems in which we are interested in the probability that an event will occur $x$ times out of $n$. For instance, we may be interested in the probability that 320 of 500 television viewers (interviewed by a television rating service) will say that they watched a certain program, the probability that 38 of 200 housewives (interviewed in a survey) will prefer an unadvertised detergent to a nationally advertised brand, the probability that 18 of 20 persons suffering from a certain disease will be cured by a new kind of "miracle" drug, the probability that 8 out of 10 workers will cash their paychecks on payday, and so on. To borrow from the language of games of chance, in each of these examples we are interested in the probability of getting $x$ **successes in $n$ trials,** or in other words, the probability of $x$ **successes** and $n - x$ **failures.** In the problems which we shall study in this section it will always be assumed that:

> There is a fixed number of trials; the probability of a success is the same for each trial; and the trials are, furthermore, all independent.

This means that the theory will *not* apply, for example, if we alternately interview men and women about their views on interior decorating, or if we are interested in the chances that a student will get a passing grade in four consecutive courses in accounting. (In the first case, men and women may well have different ideas about interior decorating, and in the second case the trials are not likely to be independent since the grades which the student will get in the later courses will depend to some extent on the grades which he got in the earlier ones.) Also, the theory would *not* apply if we were interested in the number of hats a woman will try on before she finally buys one—in this case, the number of "trials" would not be fixed.

To handle problems which *do* meet the conditions listed in the preceding paragraph, we use a special kind of probability function called the **binomial probability function,** or simply the **binomial distribution.** It includes the example of the preceding section, and its formula is fairly easily obtained. If $p$ and $1 - p$ are the probabilities for a success and a failure on any given trial, then the probability of getting $x$ successes and $n - x$ failures *in some specific order* is $p^x(1 - p)^{n-x}$; clearly, in this product of $p$'s and $(1 - p)$'s there is *one* factor $p$ for each success, *one* factor $1 - p$ for each failure, and the $x$ factors $p$ and $n - x$ factors $1 - p$ are all multiplied together by virtue of the Special Multiplication Rule which we generalized for more than two independent events on page 108. Since this probability applies to any point of the sample space which represents $x$ successes and $n - x$ failures (namely, to any specific arrangement of $x$ successes and $n - x$ failures), we have only to see how many points of this kind there are, and then multiply their number by $p^x(1 - p)^{n-x}$. Clearly, there are $\binom{n}{x}$, the number of combinations of $x$ elements selected from a set of $n$ elements (namely, the number of ways in which we can select the $x$ trials on which there is to be a success), and we have thus arrived at the result that the probability for $x$ successes in $n$ trials is given by

$$f(x) = \binom{n}{x} p^x (1 - p)^{n-x} \quad \text{for } x = 0, 1, 2, \ldots, \text{ or } n$$

This is the formula for the *binomial distribution,* and to remind the reader, $\binom{n}{x}$ is a binomial coefficient whose value can be looked up in Table VIII, or it can be calculated directly according to the formula on page 75. The reason why we refer to this probability function as the *binomial distribution* is that the values we get for $x = 0, x = 1, x = 2, \ldots,$ and $x = n$ are the successive terms of the *binomial expansion* of $[(1 - p) + p]^n$, with which the reader may be familiar from elementary algebra (see also Exercise 4 on page 130). Incidentally, since

$$[(1 - p) + p]^n = 1^n = 1$$

this demonstrates that the sum of the values of a binomial probability function always equals 1, as it should.

To illustrate the use of the formula we have just derived, let us calculate the probability that three of ten sales prospects will purchase a used car from a certain salesman, when the probability that any such prospect will purchase a used car from him is 0.20. Substituting $x = 3$, $n = 10$, $p = 0.20$, and $\binom{10}{3} = 120$ in accordance with Table VIII, we get

$$f(3) = \binom{10}{3}(0.20)^3(1 - 0.20)^{10-3}$$
$$= 120(0.20)^3(0.80)^7$$
$$= 0.2013$$

rounded to four decimals, or approximately 0.20. Similarly, to find the probability that four of eight welders on a job will be wearing safety shoes when the probability that any one of them will be wearing such shoes is 0.60, we substitute $x = 4$, $n = 8$, $p = 0.60$, and $\binom{8}{4} = 70$, and get

$$f(4) = 70(0.60)^4(0.40)^4 = 0.232$$

rounded to three decimals.

To give an example in which we calculate *all* of the values of a binomial distribution, suppose that a production engineer claims that three in ten substandard workers have physical disabilities (such as poor eyesight or hearing, or generally poor health). Assuming that these figures are correct, let us find the probabilities that among six substandard workers 0, 1, 2, 3, 4, 5, or 6 have physical disabilities. Substituting $n = 6$, $p = 0.30$, $1 - p = 0.70$, and, respectively, $x = 0, 1, 2, 3, 4, 5$, and 6, we get

$$f(0) = \binom{6}{0}(0.30)^0(0.70)^6 = 0.1176$$

$$f(1) = \binom{6}{1}(0.30)^1(0.70)^5 = 0.3025$$

$$f(2) = \binom{6}{2}(0.30)^2(0.70)^4 = 0.3241$$

$$f(3) = \binom{6}{3}(0.30)^3(0.70)^3 = 0.1852$$

$$f(4) = \binom{6}{4}(0.30)^4(0.70)^2 = 0.0595$$

$$f(5) = \binom{6}{5}(0.30)^5(0.70)^1 = 0.0102$$

$$f(6) = \binom{6}{6}(0.30)^6(0.70)^0 = 0.0007$$

Thus, it can be seen that almost 12 percent of the time none of six substandard workers have physical disabilities, about 30 percent of the time one of six substandard workers has physical disabilities, and so on. What do you think we might conclude from this about the production engineer's claim (that three in ten substandard workers have physical disabilities) if a careful examination of six substandard workers shows that as many as five of them have physical disabilities?

Finally, let us calculate the probability that three of fifteen persons will complain about the medicinal taste of a new brand of cough drop, given that the probability is 0.05 that any one of them will complain. Substituting $x = 3$, $n = 15$, and $p = 0.05$ into the formula for the binomial distribution, we get

$$f(3) = \binom{15}{3}(0.05)^3(1 - 0.05)^{15-3}$$

$$= 455(0.05)^3(0.95)^{12}$$

where the value of $\binom{15}{3}$ was obtained from Table VIII. If we performed the necessary calculations (perhaps, with the use of logarithms), we could show that the answer is 0.031 rounded to three decimals, but actually we obtained this result from Table XI, which contains the values of binomial probabilities for selected values of $p$ and for $n = 2, 3, 4, \ldots$, and 15. This table greatly simplifies work with the binomial distribution. For instance, we can immediately read off the result that the probability of getting nine *heads* and six *tails* in 15 flips of a balanced coin is 0.153; also, if a professor misplaces his office key about one day in 20, the probability that he will misplace his key in three of 14 days is 0.026; and if a stock market speculator is successful in 70 percent of his investments, the probability that he will be successful in at least ten out of 12 investments is

$$0.168 + 0.071 + 0.014 = 0.253$$

## THE HYPERGEOMETRIC DISTRIBUTION

To introduce another kind of probability function, let us suppose that an automatic washing machine manufacturing company receives its timers from a subcontractor in lots of 20. When the timers arrive at the manufacturer's warehouse, an inspector randomly selects three timers from each lot and *the whole lot is accepted if all three are in good working condition*; otherwise, each timer is inspected individually (and repaired, if necessary) at a considerable cost. It is evident that this kind of sampling inspection involves certain risks, for it is possible (though highly unlikely) that a lot will be accepted without further inspection even though 17 of the 20 timers are defective. Of course, this is an extreme case; more realistically, perhaps, it may be of interest to know the probability that a lot will be accepted without further inspection even though, say, four of the 20 timers are defective. This means that we would be interested in the probability of three *successes* (non-defective timers) in three *trials* (among the three timers inspected), and we might be tempted to argue that since 16 of the 20 timers are non-defective, the probability of getting a non-defective timer is $\frac{16}{20} = 0.80$, and, the probability of "three successes in three trials" is therefore

$$f(3) = \binom{3}{3}(0.80)^3(1 - 0.80)^0 = 0.512$$

according to the formula for the binomial distribution.

Remembering what we said on page 123, however, we find that what we have done here is not quite legitimate—we have violated the condition that *successive trials must be independent*. Although the probability that the first timer examined by the inspector is not defective is $\frac{16}{20}$, the probability that the second timer is not defective *given that the first one was in good working condition* is $\frac{15}{19}$, and the probability that the third timer is not defective *given that the first two were in good working condition* is $\frac{14}{18}$. Consequently, if we follow the multiplication rule explained in Exercise 18 on page 112, we find that the correct value of the probability (of getting three non-defective timers) is $\frac{16}{20} \cdot \frac{15}{19} \cdot \frac{14}{18} = 0.491$. The difference between 0.512 and 0.491 may not be terribly large, and it may not be of critical significance, but this does not make up for the fact that we were *wrong* in using the formula for the binomial distribution. The formula would have been correct *if each timer had been replaced before* the next one was randomly selected, but of course this is not what anyone would do in actual practice.

The correct probability function for this kind of problem (namely, for *sampling without replacement*) is that of the **hypergeometric distribution**. It applies whenever $n$ elements are randomly selected from a set containing $a$ elements of one kind (successes) and $b$ elements of another kind (failures), and we are interested in the probability of getting $x$ successes and $n - x$ failures. The formula for this probability function is

$$f(x) = \frac{\binom{a}{x} \cdot \binom{b}{n-x}}{\binom{a+b}{n}} \quad \text{for } x = 0, 1, 2, \ldots, \text{ or } n*$$

* Note that when $x$ is greater than $a$ or $n - x$ is greater than $b$, the respective binomial coefficients $\binom{a}{x}$ and $\binom{b}{n-x}$ are equal to 0. Clearly, there is *no way* in which we can form a combination of *more* elements of a set than there are in the set.

If we apply this formula to our example, where $x = 3$, $n = 3$, $a = 16$ (non-defective timers), and $b = 4$ (defective timers), we get

$$f(3) = \frac{\binom{16}{3}\binom{4}{0}}{\binom{20}{3}} = 0.491$$

which is identical with the result we obtained earlier by other means.

To prove the formula for the hypergeometric distribution, we have only to observe that the $x$ successes can be selected from among $a$ possibilities in $\binom{a}{x}$ ways, the $n - x$ failures can be selected from among $b$ possibilities in $\binom{b}{n-x}$ ways, so that the $x$ successes *and* $n - x$ failures can be selected in $\binom{a}{x} \cdot \binom{b}{n-x}$ ways in accordance with the rule on page 72. The total number of ways in which we can select $n$ elements from a set of $a + b$ elements is $\binom{a+b}{n}$, and according to the special rule for equiprobable events on page 98 the desired probability is given by the ratio of $\binom{a}{x} \cdot \binom{b}{n-x}$ to $\binom{a+b}{n}$.

To consider another example in which the formula for the hypergeometric distribution applies, suppose that a building inspector in a certain community decides to inspect eight of 16 new buildings for possible violations of the building code. On the basis of past experience, he suspects that four of the 16 buildings contain violations of the building code, and he wants to know the probability that *if his suspicion is correct*, his sample will disclose *at least* two of the four. The probability which he wants to know is given by $f(2) + f(3) + f(4)$, where each of the three terms in this sum is to be calculated by means of the formula for the hypergeometric distribution with $a = 4$, $b = 12$, and $n = 8$. Substituting these values together with $x = 2$, $x = 3$, and $x = 4$, we get

$$f(2) = \frac{\binom{4}{2}\binom{12}{6}}{\binom{16}{8}} = \frac{6 \cdot 924}{12{,}870} = 0.431$$

$$f(3) = \frac{\binom{4}{3}\binom{12}{5}}{\binom{16}{8}} = \frac{4 \cdot 792}{12{,}870} = 0.246$$

$$f(4) = \frac{\binom{4}{4}\binom{12}{4}}{\binom{16}{8}} = \frac{1 \cdot 495}{12{,}870} = 0.038$$

and the probability that the inspection will reveal at least two of the four buildings which contain violations of the building code is

$$0.431 + 0.246 + 0.038 = 0.715$$

Perhaps, this should make the building inspector check more than eight of the new buildings, and it will be left to the reader (in Exercise 16 on page 132) to find the probability that at least two of the buildings with violations will be discovered if he inspects ten of the 16 new buildings.

In the beginning of this section we introduced the hypergeometric distribution in connection with a problem in which we *erroneously* used the binomial distribution. Actually, when $n$ is small compared to $a + b$, the binomial distribution often provides a very good *approximation* to the hypergeometric distribution. It is generally agreed that this approximation can be used so long as $n$ constitutes less than 5 percent of $a + b$; this will simplify many problems because the binomial distribution has been tabulated much more extensively than the hypergeometric distribution, and it is generally easier to use (see also Exercises 17 and 18 below).

## EXERCISES

1. Check the general rules on page 123 to decide whether the following can be probability functions, and explain your answers:
   (a) $f(x) = \frac{1}{2}$   for $x = 1, 2$, or $3$;
   (b) $f(x) = \dfrac{x}{10}$   for $x = 1, 2, 3$, or $4$;
   (c) $f(x) = \dfrac{x - 4}{5}$   for $x = 1, 3, 5, 7$, or $9$.

2. An experiment consists of rolling a pair of balanced dice, one green and one red.
   (a) Letting $(2, 3)$ represent the outcome where the green die comes up 2 and the red die comes up 3, letting $(5, 1)$ represent the outcome where the green die comes up 5 and the red die comes up 1, and so on, draw a diagram similar to that of Figure 3.2 on page 61 which shows the 36 possible outcomes.

(b) Referring to the sample space of part (a) write next to each point the corresponding total rolled with the pair of dice, and, assuming that each point has the same probability of $\frac{1}{36}$, construct the probability function for the total number of points rolled with a pair of "honest" dice, that is, construct a table showing the probabilities of rolling a total of 2, 3, 4, . . ., 11, or 12.

(c) Verify that the equation of the probability function of part (b) can be written as

$$f(x) = \frac{6 - |x - 7|}{36} \quad \text{for } x = 2, 3, \ldots, \text{ or } 12$$

where the *absolute value* $|x - 7|$ equals $x - 7$ or $7 - x$, whichever is positive or zero.

3. Using the results of part (b) of Exercise 2 or the formula of part (c), find the probabilities of
   (a) rolling 7 or 11 with a pair of "honest" dice;
   (b) rolling 2, 3, or 12 with a pair of "honest" dice;
   (c) rolling 4, 5, or 6 with a pair of "honest" dice.

4. Using the *binomial expansions* $(a + b)^3 = a^3 + 3a^2b + 3ab^2 + b^3$ and $(a + b)^4 = a^4 + 4a^3b + 6a^2b^2 + 4ab^3 + b^4$ (which can be checked by performing the necessary term-by-term multiplications), show that
   (a) the successive terms in the expansion of $(\frac{9}{10} + \frac{1}{10})^3$ equal the probabilities given in Table XI for getting 0, 1, 2, and 3 successes in three trials when $p = \frac{1}{10}$;
   (b) the successive terms in the expansion of $(\frac{1}{5} + \frac{4}{5})^4$ equal the probabilities given in Table XI for getting 0, 1, 2, 3, and 4 successes in four trials when $p = \frac{4}{5}$.

5. A multiple-choice test consists of six questions, with each having five answers of which only one is correct. If a student answers each question by randomly drawing one of five slips of paper numbered 1, 2, 3, 4, and 5, and marking the corresponding alternative, calculate with the formula for the binomial distribution the probabilities that he will get
   (a) all wrong answers;
   (b) one right answer and five wrong answers;
   (c) two right answers and four wrong answers;
   (d) at least three right answers.
   [*Hint:* For part (d) use the results of parts (a), (b), and (c).]

6. In a certain large office, illness is given as the reason for 90 percent of all absences from work. Find the probability that four of five absences from

work (randomly selected from payroll records) were claimed to be due to illness, by using
(a) the formula on page 124;
(b) Table XI at the end of the book.

7. A safety engineer claims that among all workers in a certain factory who operate grinding wheels only 70 percent use safety goggles for short jobs. Find the probabilities that *exactly* three and *at least* five of six workers who operate grinding wheels will be using such goggles for short jobs
   (a) by using the formula on page 124;
   (b) by referring to Table XI at the end of the book.

8. It is known from experience that 60 percent of all persons who get a certain mail-order catalog will end up ordering something from the catalog. Find the probability that *at most two* (0, 1, or 2) of eight persons who get this mail-order catalog will end up ordering something from the catalog
   (a) by using the formula on page 124;
   (b) by referring to Table XI at the end of the book.

9. A small loan company knows that 20 percent of all its borrowers will make at least one late payment on their installment loans. Use Table XI to find the probabilities that among 12 borrowers
   (a) none will make any late payments on their installment loans;
   (b) at least six will make at least one late payment on their installment loans;
   (c) at most three will make at least one late payment on their installment loans.

10. A study has shown that one half of the families in a certain area have at least two cars. Use Table XI to find the probabilities that among 15 such families (randomly selected for further study)
    (a) exactly eight have at least two cars;
    (b) more than 11 have at least two cars;
    (c) fewer than six have at least two cars;
    (d) anywhere from ten to 12, inclusive, have at least two cars.

11. An agricultural cooperative claims that 95 percent of the watermelons that are shipped out are ripe and ready to eat. Use Table XI to find the probabilities that among 12 watermelons that are shipped out, 0, 1, 2, 3, ..., 11, and 12 are ripe and ready to eat, and draw a *histogram* of this probability distribution.

12. On page 127 we pointed out that it is possible (though highly unlikely) that three non-defective timers will be randomly selected from a lot of

20 timers among which 17 are defective. Show that the probability of this happening is $\frac{1}{1,140}$ by using
(a) the method of Exercise 18 on page 112;
(b) the formula for the hypergeometric distribution.

13. Verify the answers obtained for the three parts of Exercise 13 on page 111 with the use of the formula for the hypergeometric distribution. Also find the probability that if five of the 20 invoices are chosen at random, three will contain errors while the other two will not.

14. With reference to Exercise 12 on page 111, find the probability that if four of the construction workers are randomly selected and then interviewed, two of them will say that they like the new kind of helmet, while the other two will say that they do not like it.

15. Of 15 secretaries in a business office, six are graduates of a secretarial school. If four of these secretaries are randomly selected for a certain task, what are the probabilities that
    (a) none of those who are chosen are graduates of a secretarial school;
    (b) only one of those who are chosen is a graduate of a secretarial school;
    (c) at least two of those who are chosen are graduates of a secretarial school.
    [*Hint:* For part (c) use the results of parts (a) and (b).]

16. With reference to the illustration on page 129, find the probability that the inspector will discover at least two of the new buildings which contain violations of the building code if he increases his sample to ten of the new buildings.

17. A mailing list of 400 names and addresses includes 40 old addresses of persons who have moved. If a mail clerk randomly selects four names from this list, what is the probability that all of the addresses will be current (that is, none of them will be old)? By how much would we be off if we *approximated* this probability by calculating the binomial probability of zero successes in four trials with $p = 0.10$?

18. A fleet of 100 trucks includes five trucks with faulty brakes. If three of these trucks, randomly selected, are assigned to a job, find the probability that exactly one of them will have faulty brakes by
    (a) using the formula for the hypergeometric distribution;
    (b) approximating the probability with the use of the formula for the binomial distribution having $n = 3$ and $p = 0.05$.

## THE MEAN OF A PROBABILITY DISTRIBUTION

On page 87 we stated that in the United States a household can expect to have 3.19 members, and we pointed out that this figure is the sum of the products obtained by multiplying 0, 1, 2, 3, 4, ..., by the respective probabilities that a household in the United States actually has that many members. To give a similar example in which we already know all of the probabilities, let us find the expected number of heads for four flips of a balanced coin. Multiplying 0, 1, 2, 3, and 4 by the corresponding probabilities shown in the table on page 121, we find that the sum of these products is

$$0(\tfrac{1}{16}) + 1(\tfrac{4}{16}) + 2(\tfrac{6}{16}) + 3(\tfrac{4}{16}) + 4(\tfrac{1}{16}) = 2$$

and this result should not come as a surprise—as we pointed out on page 85, a mathematical expectation represents an *average*, or as we referred to it in Chapter 2, a *mean*. Thus, let us state formally that

> The mean of a probability distribution is the mathematical expectation (also called the expected value) of a random variable having the particular distribution.

Symbolically, the mean of a probability distribution is given by the formula

$$\mu = \Sigma \, x \cdot f(x)$$

where the summation extends over all values of $x$ which the random variable can take on, and the quantities $f(x)$ are the corresponding probabilities. The symbol $\mu$ which we use to denote the mean of a probability distribution is the Greek letter *mu*, and it should be noted that the formula for $\mu$ is very similar to the one for the weighted mean on page 33. The weights which we assign to the $x$'s are simply their probabilities, and the denominator in the formula on page 33 can be omitted since the sum of the probabilities (of a probability function) must always be equal to 1.

To give another example, let us calculate the mean of the probability distribution on page 125, which pertained to the number of substandard workers who have physical disabilities. Substituting the probabilities which we calculated on page 125 into the formula for $\mu$, we get

$$\mu = 0(0.1176) + 1(0.3025) + 2(0.3241) + 3(0.1852)$$
$$+ 4(0.0595) + 5(0.0102) + 6(0.0007)$$
$$= 1.7995$$

or approximately 1.80. Thus, if the production engineer's estimate of $p = 0.30$ is correct, namely, if 30 percent of all substandard workers have physical disabilities, then *on the average* 1.8 of six substandard workers should have physical disabilities.

When a random variable can take on many different values, the calculation of $\mu$ becomes very laborious unless we use simplifications. For instance, if we wanted to determine how many of 380 patrons of an expensive restaurant can be expected to charge their meals using credit cards (when the probability that any one of them will use his credit card is 0.20), we would have to calculate the 381 probabilities corresponding to 0, 1, 2, 3, ..., and 380 of the patrons using their credit cards, multiply each of these probabilities by the corresponding value of the random variable (the number of patrons using their credit cards), and then add the 381 products. However, if we think for a moment, we might argue that in the long run 20 percent of the patrons will charge their meals using credit cards, 20 percent of 380 is $0.20(380) = 76$, and, hence, we might conclude that 76 of the 380 patrons can be expected to charge their meals using credit cards. Similarly, if a balanced coin is flipped 1,000 times, we can argue that *heads* should come up about 50 percent of the time, and, hence, that we can expect to get $0.50(1,000) = 500$ heads in 1,000 flips of a balanced coin. Without this argument, we would have had to determine the 1,001 probabilities of getting 0, 1, 2, 3, ..., and 1,000 heads, and it is fortunate, indeed, that there is the special formula

$$\mu = n \cdot p$$

for the mean of a binomial distribution. In words, this formula expresses the fact that

**The mean of a binomial distribution is given by the product of the number of trials and the probability of success on each individual trial.**

Proofs of this result may be found in most textbooks on mathematical statistics; see for instance, those listed in the Bibliography at the end of the book.

Using the formula $\mu = n \cdot p$, we can now verify the results obtained in the first two examples of this section. In the example dealing with the number of heads which we obtain in four flips of a balanced coin, we have $n = 4$, $p = 0.50$, and, hence

$$\mu = 4(0.50) = 2$$

and in the example dealing with substandard workers having physical disabilities, we have $n = 6$, $p = 0.30$, and, hence,

$$\mu = 6(0.30) = 1.80$$

This agrees with the results which we obtained before (except for the rounding error of 0.0005 in the second of the two examples).

It is important to remember, of course, that the formula $\mu = n \cdot p$ applies only to binomial distributions. Fortunately, there are other special formulas for other special distributions; for the *hypergeometric distribution*, for example, the formula is

$$\mu = \frac{n \cdot a}{a + b}$$

so that in the example on page 128 the building inspector can *expect* to catch

$$\mu = \frac{8 \cdot 4}{4 + 12} = 2$$

of the four buildings containing violations of the building code.

## MEASURING CHANCE VARIATION

In the introduction to this chapter we claimed that it is possible to *predict, measure,* and sometimes even *control* chance variations. This is tremendously important, for we are often faced with situations in which we must decide whether differences (discrepancies) between *what we expect* and *what we get* can reasonably be attributed to chance. For instance, when a balanced coin is flipped 100 times, we would expect to get 50 heads and 50 tails, but what can we say about the coin (or the manner in which it is flipped) if we get 42 heads and 58 tails, or perhaps 25 heads and 75 tails? Similarly, if a spray painting operation is considered to be "under control" if only half of one percent of the furniture being painted has runs or streaks, what can we conclude about the spray painting of 5,000 pieces of furniture if 42 items (which is more than half of one percent) have runs or streaks? Is the difference small enough to be attributed to chance, or is it big enough to blame the whole thing on a faulty spray machine setting, inferior paint, or perhaps poor workmanship?

To answer questions of this kind we shall first have to develop a way of measuring the fluctuations of random variables, or as we put it on page 121, we need a "yardstick" for measuring variations that are due to chance. Now, if a random variable takes on the value $x$ and the mean of its probability distribution is $\mu$, then the difference $x - \mu$ is called the **deviation from the mean,** and it measures by how much the value of the random variable differs from what we would expect. For instance, if we get $x = 42$ heads in 100 flips of a balanced coin, then $\mu = 50$ and the amount by which we are "off" is

$$x - \mu = 42 - 50 = -8$$

In other words, we got eight fewer heads than expected, and if the coin is properly balanced and there is no other kind of chicanery, this difference will have to be attributed to chance. This suggests that we measure and judge chance variations in terms of the amounts by which we can *expect* to be off. For instance, for four flips of a balanced coin we can expect $\mu = 2$ heads, and if we get only 0 heads we are off by $0 - 2 = -2$; similarly, if we get one head we are off by $1 - 2 = -1$, if we get two heads we are off by $2 - 2 = 0$, if we get three heads we are off by $3 - 2 = 1$, and if we get four heads we are off by $4 - 2 = 2$. Since the corresponding probabilities are $\frac{1}{16}$, $\frac{4}{16}$, $\frac{6}{16}$, $\frac{4}{16}$, and $\frac{1}{16}$ according to the table on page 121, we find that the (average) amount by which we can *expect* to be off is

$$(-2)(\tfrac{1}{16}) + (-1)(\tfrac{4}{16}) + 0(\tfrac{6}{16}) + 1(\tfrac{4}{16}) + 2(\tfrac{1}{16}) = 0$$

The trouble with this answer is that some of the amounts by which we are off are *positive* while others are *negative*, so that the amount by which we can *expect* to be off is zero, and, hence, *not indicative of the size of the chance fluctuations in the number of heads we might get in four flips of a coin.*

There are several ways in which we can get rid of the minus signs in which we are really not interested. One possibility is to ignore the signs of the amounts by which we are off (namely, take their *absolute values*), but mathematically more convenient is to *square* the amounts by which we are off—after all, the squares of real numbers are always positive or zero. Referring again to the distribution of the number of heads which we get in four flips of a balanced coin, we can thus say that the *average* (namely, the *mathematical expectation*) of the squares of the amounts by which we are off is

$$(-2)^2(\tfrac{1}{16}) + (-1)^2(\tfrac{4}{16}) + 0^2(\tfrac{6}{16}) + 1^2(\tfrac{4}{16}) + 2^2(\tfrac{1}{16}) = 1$$

and this defines the **variance** of this probability distribution. In general,

> The variance of a probability distribution is the mathematical expectation of the squared deviation from the mean; symbolically, it is given by the formula
>
> $$\sigma^2 = \Sigma\, (x - \mu)^2 \cdot f(x)$$

where the summation extends over all values of $x$ for which the probability function is defined.

The symbol which we use to denote the variance of a probability distribution is $\sigma^2$, where $\sigma$ is the lower case Greek letter *sigma*. If we want to compensate for the fact that the variance is the expected value of the *squared* deviations from the mean, we can simply take the square root of the variance and we will get another widely used measure of chance variation, the **standard deviation**

$$\sigma = \sqrt{\Sigma (x - \mu)^2 \cdot f(x)}$$

To give the reader some idea how the variance reflects the average size of chance fluctuations, let us refer to Figure 5.1, which contains the histograms

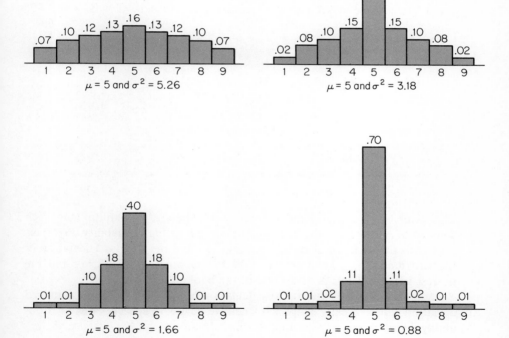

**FIGURE 5.1** Histograms of four probability distributions.

of four probability distributions defined for $x = 1, 2, 3, \ldots,$ and 9. They all have the mean $\mu = 5$ as can easily be verified using the formula on page 133, but their respective variances are $\sigma^2 = 5.26$, $\sigma^2 = 3.18$, $\sigma^2 = 1.66$, and $\sigma^2 = 0.88$. Aside from the obvious differences in the "spread" of these four distributions, the reader can verify for himself that the probabilities of getting a value which differs from $\mu = 5$ by *two or more* are, respectively, 0.58, 0.40, 0.24, and 0.08. Similarly, the probabilities of getting a value which differs from $\mu = 5$ by *three or more* are, respectively, 0.34, 0.20, 0.04, and 0.04. Thus, a small value of $\sigma^2$ implies that we are more likely to get a value close to the mean, while a large value of $\sigma^2$ implies that we are more likely to get a value far away from the mean. This will be discussed further in the section beginning on page 139.

To give another illustration of the calculation of the variance of a probability distribution, let us refer again to the example on page 125, which dealt with the number of substandard workers (among $n = 6$) who have physical disabilities. Since we have already shown that the mean of this distribution is $\mu = 1.80$, we can arrange the necessary calculations as in the following table, where $x$ is the number of substandard workers having physical disabilities and $x - \mu$ is the corresponding deviation from the mean, namely, $x - 1.80$:

| $x$ | $x - \mu$ | $(x - \mu)^2$ | Probability $f(x)$ | $(x - \mu)^2 f(x)$ |
|---|---|---|---|---|
| 0 | −1.80 | 3.24 | 0.1176 | 0.3810 |
| 1 | −0.80 | 0.64 | 0.3025 | 0.1936 |
| 2 | 0.20 | 0.04 | 0.3241 | 0.0130 |
| 3 | 1.20 | 1.44 | 0.1852 | 0.2667 |
| 4 | 2.20 | 4.84 | 0.0595 | 0.2880 |
| 5 | 3.20 | 10.24 | 0.0102 | 0.1044 |
| 6 | 4.20 | 17.64 | 0.0007 | 0.0123 |

$$\sigma^2 = 1.2590$$

The values in the right-hand column were obtained by multiplying each squared deviation from the mean by the corresponding probability, and their sum gives the variance of the distribution. To find the standard deviation of this distribution, we have only to take the square root of 1.2590, and the nearest value in Table X at the end of the book is $\sqrt{1.3} = 1.14$.

As in the case of the mean, the calculation of the variance or the standard deviation can be greatly simplified when we deal with special kinds of probability distributions. For instance, for the variance of the *binomial distribution* we have the formula

$$\sigma^2 = np(1 - p)$$

which we shall not prove, but which can easily be verified for our two examples. For the distribution of the number of heads which we obtain in four flips of a coin we have $n = 4$ and $p = 0.50$, and the formula yields

$$\sigma^2 = 4(0.50)(0.50) = 1$$

which agrees with the result on page 136. Similarly, for the distribution of the number of substandard workers with physical disabilities we have $n = 6$ and $p = 0.30$, and the formula yields

$$\sigma^2 = 6(0.30)(0.70) = 1.26$$

which agrees with the result obtained above (except for the rounding error of 0.001).

## CHEBYSHEV'S THEOREM

The variance and the standard deviation of a probability distribution were introduced as measures of the *expected chance fluctuations* of a random variable having that distribution, and as we already pointed out on page 138, a small value of $\sigma$ or $\sigma^2$ implies that we are likely to get a value which is very close to the mean, while a large value of $\sigma$ or $\sigma^2$ implies that we are more likely to get a value far away from the mean. Formally, this important idea is expressed by the following rule, called **Chebyshev's Theorem**:

> The probability that a random variable will take on a value within $k$ standard deviations of the mean is always at least $1 - (1/k^2)$, where $k$ can be any positive number.

Thus, the probability of getting a value within *two* standard deviations of the mean (namely, a value on the interval from $\mu - 2\sigma$ to $\mu + 2\sigma$) is *always* at least $1 - (1/2^2) = 0.75$. Similarly, the probability of getting a value within *five* standard deviations of the mean is *always* at least $1 - (1/5^2) = 0.96$, and the probability of getting a value within *ten* standard deviations of the mean is *always* at least $1 - (1/10^2) = 0.99$. If we change the argument around, Chebyshev's Theorem can also be stated as follows:

> The probability that a random variable will take on a value which differs from the mean by at least $k$ standard deviations is always at most $1/k^2$.

Thus, the probability is *always* at most $1/2^2 = 0.25$ that we will get a value of a random variable which differs from the mean of its distribution by at least *two* standard deviations, and it is *always* at most $1/10^2 = 0.01$ that we will get a value of a random variable which differs from the mean of its distribution by at least *ten* standard deviations.

To give a concrete example, suppose that we have won only 138 times betting on *heads* in 400 flips of a coin, and that we are beginning to wonder whether we should raise any questions about the conduct of the game. If the game is honest, we should have a fifty-fifty chance of winning on each flip of the coin, and the number of heads in 400 flips of a coin is a random variable having a binomial distribution with $n = 400$, $p = 0.50$, and, hence,

$$\mu = np = 400(0.50) = 200$$

and

$$\sigma = \sqrt{np(1-p)} = \sqrt{400(0.50)(0.50)} = 10$$

Since the 138 heads we got differs from $\mu = 200$ by 62, which is *more than six standard deviations*, we find that the probability of this happening by chance is at most $1/6^2 = \frac{1}{36}$ or approximately 0.028. Thus, it would probably be advisable to be a bit wary and stop playing the game.

In general, when Chebyshev's Theorem is applied to random variables having binomial distributions, it leads to the **Law of Large Numbers,** which laymen often refer to as the **Law of Averages.** For instance, it follows from the preceding example that for 400 flips of a coin, the probability is at least $1 - 0.028 = 0.972$ that the number of heads will differ from 200 by less than 60. If we express this result in terms of proportions, we can assert with a probability of at least 0.972 that in 400 flips of a coin the *proportion* of heads will differ from $\frac{200}{400} = 0.50$ by less than $\frac{60}{400} = 0.15$. To continue this argument, the reader will be asked to show in Exercise 17 on page 143 that we can assert with a probability of at least 0.972 that in 10,000 flips of a coin the *proportion* of heads will differ from 0.50 by less than 0.03, and that in 1,000,000 flips of a coin the *proportion* of heads will differ from 0.50 by less than 0.003. Correspondingly, the Law of Large Numbers states that:

> When the number of trials grows larger and larger, the proportion of successes will tend to come closer and closer to the probability of a success for each individual trial.

This is true, incidentally, regardless of whether or not the probability of a success on each individual trial happens to be 0.50. A very important aspect of the Law of Large Numbers is that it justifies the *frequency interpretation of probability*, which we studied in Chapter 3—after all, it was there that we talked about what happens to a proportion "in the long run."

In actual practice, Chebyshev's Theorem is very rarely used; the very fact that it applies to *any* probability distribution has the unfortunate side-effect that the probability "at least $1 - (1/k^2)$" is often *unnecessarily small,* and the probability "at most $1/k^2$" is often *unnecessarily large.* For instance, in the example at the top of this page we stated that the probability of getting a value more than $k = 6$ standard deviations above or below the mean is *at most*

$1/6^2 = 0.028$, while the *actual* probability that this will happen for a binomial distribution with $n = 400$ and $p = 0.50$ is about 0.000000002. "At most 0.28" is, of course, correct, but it does not tell us very much.

## EXERCISES

1. The following table gives the probabilities that a residential real estate salesman will sell 0, 1, 2, 3, 4, or 5 residences in a week:

| Number of Residences $x$ | 0 | 1 | 2 | 3 | 4 | 5 |
|---|---|---|---|---|---|---|
| $f(x)$ | 0.33 | 0.41 | 0.20 | 0.05 | 0.01 | 0.00 |

Find the mean of this probability distribution.

2. The owner of a furniture store figures that one-half of his sales prospects make a purchase. Use Table XI to find the probabilities that among ten of his sales prospects 0, 1, 2, 3, ..., or 10 will make a purchase, and use these probabilities to calculate the mean of this probability distribution. Verify that the result equals the *product* of the number of "trials" and the probability of "success" for each trial.

3. A new-car dealer sent invitations to 15 of his best customers to look at a new model of a luxury car, and he figures that for each one the odds are 3 to 2 that he will *not* come. Use Table XI to find the probabilities that 0, 1, 2, 3, ..., or 15 of the customers *will* come to see the new model and use these probabilities to calculate the mean (namely, the average number he can expect to come). Verify that the result equals the *product* of the number of "trials" and the probability of "success" for each trial.

4. Use the probabilities of Exercise 11 on page 131 to calculate the mean of the probability distribution of the number of watermelons that are ripe (among the 12 that are shipped). Verify the result with the use of the special formula on page 134.

5. Find the mean of the probability distribution of part (b) of Exercise 1 on page 129.

6. Find the mean and the standard deviation of the distribution of each of the following *binomial* random variables:
   (a) the number of heads obtained in 1,600 tosses of a balanced coin;
   (b) the number of *sixes* obtained in 720 rolls of an "honest" die;

(c) the number of business loans a banker will make to the 150 businesses in a given city if the probability that the banker will make a loan to any one of these businesses is 0.60;

(d) the number of applicants for life insurance (among 900 applicants) who will fail to meet medical requirements if the probability that any one applicant will fail to meet the medical requirements is 0.10.

**7.** With reference to the illustration on page 127, find the probabilities that 0, 1, and 2 of the three timers are non-defective, and use *all* of the probabilities to calculate the mean of this probability distribution. Verify the result by substituting the appropriate values of $a$, $b$, and $n$ into the formula on page 135.

**8.** Use the formula for the mean of the hypergeometric distribution to find
(a) the number of graduates of secretarial schools one can expect to find in the sample referred to in Exercise 15 on page 132;
(b) the number of buildings containing violations of the building code which the inspector (of Exercise 16 on page 132) can expect to discover.

**9.** As can readily be verified by means of the formula for the binomial distribution (or by listing all 32 possibilities), the probabilities of getting 0, 1, 2, 3, 4, or 5 heads in five flips of a balanced coin are, respectively, $\frac{1}{32}$, $\frac{5}{32}$, $\frac{10}{32}$, $\frac{10}{32}$, $\frac{5}{32}$, and $\frac{1}{32}$.
(a) Use these probabilities to show that the mean of this probability distribution is $\mu = 2.5$.
(b) Use the given probabilities and the result of part (a) to show that the standard deviation of this probability distribution is $\sigma = \sqrt{5}/2$ (or approximately 1.12).
(c) According to Chebyshev's Theorem, the probability of getting a value which differs from the mean by at least $k = 2$ standard deviations is always at most $1/k^2 = \frac{1}{4}$. What is the *exact* value of this probability for the given distribution?

**10.** Referring to the probabilities looked up in Exercise 3 and making use of the fact that $\mu = 6$, calculate the variance of the distribution (of the number of customers who will come to look at the new model). Verify the result by substituting $n = 15$ and $p = 0.40$ into the special formula on page 138.

**11.** According to Chebyshev's Theorem, the probability of getting a value within $k = 3$ standard deviations of the mean is always at least $1 - (1/k^2) = 1 - (1/3^2) = \frac{8}{9}$. Use the probabilities of Exercise 3 and the results of Exercise 10 to find the *exact* probability that the number of customers who will come to look at the car will be within three standard deviations of the mean.

12. For each of the 64 customers who patronize a given barber shop the probability is 0.50 that he will get immediate service.
    (a) Use the special formulas for the mean and the standard deviation of a binomial distribution to calculate $\mu$ and $\sigma$ for the distribution of the number of customers (among the 64) who will get immediate service.
    (b) What does Chebyshev's Theorem tell us about the probability that between 24 and 40 of the customers will get immediate service?
    (c) What does Chebyshev's Theorem tell us about the probability that between 12 and 52 of the customers will get immediate service?

13. Referring to part (b) of Exercise 6, what does Chebyshev's Theorem tell us about the probability of getting between 70 and 170 sixes in 720 rolls of an "honest" die.

14. Referring to part (c) of Exercise 6, what does Chebyshev's Theorem tell us about the probability that the banker will make loans to at most 66 or at least 114 of the businesses?

15. Referring to part (d) of Exercise 6, what does Chebyshev's Theorem tell us about the probability that between 18 and 162 of the applicants for life insurance will fail to meet medical requirements?

16. Suppose that Mr. Jones flips a coin 100 times and that he gets 30 heads, which is 20 less than the 50 he can expect. Then he flips the coin another 100 times and gets 44 heads, so that altogether he has 74 heads, which is 26 less than the 100 he can expect. Discuss his complaint that the "Law of Averages," namely, the Law of Large Numbers, is letting him down.

17. Duplicating the work on page 140, show that
    (a) we can assert with a probability of at least $\frac{35}{36} = 0.972$ that in 10,000 flips of a balanced coin the *proportion* of heads will be between 0.47 and 0.53;
    (b) we can assert with a probability of at least $\frac{35}{36} = 0.972$ that in 1,000,000 flips of a balanced coin the *proportion* of heads will be between 0.497 and 0.503.

# 6

# THE NORMAL DISTRIBUTION

**INTRODUCTION**

Continuous sample spaces arise whenever we deal with quantities that are measured on continuous scales—when we measure the speed of a car, when we measure the weight of a jar of instant coffee, when we measure the amount of tar in cigarettes, when we measure the floor space in an office, and so forth. It is true, of course, that we always round our answers to the nearest whole unit or to a few decimals, but there *is* a continuum of possibilities in each of these examples, and we *could* ask for probabilities associated with individual points of the sample space. If we did this, however, we would find that (for all practical purposes, at least) our answers would always be equal to zero. Surely, we should be willing to give *any odds* that a car will *not* be traveling at *exactly* $20\pi$ miles per hours (where $\pi$ is the irrational number 3.1415926 ... which arises in connection with the area of a circle), and we should be willing to give *any odds* that a jar of instant coffee will *not* contain exactly $\sqrt{63} = 7.937253$ ... ounces of coffee. *In the continuous case we are thus not really interested in probabilities associated with individual points of the sample space, but in probabilities associated with intervals or regions.* For instance, we might ask for the probability that a car will be traveling anywhere from 70 to 75 miles per hour, for the probability that a jar of instant coffee contains anywhere from 7.9 to 8.1 ounces of coffee, or for the probability that the floor space of an office is anywhere from 4,000 to 4,200 square feet.

When we first discussed histograms in Chapter 2, we pointed out that the frequencies, percentages, or proportions associated with the various classes are given by the *areas* of the rectangles, and with reference to Figure 5.1 on page 137 we might add that this is true also for the probabilities associated with the values of a random variable. In the continuous case, we also represent probabilities by means of areas as is illustrated in Figure 6.1, but instead of the areas of rectangles we now use areas under continuous curves. The first diagram of Figure 6.1 represents the probability distribution of a random variable which takes on the values 0, 1, 2, ..., 9, and 10, and the probability of getting a 3 is given by the area of the shaded rectangle; the second diagram refers to a random variable which can take on *any value* on the interval from 0 to 10, and the probability of getting a value between 2.5 and 3.5 is given by the medium blue area under the curve. Similarly, the light blue area under the

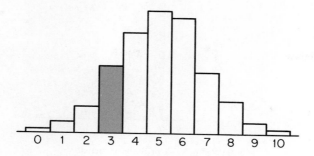

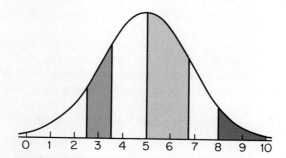

**FIGURE 6.1** Histogram of probability function and graph of continuous distribution.

curve gives the probability of getting a value between 5 and 6.8, while the dark blue area gives the probability of getting a value greater than 8.

Curves like the one shown in the second diagram of Figure 6.1 are the graphs of functions which we refer to as **probability densities,** or more informally as **continuous distributions.** (The first of these terms is borrowed from the language of physics, where the terms "weight" and "density" are used in very much the same way in which we use the terms "probability" and "probability density" in statistics.) What characterizes a probability density is the fact that

> The area under the curve between any two values $a$ and $b$ (for instance, those of Figure 6.2) gives the probability that a random variable having this "continuous distribution" will take on a value on the interval from $a$ to $b$.

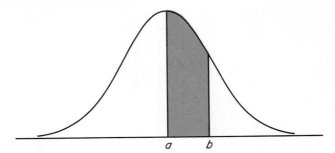

**FIGURE 6.2** A probability density.

It follows from this that the values of a probability density function *cannot be negative*, and that *the total area under the curve (representing the certainty that a corresponding random variable must take on one of its values) is always equal to 1*.

It is impossible to define the mean and the standard deviation (or variance) of a continuous distribution without the use of calculus, but we can always picture continuous distributions as approximated by means of histograms of probability functions for which the mean and the standard deviation can be calculated in accordance with the formulas on pages 133 and 137 (see Figure 6.3). Then, if we choose histograms with narrower and narrower classes, the

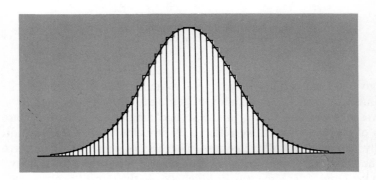

**FIGURE 6.3** Continuous distribution approximated by means of histogram of probability function.

means and the standard deviations of the corresponding probability functions will approach the mean and the standard deviation of the continuous distribution. Actually, the mean and the standard deviation of a continuous distribution measure the same features as the mean and the standard deviation of a probability function—the *expected value* of a random variable having the given distribution, and the *average* (that is, the mathematical expectation) *of its squared deviations from the mean*. More intuitively, the mean $\mu$ of a continuous distribution is indicative of its "center" or "middle," while the standard deviation $\sigma$ of a continuous distribution measures its "dispersion" or "spread."

## THE STANDARD NORMAL DISTRIBUTION

Among the many special kinds of continuous distributions that are used in statistics, **normal distributions** are by far the most important. Their study dates back to the eighteenth century and investigations into the nature of experimental errors. It was observed that variations among repeated measurements of the same physical quantity displayed a surprising degree of regularity, and it was found that their pattern (distribution) could be closely approximated by a certain kind of continuous distribution. This distribution was referred to as the "normal curve of errors" and it was attributed to the laws of chance.

The graph of a normal distribution is a bell-shaped curve that extends indefinitely in both directions, coming closer and closer to the horizontal axis without ever reaching it. Actually, it is seldom necessary to extend the "tails" of the curve very far, since the area under the curve becomes negligible when we go more than four or five standard deviations away from the mean. A very important feature of any normal distribution is that it is *symmetrical about its mean;* with reference to Figure 6.4 this means that if we folded the diagram along the dashed line, the two halves of the curve would coincide. Another important feature is that *there is one and only one normal distribution which has a given mean $\mu$ and a given standard deviation $\sigma$.* For instance, the

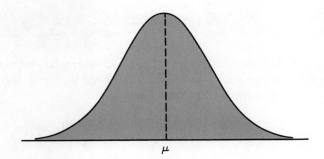

**FIGURE 6.4** Graph of normal distribution.

normal distribution which has $\mu = 20$ and $\sigma = 5$ is shown in Figure 6.5 together with the normal distribution which has $\mu = 30$ and $\sigma = 10$. Thus,

Normal distributions can have different shapes and they can be moved to the left or to the right, but there is one and only one normal distribution for any given pair of values of $\mu$ and $\sigma$.

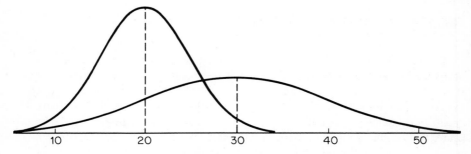

**FIGURE 6.5** Two normal distributions.

In actual practice, probabilities related to normal distributions (that is, areas under the corresponding curves) are obtained from special tables, such as Table I at the end of this book. As is apparent from Figure 6.5, areas under normal distributions (and, hence, probabilities related to these distributions) will differ depending on their means and standard deviations, but this difficulty is resolved by always working with the **standard normal distribution,** namely, with the normal distribution which has $\mu = 0$ and $\sigma = 1$. Areas under any other normal curve can then be obtained by performing the change of scale illustrated in Figure 6.6. All we are really doing here is converting the units of measurement into so-called **standard units** by means of the formula

$$z = \frac{x - \mu}{\sigma}$$

For instance to find the area between 22 and 25 under the *first* normal curve of Figure 6.5 (the one with $\mu = 20$ and $\sigma = 5$), we simply look for the area under the standard normal distribution between

$$z = \frac{22 - 20}{5} = 0.40 \quad \text{and} \quad z = \frac{25 - 20}{5} = 1.00$$

Similarly, to find the corresponding area under the *second* normal curve of

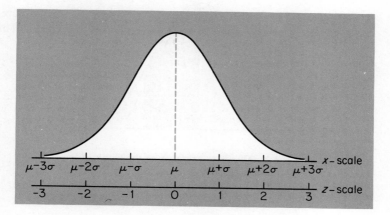

**FIGURE 6.6** Change of scale.

Figure 6.5 (the one with $\mu = 30$ and $\sigma = 10$), we look for the area under the standard normal distribution between

$$z = \frac{22 - 30}{10} = -0.80 \quad \text{and} \quad z = \frac{25 - 30}{10} = -0.50$$

Note that in the first case the two z-values are *positive* because 22 and 25 exceed the mean $\mu = 20$, while in the second case they are *negative* because 22 and 25 are both less than $\mu = 30$. The two areas, themselves, which we shall determine on page 152, will be positive, of course, because they represent probabilities.

The table to which we shall have to refer in problems like this is Table I at the end of the book, *whose entries are the areas under the standard normal distribution between the mean ($z = 0$) and $z = 0.00, 0.01, 0.02, 0.03, \ldots$, 3.08, and 3.09*. In other words, the entries of Table I are areas like the one shaded in Figure 6.7. Observe that Table I has no entries corresponding to *negative* values of z, for these are not needed by virtue of the *symmetry* of the standard normal distribution about its mean, namely, about $z = 0$. To find the area between $z = -1.50$ and $z = 0$, for example, we simply look up the area between $z = 0$ and $z = 1.50$ (see Figure 6.8), and to determine the area between $z = -0.75$ and $z = -0.25$, we simply find the area between $z = 0.25$ and $z = 0.75$. As can easily be verified, the answer to the first problem is 0.4332, namely, the entry of Table I which corresponds to $z = 1.50$. So far as the second problem is concerned, the area cannot be looked up directly, but we can look up the area under the curve between $z = 0$ and $z = 0.75$, the area between $z = 0$ and $z = 0.25$, and then take the *difference* between the

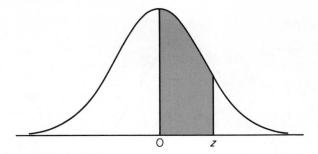

**FIGURE 6.7** Tabulated areas under the graph of the standard normal distribution.

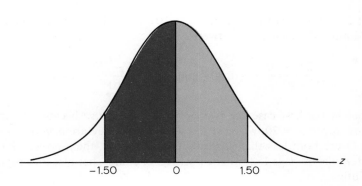

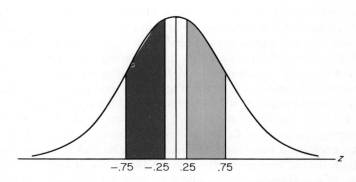

**FIGURE 6.8** Areas under standard normal distributions.

two. We thus get $0.2734 - 0.0987 = 0.1747$, as the reader will be asked to verify in Exercise 11 on page 157.

Questions concerning areas under normal distributions arise in various ways, and the ability to find any desired area quickly can be a big help. Although the table gives only areas between the mean $z = 0$ and selected positive values of $z$, we often have to find areas to the left or to the right of given positive or negative values of $z$, or areas between two given values of $z$ (as in the second example of the preceding paragraph). Finding any of these areas is easy, provided we remember exactly what areas are represented by the entries of Table I, and make use of the fact that the standard normal distribution is symmetrical about $z = 0$, so that the area to the left of $z = 0$ and the area to the right of $z = 0$ are both equal to 0.5000. For instance, we find that the probability of getting a $z$ less than 0.81 (namely, the area under the curve to the left of $z = 0.81$ which is shaded in the first diagram of Figure 6.9)

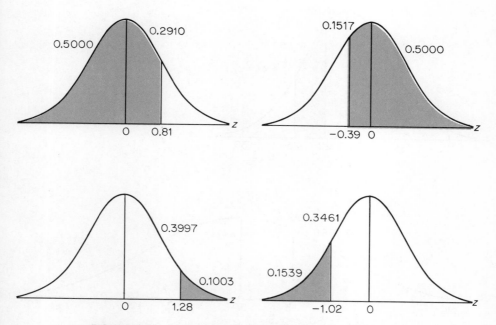

**FIGURE 6.9** Areas under standard normal distributions.

is $0.5000 + 0.2910 = 0.7910$, and the probability of getting a $z$ greater than $-0.39$ (namely, the area under the curve to the right of $z = -0.39$ which is shaded in the second diagram of Figure 6.9) is $0.5000 + 0.1517 = 0.6517$. Similarly, we find *by subtraction* that the probability of getting a $z$ greater than 1.28 (namely, the area under the curve to the right of 1.28 which is shaded in the third diagram of Figure 6.9) is $0.5000 - 0.3997 = 0.1003$, and that the probability of getting a $z$ less than $-1.02$ (namely, the area under the curve to the left of $z = -1.02$ which is shaded in the fourth diagram of Figure 6.9) is $0.5000 - 0.3461 = 0.1539$.

Returning now to the example on page 148, we find that for a normal distribution with $\mu = 20$ and $\sigma = 5$ the probability of getting a value between 22 and 25 is given by the area of the shaded region of the *first* diagram of Figure 6.10. Looking up the area between $z = 0$ and $z = 1.00$ as well as

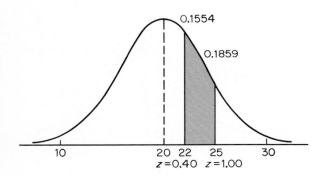

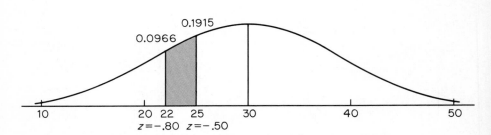

**FIGURE 6.10** Areas under normal distributions.

the area between $z = 0$ and $z = 0.40$, we find *by subtraction* that the area between $z = 0.40$ and $z = 1.00$ is $0.3413 - 0.1554 = 0.1859$. Similarly, for a normal distribution with $\mu = 30$ and $\sigma = 10$ the probability of getting a value between 22 and 25 is given by the area of the shaded region of the *second* diagram of Figure 6.10. Looking up the area between $z = 0.50$ and $z = 0.80$ (instead of $z = -0.80$ and $z = -0.50$), we find *by subtraction* that the area between $z = -0.80$ and $z = -0.50$ is $0.2881 - 0.1915 = 0.0966$.

When one $z$-value is positive and the other is negative, the normal curve area between the two is always given by the *sum* of the corresponding entries in Table I; for instance, the probability of getting a $z$ between $-0.51$ and $0.80$ (namely, the area under the curve shaded in Figure 6.11) is $0.1950 + 0.2881 = 0.4831$.

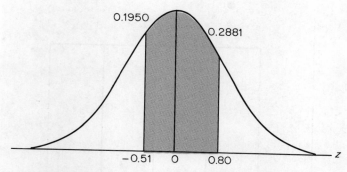

**FIGURE 6.11** Area under standard normal distribution.

There are also problems in which we are given areas under normal curves and are asked to find the corresponding values of $z$. For instance, if we want to find a $z$ which is such that the total area to its right equals $0.1900$ (see Figure 6.12), it is apparent that it must correspond to an entry of $0.5000 - 0.1900 = 0.3100$ in Table I; hence, the result is approximately $0.88$ (which corresponds to an entry of $0.3106$).

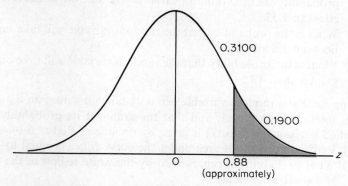

**FIGURE 6.12** Area under standard normal distribution.

## EXERCISES

1. Suppose that a random variable can only take on values on the continuous interval from 1 to 5, and that the graph of its probability density (called a **uniform density**) is given by the horizontal line of Figure 6.13.
   (a) What probability is represented by the white region of the diagram and what is its value?

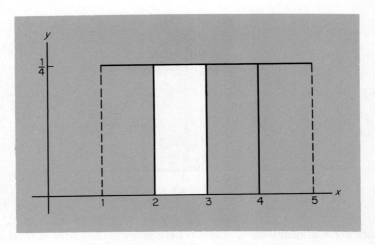

**FIGURE 6.13** A uniform probability density.

   (b) What is the probability that the random variable will take on a value less than 1.5? Would the answer be the same if we asked for the probability that the random variable will take on values less than or equal to 1.5?
   (c) What is the probability that the random variable will take on a value between 1.6 and 4.2?
   (d) What is the probability that the random variable will take on a value greater than 4.2?

2. Suppose that a random variable can only take on values on the continuous interval from 0 to 1, and that the graph of its probability density (called a **triangular density**) is given by the line of Figure 6.14.
   (a) Verify that the total area under the curve (line) is equal to 1.
   (b) What probability is represented by the white region of the diagram and what is its value?
   (c) What is the probability that the random variable will take on a value between 0.50 and 0.75?

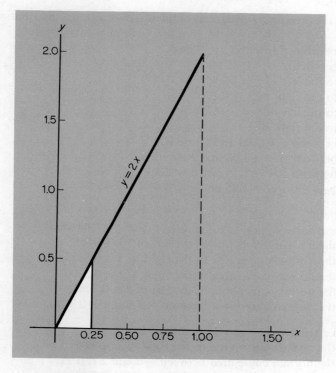

**FIGURE 6.14** A triangular probability density.

(d) What is the probability that the random variable will take on a value greater than 0.60?

[*Hint:* In parts (c) and (d) subtract the areas of appropriate triangles or use the formula for the area of a trapezoid; incidentally, the area of a triangle is one half times the product of its base and its height.]

3. Find the area under the graph of the standard normal distribution which lies
   (a) between $z = 0$ and $z = 0.76$;
   (b) between $z = -0.48$ and $z = 0$;
   (c) to the left of $z = 1.54$;
   (d) to the right of $z = 0.12$;
   (e) to the left of $z = -1.23$;
   (f) to the right of $z = -0.91$;
   (g) between $z = 0.88$ and $z = 1.88$;
   (h) between $z = -2.04$ and $z = -0.59$;
   (i) between $z = -0.31$ and $z = 1.50$;
   (j) between $z = -1.25$ and $z = 1.25$.

**4.** Find the area under the graph of the standard normal distribution which lies
   (a) between $z = 0$ and $z = 1.96$;
   (b) between $z = -0.85$ and $z = 0$;
   (c) to the left of $z = 0.75$;
   (d) to the right of $z = 0.10$;
   (e) to the left of $z = -0.75$;
   (f) to the right of $z = -1.11$;
   (g) between $z = 2.00$ and $z = 3.00$;
   (h) between $z = -3.00$ and $z = -3.09$;
   (i) between $z = -1.15$ and $z = 2.50$;
   (j) between $z = -1.45$ and $z = 1.45$.

**5.** Find $z$ if
   (a) the normal curve area between 0 and $z$ is 0.4484;
   (b) the normal curve area to the left of $z$ is 0.9868;
   (c) the normal curve area to the right of $z$ is 0.8413;
   (d) the normal curve area to the right of $z$ is 0.3300;
   (e) the normal curve area to the left of $z$ is 0.3085;
   (f) the normal curve area between $-z$ and $z$ is 0.9700.

**6.** Find $z$ (rounded to two decimals) so that
   (a) the normal curve area to the right of $z$ is 0.05;
   (b) the normal curve area between $-z$ and $z$ is 0.95;
   (c) the normal curve area to the right of $z$ is 0.02;
   (d) the normal curve area between $-z$ and $z$ is 0.98;
   (e) the normal curve area to the right of $z$ is 0.01;
   (f) the normal curve area between $-z$ and $z$ is 0.99.

**7.** A random variable has a normal distribution with the mean $\mu = 70.0$ and the standard deviation $\sigma = 8.4$. What are the probabilities that the random variable will take on
   (a) a value less than 82.6;
   (b) a value less than 63.7;
   (c) a value between 72.1 and 88.9;
   (d) a value between 57.4 and 86.8?

**8.** A random variable has a normal distribution with the mean $\mu = 65.3$ and the standard deviation $\sigma = 5.0$. What are the probabilities that this random variable will take on
   (a) a value greater than 62.8;
   (b) a value greater than 74.3;
   (c) a value between 58.3 and 63.3;
   (d) a value between 61.5 and 67.5?

CHAP. 6   SOME APPLICATIONS   157

9. A normal distribution has the mean $\mu = 51.6$. Find its standard deviation if 20 percent of the area under the curve lies to the right of 60.0.

10. A random variable has a normal distribution with the standard deviation $\sigma = 4.0$. Find its mean if the probability that the random variable will take on a value less than 87.6 is 0.9713.

11. Verify the result given on page 151, namely, that the area under the standard normal curve between $z = 0.25$ and $z = 0.75$ equals 0.1747.

## SOME APPLICATIONS

Let us now consider some examples dealing with random variables having at least approximately normal distributions. Suppose, for instance, that an analysis of the length of long distance telephone calls made from a large business office shows that the length of these telephone calls can be looked upon as a random variable having a normal distribution with a mean of 129.6 seconds and a standard deviation of 30 seconds. *With this information we can determine all sorts of probabilities, or percentages, about the length of long distance calls made from this office.* For instance, let us ask what percentage of these calls exceed 3 minutes or 180 seconds (and are, therefore, subject to overtime charges). The answer to this question is given by the shaded region of Figure 6.15, namely, that to the right of

$$z = \frac{180 - 129.6}{30} = 1.68$$

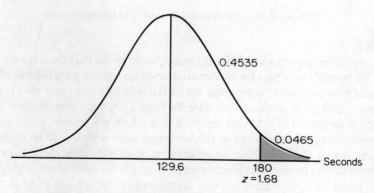

**FIGURE 6.15**  Distribution of lengths of telephone calls.

## 158 THE NORMAL DISTRIBUTION

and since the corresponding entry in Table I is 0.4535, we find that 0.5000 − 0.4535 = 0.0465 or approximately 5 percent of the long distance calls exceed 180 seconds (and are, therefore, subject to overtime charges).

Continuing with the same example, let us also find the probability that a long distance call (made at the given office) will last anywhere from 80 to 140 seconds. This probability is given by the area of the shaded region of Figure 6.16, namely, that between

$$z = \frac{80 - 129.6}{30} = -1.65 \quad \text{and} \quad z = \frac{140 - 129.6}{30} = 0.35$$

and since the corresponding entries in Table I are 0.4505 for $z = 1.65$ and 0.1368 for $z = 0.35$, we find that the answer is $0.4505 + 0.1368 = 0.5873$. Thus, the odds are just about 7 to 5 (0.5873 to 0.4127 to be more exact) that a long distance call made at the given office will last anywhere from 80 to 140 seconds.

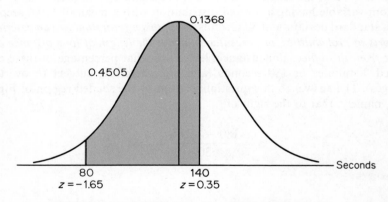

**FIGURE 6.16** Distribution of lengths of telephone calls.

To consider a somewhat different example, suppose that the actual amount of a "6 ounce" serving of a carbonated beverage which a soft-drink dispensing machine puts into paper cups varies from cup to cup, and that it can be looked upon as a random variable having a normal distribution with a standard deviation of 0.12 ounces. Well then, if only 4 percent of the cups are to contain less than 6 ounces of the beverage, *what will have to be the average amount which the soft-drink dispensing machine puts into these cups?* This example differs from the preceding ones in which we were given $\mu$ and $\sigma$, and were asked to determine a normal curve area. Now we are given a normal curve area (see Figure 6.17) and $\sigma$, and are asked to find $\mu$. Since the value of $z$

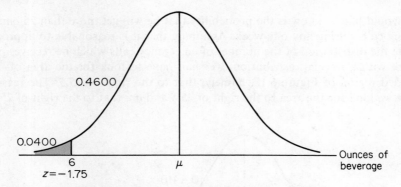

**FIGURE 6.17** Distribution of actual amount of beverage in "6-ounce" servings.

which corresponds to $0.5000 - 0.0400 = 0.4600$ is approximately 1.75 (according to Table I), we have

$$-1.75 = \frac{6.00 - \mu}{0.12}$$

and, to solve for $\mu$, we get

$$6.00 - \mu = -1.75(0.12) = -0.21$$

and then

$$\mu = 6.00 + 0.21 = 6.21 \text{ ounces}$$

So far as the owner of the soft-drink dispensing machine is concerned this is not very satisfactory, for he has to put quite a bit more of the beverage into the cups than the designated 6 ounces. If he wanted to *lower the average* yet keep 96 percent of the cups above 6 ounces, he would have to get a dispensing machine with *less variability*—for instance, in Exercise 4 on page 165 the reader will be asked to show that if the variability of the machine is reduced so that $\sigma = 0.04$ ounces, this will lower the required average amount of beverage per cup to $\mu = 6.07$ ounces, yet keep 96 percent of the cups above 6 ounces.

Although, strictly speaking, the normal distribution applies to *continuous measurements*, it is often used to approximate distributions of random variables which can take on only a finite number of values. This is perfectly all right in many situations, but we have to be careful to make the **continuity correction** illustrated in the following example. Suppose that the number of emergency calls which the operator of a tow truck gets per week is a random variable having the mean $\mu = 18.3$ and the standard deviation $\sigma = 4.6$. What we would like to know is the probability that he will get more than 25 emergency calls during any one week. Assuming that it is reasonable to approximate the distribution of the number of emergency calls which he receives per week with a normal distribution, we shall have to look for the area of the shaded region of Figure 6.18, namely, that to the right of 25.5. The reason why we look for the area to the right of 25.5 and *not* that to the right of 25 is

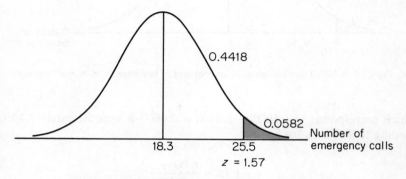

**FIGURE 6.18** Distribution of number of emergency calls.

that *the number of emergency calls which the operator receives per week can only be a whole number.* Thus, if we want to approximate the distribution of the number of calls with a normal curve, we have to "spread" the values of this random variable over a continuous scale, and we can do this by representing each whole number $k$ by the interval from $k - \frac{1}{2}$ to $k + \frac{1}{2}$. In particular, 5 is thus represented by the interval from 4.5 to 5.5, 12 is represented by the interval from 11.5 to 12.5, 25 is represented by the interval from 24.5 to 25.5, 26 is represented by the interval from 25.5 to 26.5, 27 is represented by the interval from 26.5 to 27.5, and the probability of getting a value greater than 25 is given by the area under the curve to the right of 25.5.* We thus get

$$z = \frac{25.5 - 18.3}{4.6} = 1.57 \text{ (approximately)}$$

and the area of the shaded region of Figure 6.18 turns out to be 0.5000 − 0.4418 = 0.0582. This means that the probability of the operator getting more

\* It does not matter in situations like this where we include the dividing lines between successive intervals; for instance, it does not matter whether we include 25.5 in the interval from 24.5 to 25.5 or in the interval from 25.5 to 26.5. As we indicated on page 144, *in the continuous case the probability of a point (that is, a single number) is for all practical purposes always equal to zero.*

than 25 emergency calls during any given week is approximately 0.06. The continuity correction which we have introduced in this example will be discussed further on page 167.

In Chapter 5 we showed that the probability of a random variable taking on a value within two standard deviations of the mean is always *at least* 0.75. This result was based on Chebyshev's Theorem (which, incidentally, applies also to continuous distributions), but if we know for a fact that we are dealing with a random variable having a normal distribution, we can make the much *stronger* statement that the probability of getting a value within two standard deviations of the mean actually *equals* 0.9544. To demonstrate this we have only to determine the area of the region shaded in the second diagram of Figure 6.19, namely, that between $z = -2$ and $z = 2$. Thus, if a random variable has a normal distribution, *approximately 95 percent of the time its values will fall within two standard deviations of the mean.* Correspondingly, if we look for the normal curve area between $z = -1$ and $z = 1$, and that between $z = -3$ and $z = 3$ (as in the first and third diagrams of Figure 6.19),

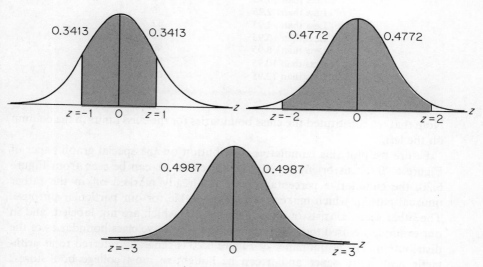

**FIGURE 6.19** Areas under normal distributions.

we find that *approximately 68 percent of the time the values of a normally distributed random variable will fall within one standard deviation of the mean and approximately 99.7 percent of the time its values will fall within three standard deviations of the mean.* It will be important to remember these percentages for future applications.

## PROBABILITY GRAPH PAPER

As in the examples of the preceding section, we often assume that observed data follow the over-all pattern of a normal distribution, and this raises the question "How can we tell whether this is actually the case?" Among the various methods that can be used to test whether the assumption is reasonable, or justifiable, the one we shall give here is *not* the best—it is largely subjective, but it has the advantage that it is very easy to perform. To illustrate this method, let us refer again to the telephone-order distribution, which we used as an example throughout Chapter 2. If we divide each of the cumulative frequencies on page 17 by 150 (the total number of telephone orders), and then multiply by 100 to express the figures as percentages, we obtain the following *cumulative percentage distribution:*

| Time Required (minutes) | Cumulative Percentage |
|---|---|
| Less than  0.95 | 0 |
| Less than  2.95 | 6 |
| Less than  4.95 | 32 |
| Less than  6.95 | 62 |
| Less than  8.95 | 90 |
| Less than 10.95 | $97\frac{1}{3}$ |
| Less than 12.95 | 100 |

Note that we substituted the class boundaries for the class limits in the column on the left.

Before we plot this cumulative distribution on the special graph paper of Figure 6.20, let us briefly investigate its scales. As can be seen from Figure 6.20, the cumulative percentage scale is already marked off in the rather unusual pattern which makes the paper suitable for our particular purpose. The other scale consists of equal subdivisions which are not labeled, and in our example we used them to represent the successive class boundaries of the distribution. The graph paper of Figure 6.20 is generally referred to as **arithmetic probability paper** and it can be bought in most college bookstores; actually, it would be much more appropriate to refer to it as **normal probability paper.** *

Now, if we plot the cumulative "less than" percentages which correspond to 2.95, 4.95, 6.95, 8.95, and 10.95 on this special kind of graph paper, we

---

* The paper is referred to as *arithmetic* in order to distinguish it from **logarithmic probability paper** in which the equal subdivisions are replaced by subdivisions like those on the logarithmic scale of a slide rule.

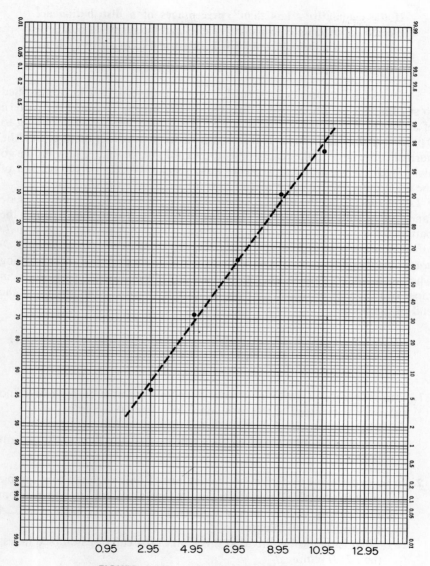

**FIGURE 6.20** Arithmetic probability paper.

**164** THE NORMAL DISTRIBUTION CHAP. 6

obtain the five points shown in Figure 6.20, and it should be observed that they almost fall on a straight line. This, in fact, is the criterion:

> If the cumulative "less than" percentages of a distribution are plotted on arithmetic probability paper and the resulting points lie very close to a straight line, this is construed as positive evidence that the distribution follows the general pattern of a normal curve.

So far as our example is concerned, it would thus seem that we are justified to say that the telephone-order distribution follows quite closely the pattern of a normal distribution. Note that we did not plot points corresponding to 0 percent at 0.95 and 100 percent at 12.95. As we pointed out on page 147, the normal curve never quite reaches the horizontal axis no matter how far we go away from the mean; thus, we never quite reach 0 percent or 100 percent of the area under the curve no matter how far we go in either direction.

Let us repeat that the method which we have described here is only a crude (and highly subjective) way of checking whether a distribution follows the pattern of a normal curve. Also, in a graph like that of Figure 6.20 *only pronounced departures from a straight line are real evidence that the data do not follow the pattern of a normal distribution.* A more objective way of testing whether observed data follow the pattern of a normal curve is discussed in most of the more advanced statistics texts listed in the Bibliography at the end of the book.

## EXERCISES

1. With reference to the long distance telephone example on page 157, find
   (a) what percentage of the calls lasted less than 60 seconds;
   (b) what percentage of the calls lasted more than 150 seconds;
   (c) above what number of seconds will we find the longest 30 percent of the calls.

2. In an experiment to determine the amount of time required to assemble an "easy to assemble" toy, the assembly time was found to be a random variable having approximately a normal distribution with $\mu = 32.4$ minutes and $\sigma = 4.0$ minutes.
   (a) What is the probability that this kind of toy can be assembled in less than 21.0 minutes?
   (b) What is the probability that the assembly of this kind of toy will take more than 36.0 minutes?
   (c) What is the probability that this kind of toy can be assembled in anywhere from 25.4 to 39.4 minutes?

3. The lengths of the sardines received by a certain cannery have a mean of 4.50 inches and a standard deviation of 0.20 inches. Assuming that the distribution of the length of these sardines can be treated as if it were a normal distribution, find
   (a) what percentage of these sardines are longer than 5.00 inches;
   (b) what percentage of these sardines are shorter than 4.20 inches;
   (c) the probability that if we randomly select one of the sardines its length will be between 4.40 and 4.60 inches;
   (d) the value below which we will find the shortest 10 percent of these sardines.

4. With reference to the example on page 159, show that if the soft-drink dispensing machine is adjusted so that $\mu = 6.07$ ounces and $\sigma = 0.04$ ounces (for the amount of the beverage it puts into the cups designed to serve 6 ounces), then 96 percent of the cups will contain at least 6 ounces.

5. In a very large class in business law, the final examination grades have a mean of 75.2 and a standard deviation of 9.7. Assuming that it is reasonable to treat the distribution of these grades (which, incidentally, are all *whole numbers*) as if it were a normal distribution, find
   (a) what percentage of the grades should exceed 89;
   (b) what percentage of the grades should be 50 or less;
   (c) the lowest A, if the highest 10 percent of the grades are to be regarded as A's;
   (d) the highest grade a student can get yet fail the test, if the lowest 15 percent of the grades are to be regarded as failing grades.

6. The number of days which guests stay at a large resort is a random variable having approximately a normal distribution with the mean $\mu = 8.5$ and the standard deviation $\sigma = 2.2$. Among 1,000 guests, how many can be expected to stay at the resort anywhere from 5 to 10 days, inclusive? (Use the continuity correction.)

7. A taxicab driver knows from experience that the number of "fares" he will pick up during an evening is a random variable which has the mean $\mu = 25.6$ and the standard deviation $\sigma = 4.2$. Assuming that the distribution of the number of "fares" picked up during an evening can be approximated with a normal distribution, find the probabilities that the driver will pick up (a) more than 30 "fares", and (b) fewer than 20 "fares." (Use the continuity correction.)

8. Convert the distribution of average hourly earnings on page 10 into a cumulative "less than" percentage distribution, and use arithmetic probability paper to judge whether it is reasonable to treat the data as having roughly a normal distribution.

**166** THE NORMAL DISTRIBUTION  CHAP. 6

9. Convert the distribution of finance charges of Exercise 7 on page 25 into a cumulative "less than" percentage distribution, and use arithmetic probability paper to judge whether this distribution has roughly the shape of a normal distribution.

10. Plot the cumulative "less than" percentage distribution of whichever data you grouped among those of Exercises 8 through 10 on pages 26 and 27 on arithmetic probability paper, and judge whether it is reasonable to treat the data as having roughly a normal distribution.

## APPROXIMATING THE BINOMIAL DISTRIBUTION

The normal distribution is often introduced as a continuous distribution which provides a very close approximation to the binomial distribution when $n$ (the number of trials) is large, and $p$ (the probability of a success on an individual trial) is close to 0.50. This is illustrated in Figure 6.21, which contains

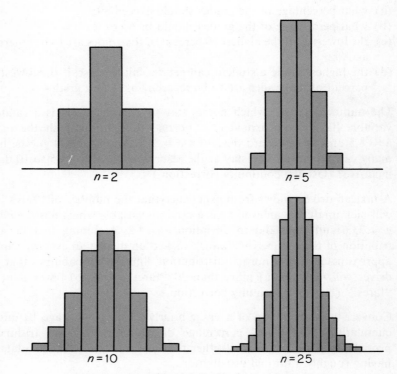

**FIGURE 6.21**   Binomial distributions with $p = 0.50$.

the histograms of binomial distributions having $p = 0.50$ and $n = 2, 5, 10,$ and 25—evidently, when $n$ becomes large, the distributions rapidly approach the bell-shaped pattern of a normal curve. In reality, normal distributions with

$$\mu = np \quad \text{and} \quad \sigma = \sqrt{np(1-p)}$$

are often used to approximate corresponding binomial distributions even when $n$ is "not too large" and $p$ differs from 0.50, but is *not too close* to either 0 or 1. It is a good rule of thumb to use this approximation only when $np$ and $n(1 - p)$ are *both* greater than 5.

To illustrate the normal curve approximation of a binomial distribution, let us first determine the probability of getting 6 *heads* and 12 *tails* in 18 tosses of a balanced coin. Substituting $n = 18, p = 0.50, x = 6$ (for the six heads), and $\binom{18}{6} = 18{,}564$ into the formula on page 124, we get

$$18{,}564(0.50)^6(1 - 0.50)^{12} = 0.0708$$

after fairly tedious calculations. A good deal of this work can be avoided by using the normal curve approximation, namely, by determining the area of the shaded region of Figure 6.22, where the interval from 5.5 to 6.5 represents *six* heads in accordance with the *continuity correction* mentioned on page 159.

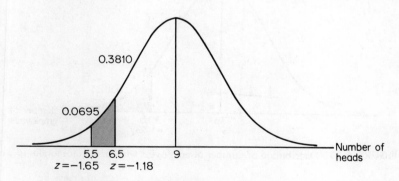

**FIGURE 6.22** Distribution of number of heads ($n = 18$).

Since $\mu = 18(0.50) = 9$ and $\sigma = \sqrt{18(0.50)(0.50)} = 2.12$ in this case, we shall have to find the area between

$$z = \frac{5.5 - 9}{2.12} = -1.65 \quad \text{and} \quad z = \frac{6.5 - 9}{2.12} = -1.18$$

for which the corresponding entries in Table I are 0.4505 and 0.3810. Thus, the probability we are trying to determine is $0.4505 - 0.3810 = 0.0695$, and it should be observed that the difference between the original result and the value obtained by means of the normal curve approximation is only $0.0708 - 0.0695 = 0.0013$ (which is negligible for most practical purposes).

The normal curve approximation of binomial distributions is especially useful in problems in which we would otherwise have to calculate the probabilities of many different values of a random variable. Suppose, for example, we want to determine the probability that *at least* 40 of the 50 employees of a business will drive an automobile to work on a certain day, when the probability that any one of the employees will drive to work is 0.70. (In other words, we want to know the probability of getting *at least* 40 successes in 50 trials, when the probability of a success on any one trial is 0.70.) If we tried to solve this problem by using the formula for the binomial distribution, we would have to find the *sum* of the 11 probabilities which correspond to 40, 41, 42, . . ., 49, and 50 successes, and it would hardly seem necessary to point out that this would require an enormous amount of work. However, since $np = 50(0.70) = 35$ and $n(1 - p) = 50(0.30) = 15$ both exceed 5 (the minimum suggested on page 167), we can use the normal curve approximation and find the answer by determining the area of the shaded region of Figure 6.23. Note that by looking for the area to the right of 39.5, we are again using

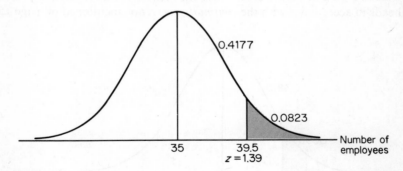

**FIGURE 6.23** Distribution of number of employees who drive automobiles to work.

the continuity correction discussed on page 160—40 is represented by the interval from 39.5 to 40.5, and "at least 40" is given by the interval *to the right of 39.5*. Since $\mu = 50(0.70) = 35$ and $\sigma = \sqrt{50(0.70)(0.30)} = 3.24$ in this example, we find that corresponding to 39.5 we get

$$z = \frac{39.5 - 35}{3.24} = 1.39$$

and it follows (according to Table I) that the desired probability is $0.5000 - 0.4177 = 0.0823$. As can easily be verified, the odds are thus about 11 to 1 *against* the possibility that 40 or more of the 50 employees of the given business will drive an automobile to work. As a matter of interest, let us point out that the *actual value* of the probability (looked up in a more extensive table than Table XI and rounded to three decimals) is 0.079, so that both answers are 0.08 rounded to two decimals.

## EXERCISES

1. Use the normal curve approximation to find the probability of getting five heads and seven tails in twelve flips of a balanced coin, and compare the result with the value given for this probability in Table XI.

2. Use the normal curve approximation to find the probability that a student will get at least eight correct answers in a true-false test, if he decides how to mark each of the 14 questions by flipping a balanced coin. Compare the result with the answer which we would get with the use of Table XI.

3. A multiple-choice test consists of 25 questions, each with four possible answers. If a student answers the test by guessing the answers before reading the questions (that is, by randomly checking the first, second, third, or fourth answer to each question), find with the use of the normal curve approximation the probability that he will get
   (a) exactly six correct answers;
   (b) at least seven correct answers;
   (c) fewer than four correct answers;
   (d) more than two but fewer than nine correct answers.

4. If 60 percent of the customers of a very large department store charge all of their purchases, what is the probability that among 600 customers (randomly selected) at least 366 charge all their purchases.

5. An electrical appliance manufacturer claims that 20 percent of all appliance breakdowns are caused by the failure of customers to follow operating instructions. If this claim is correct, what is the probability that among 100 breakdowns, more than 25 were caused by the failure of customers to follow operating instructions?

6. What is the probability that fewer than 575 of 1,000 patrons of a drive-in theater will patronize its refreshment stand, when the probability that any one of them will patronize the refreshment stand is 0.60?

7. If 80 percent of the loan applications received by a finance company are approved, what is the probability that among 200 loan applications more than 150 will be approved?

8. A department store manager knows that on the average 3 percent of the sales slips filled out by the department store's clerks contain mistakes. What is the probability that among 1,500 of these sales slips at least 50 will contain mistakes?

9. To illustrate the Law of Large Numbers which we discussed on page 140, suppose that among all adults living in a very large city there are as many men as there are women. Using the normal curve approximation, find the probabilities that in a random sample of adults living in this city the *proportion* of men will be anywhere from 0.49 to 0.51 when the number of persons in the sample is (a) 100, (b) 1,000, and (c) 10,000.

# 7

# CHANCE VARIATION: SAMPLING

**INTRODUCTION**

So far we have often referred to sets of data as **samples** without defining this term in a rigorous way. Intuitively, most persons know what the word "sample" means, and if we look it up we find that a sample is "a part to show what the rest is like" according to one dictionary, and "a portion, part, or piece taken or shown as representative of the whole" according to another. All this will probably make sense to most readers, but difficulties *can* arise when we ask ourselves what is meant here by "the rest" and "the whole." Suppose, for instance, that after hearing two recordings out of a large album of recordings made by a certain musical group, you intensely dislike and criticize these recordings. Well then, *is it your intention to criticize the two recordings which you heard, all of the recordings in the album made by the musical group, or all of the recordings which the musical group has ever made or will make?* In the first case, in which you criticize only the two recordings which you heard, you would have a **100 percent sample,** namely, *complete information* about the recordings to which you refer. In the other two cases, however, you would have only *partial information*, and it should be evident that in the second case you would have a *better sample* than in the third, namely, *relatively* more information for making any kind of generalization.

The whole question "What is a sample a sample of?" is of critical importance in statistics. For instance, if a buyer for a chain of food stores inspects the contents of a few baskets of fresh strawberries at a certain farm in California, are these strawberries to be looked upon as a sample of all the strawberries for sale at that farm on that particular day, all the strawberries which are regularly for sale at that farm, all the strawberries for sale at all farms in that locality, or perhaps even all the strawberries for sale at all farms in the state of California? Similarly, if we observe the hair styles of some of the women who visit an art museum on a given day, do our observations reflect the hair styles of all the women who visit that museum, all women who enjoy art, or all women in the United States? Finally, suppose that a person working for a market research organization stands on a street corner at a college campus and records how many passers-by wear shoes, how many wear sneakers, how many wear some other kind of footwear, and how many are barefooted. Do his findings reflect the types of footwear worn (or not worn) on that particular college campus, in the community where the college is located, in the United States, or in the entire world of fashion? As may be apparent, each of these questions can be answered by "Take your pick!", but it should be understood that in some instances we would have *good* samples which lend themselves to meaningful generalizations, while in other instances it would be utter folly to make any generalization whatsoever.

In statistics, we refer to the "whole" of which a sample is a part as the **population** from which we are sampling, and we do this even when we are talking about the diameters of oranges packed in a crate, the weights of the loaves of bread sold by a certain bakery, the speeds of aircraft, the number of bankruptcies recorded annually in the United States, and so on. As we explained in the beginning of Chapter 1, statistics used to concern itself mostly with the descriptions of *human populations* (we now call this *demography*), and as the scope of the subject grew, the term "population" has assumed this much wider connotation. Whether or not it sounds strange to refer to such things as the diameters of oranges packed in a crate as a population is beside the point—in statistics, "population" is a technical term with a meaning of its own.

A population is referred to as **finite** if it consists of a *fixed number* (that is, some definite whole number) of elements, items, objects, measurements, or observations. For instance, if we are analyzing the wages paid to all of the welders presently employed by a shipyard without any intention of making any generalizations, these wages constitute a finite population. Similarly, if we are studying the amounts of cash held by all commercial banks in a certain city on January 1, 1974 without any intention of making any generalizations, these amounts constitute a finite population. Finally, we would be sampling from a finite population if we randomly chose a number from 1 to 10, if we selected three of the 20 trucks in a fleet for a routine safety check, or if an auditor verified the accuracy of five invoices taken in some way from all the invoices which a large business firm received on a given day. We have made it a point to refer to these populations as *finite*, because there are also situations where there is (hypothetically, at least) no limit to the number of observations that could be made. This will be discussed further on page 174.

It may have occurred to the reader from the preceding discussion that *whether a set of data constitutes a finite population or a sample depends entirely on what we intend to do with it.* For instance, if we are given complete information about the sales made by drug stores in Buffalo, New York, in 1973, these data constitute a finite population if we are not going to generalize, say, to corresponding figures for the whole state of New York or for other years; the moment we make any kind of generalization, however, the data must be looked upon as a sample. Thus, it is always important to state whether a given set of data is to be regarded as a sample or a population, and to emphasize the distinction we even use different symbols for the statistical descriptions of samples and populations. For instance, whereas $\bar{x}$ denotes the mean of a sample, the Greek letter $\mu$ (*mu*) is used to denote the mean of a population, and it is not a coincidence that this is the same symbol which we used in Chapter 5 for the mean of a probability distribution. To illustrate, suppose that ten slips of paper are numbered 1, 2, 3, ..., 9, and 10, and that we are asked to draw one of these slips after they have been thoroughly mixed in a hat. Since the probability for each number is thus equal to $\frac{1}{10}$, the mean of this probability distribution is

$$\mu = 1(\tfrac{1}{10}) + 2(\tfrac{1}{10}) + 3(\tfrac{1}{10}) + \cdots + 9(\tfrac{1}{10}) + 10(\tfrac{1}{10})$$
$$= 5.5$$

according to the formula on page 133, and if we factor out $\frac{1}{10}$ it can easily be seen that the result which we have obtained is the "ordinary" mean of the finite population which consists of the numbers 1, 2, 3, ..., 9, and 10, namely,

$$\frac{1 + 2 + 3 + \cdots + 9 + 10}{10} = 5.5$$

This kind of argument holds whenever each element of a finite population has the same chance of being selected (as part of a sample).

To distinguish between samples and populations we not only use different symbols for their statistical descriptions, but we also refer to these descriptions by different names:

> **Descriptions of samples are called *statistics*** (as we have already indicated on page 29), **and descriptions of populations are called *parameters*.**

Hence, we say that $\bar{x}$ is a **statistic** whereas $\mu$ is a **parameter**. This terminology has the advantage that it will prevent a good deal of the confusion which might otherwise arise in subsequent chapters: *We will always use statistics to reach decisions about parameters* (namely, generalize from descriptions of samples to corresponding descriptions of populations).

To introduce the concept of an **infinite population,** let us refer again to the example where ten slips of paper, numbered from 1 to 10, are thoroughly mixed in a hat. Suppose now that we repeatedly draw one of the slips of paper, but that each slip is replaced before the next one is drawn. This is called **sampling with replacement,** and *except for the element of time* there is

**174** CHANCE VARIATION: SAMPLING CHAP. 7

no limit to the number of drawings that could be made. We could make 100 drawings, even a million or more, and by replacing each slip before the next one is drawn we create the illusion of an *endless supply*, namely, an *infinite population*.

Generally speaking, we say that we are sampling from an infinite population if there is no limit to the number of observations that *could* be made. For instance, we might say that five flips of a coin constitute a sample from the infinite population of all (hypothetically possible) flips of the coin; we might say that a dozen double-edge razor blades constitute a sample from the (hypothetically infinite) population of *all* the past, present, and future double-edge razor blades that will ever be made by a given firm; and we often look at the results of experiments as a sample from the (hypothetically infinite) population which consists of the results we would, or could, obtain if the experiment were repeated over and over again.

Infinite populations are not easy to visualize. Suppose, for instance, we are interested in the amounts of time that are required by cooks in a chain of restaurants to prepare the individual orders for food. If we tried to picture the "population" of *all* past, present, and future food orders prepared by the cooks, we would be confronted with a mountain of food which would exceed the height of Mount Everest. Or, if we considered the amounts of time required to prepare all these orders, we would be faced with millions upon millions of numbers, most of which could not even be determined because the food has not yet been ordered or prepared. Fortunately, there is a much easier way of looking at a situation like this—instead of trying to visualize an endless pile of food orders or an endless list of numbers (representing the times it takes to prepare the orders), we picture the times as *grouped into a distribution*, say, like the one of Figure 7.1, *which tells us the probability that, to the nearest minute, the time required to prepare any one of the orders will be 5 minutes, 6 minutes, 7 minutes, . . ., 15 minutes, . . ., and so on.* An infinite population of times (or other kinds of measurements) can thus be pictured as a probability distribution which we call the **population distribution.** If we wanted to look at the preparation times of the food orders as values of a continuous random variable (and not as rounded to the nearest minute), we would picture the population distribution as a probability density, perhaps, like the one shown in Figure 7.2.

To most statisticians, the terms "population" and "population distribution" have become practically synonymous. For instance, if an airline weighs the suitcases of 100 passengers and these weights are values of a random variable having a normal distribution, they may well say that they are *sampling from a normal population*. Similarly, a wholesale buyer of peaches might say that he is *sampling from a normal population* when, in fact, he is measuring the diameters of peaches offered for sale by several growers (and it is reasonable for him to assume that his measurements are values of a random

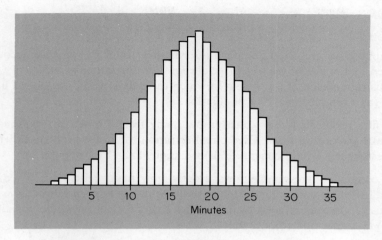

**FIGURE 7.1** Population distribution of food preparation times.

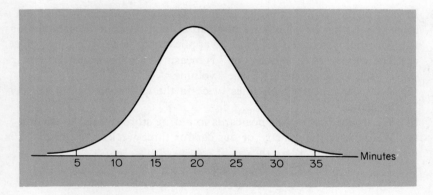

**FIGURE 7.2** Population distribution of food preparation times.

variable having a normal distribution). Statisticians even say that they are *sampling from a binomial population* when they observe the number of heads obtained in 100 flips of a coin, when they count the number of customers who exchange their coupons for free tubes of toothpaste, or when they determine the number of charge-account customers who make late payments on their debts. All this really means that in each case they are dealing with a random variable (the number of heads, the number of customers who exchange their coupons, and the number of charge-account customers who make late payments) which has a binomial distribution.

## EXERCISES

1. Suppose we are given complete information about the expense accounts filed by the executives of a large corporation for January, 1974. Give one illustration each of a problem in which these data would be looked upon as (a) a sample, and (b) a population.

2. A film producer knows exactly how many feet of film were required to make each of his films. Give one illustration each of a problem in which these figures would constitute (a) a sample, and (b) a population.

3. A certain congressman knows exactly how many letters his office has mailed to his constituents each month during his term of office. Give one illustration each of a problem in which this information would constitute (a) a sample, and (b) a population.

4. A college athletic director has a complete record of the number of wins and losses of each of his varsity teams. Give one illustration each of a problem in which these figures would be (a) a sample, and (b) a population.

5. In each of the following examples state whether we are sampling from a finite population or an infinite population, and describe the population:
   (a) An employer selects seven of 30 applicants to work in his new office.
   (b) The volume of a "moon rock" is measured six times and the average is used to determine its "true" volume.
   (c) An actor rehearsed his lines twice in the afternoon before the performance.
   (d) We selected two of 14 postcards in a drug store to mail to friends.
   (e) We observe the mileage of our car for three weeks to determine the "miles per gallon" for the car.

## RANDOM SAMPLING

The purpose of most statistical studies is to make generalizations from samples to finite or infinite populations, and there are certain rules that must be observed in order to avoid results which are obviously incorrect, irrelevant, or misleading. In other words, not all samples lend themselves to making valid generalizations, and as we pointed out in Chapter 1, there are many pitfalls. To give a few examples, suppose, for instance, that we want to determine the average income of individuals in a certain city. It is very unlikely that we would arrive at anything even remotely close to the correct figure if we sampled only the income of corporation executives, or if we sampled only the income of persons who are presently on welfare. Similarly, we can hardly expect to get a reasonable generalization about college grades from data pertaining only to students who are on academic probation, and we can hardly expect to be able to infer much about the level of air pollution in a city if we sample only the air in the city dump. Of course, these examples are extreme, but they serve to illustrate some of the many things we have to watch out for if we want to make meaningful and useful generalizations on the basis of sample data.

The whole problem of when and under what conditions samples permit reasonable generalizations is not easily answered. So far as the work in this book is concerned, we shall partly skirt the issue by limiting our discussion to so-called **random samples**—some other kinds of samples will be mentioned very briefly in the section which follows. To illustrate the idea of a random sample, let us begin with samples from *finite populations*. Suppose, for instance, that a finite population is of **size** 6, namely, that it consists of six elements which we shall label $a$, $b$, $c$, $d$, $e$, and $f$. (These elements of the finite population might be the commissions earned by a salesman in six consecutive weeks, the dividends paid by six different stocks, the weights of six football players, the ages of six management trainees, and so on.) Suppose, furthermore, that we want to select four of the six elements of this population and that we are curious to know in how many different ways this might be done. In other words, we want to find out *how many different samples of* **size** *4 can be selected from a finite population of size 6*. The answer to this question is easily determined; in fact, using the rule on page 75, we immediately get

$$\binom{6}{4} = \frac{6 \cdot 5 \cdot 4 \cdot 3}{4!} = 15$$

One of these samples consists of the elements $a$, $b$, $c$, and $d$; another consists of the elements $b$, $c$, $d$, and $e$; a third consists of the elements $a$, $c$, $d$, and $f$; and the reader should not find it too difficult to list all 15 possibilities.

Now if we select one of the 15 possible samples in such a way that each has the same probability of $\frac{1}{15}$ of being chosen, we say that the sample we get is a **simple random sample** or, more briefly, a **random sample**. One way in which we could obtain such a sample would be to write each possible combination of four of the letters $a$, $b$, $c$, $d$, $e$, and $f$ on a separate slip of paper, mix the 15 slips of paper thoroughly, and then draw one. This could easily be done, and it clearly conveys the idea that *the selection of a random sample from a finite population must somehow be left to chance, with each possible sample having the same probability.*

More generally, if we deal with a finite population of size $N$ from which we want to select a sample of size $n$, we say that

The sample we get is random if it is selected in such a way that each of the $\binom{N}{n}$ possible samples has the same probability of $\dfrac{1}{\binom{N}{n}}$ of being chosen.

How this is done in actual practice is another matter. We could, of course, use slips of paper as in the preceding example, but in most practical situations it is virtually impossible, or at least highly impractical, to list all possible samples. For instance, if a random sample of size 3 has to be drawn from a finite population of size 500, there are

$$\binom{500}{3} = \frac{500 \cdot 499 \cdot 498}{3!} = 20{,}708{,}500$$

possible samples, and it would hardly seem like a pleasant prospect to have to label that many slips of paper. Fortunately, all this work is not really necessary, for we could also obtain a random sample by listing the elements of the population on 500 slips of paper and then drawing three at once or one after the other (without replacing the slip or slips which have already been drawn). As the reader will be asked to show in Exercise 7 on page 184, this will also yield a random sample provided that in each successive drawing all of the remaining elements of the population have the same chance of being selected.

The job of numbering slips of paper can be avoided altogether if we use instead a table of **random numbers,** or **random digits,** which consists of pages on which the digits 0, 1, 2, 3, 4, 5, 6, 7, 8, and 9 are recorded in a "random" fashion, much as they would appear if they had been generated by a gambling device. We could construct a table of random numbers by using a spinner like that of Figure 7.3, but in actual practice this is usually done with electronic computers. Several commercially-published tables of random numbers are listed in the Bibliography at the end of the book, and it should be observed that before such tables are published they are generally required to "pass" various statistical tests intended (insofar as this is possible) to ensure their randomness. Another point which must not be overlooked is that *tables of random numbers will yield random samples only if they are properly used.* For *instance*, if someone presents us with the numbers 7, 7, 7, 7, 7, 7, 7, 7, 7 and refers to them as a random sample because he copied them from a table of random numbers, this is merely a bad joke if he *intentionally* picked one 7 here, one 7 there, another 7 there, and so on.

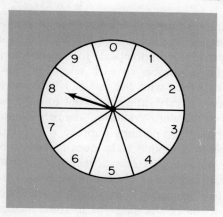

**FIGURE 7.3** Spinner.

To illustrate the proper use of a table of random numbers (in this case Table IX at the end of the book), suppose that a tax collector wants to select, at random, ten of 300 income tax returns for a careful audit. Numbering the income tax returns from 1 to 300, or better, numbering them 001, 002, 003, 004, . . ., 299, and 300 (so that each of them is represented by a sequence of three digits), he arbitrarily picks a page in a table of random numbers, he arbitrarily picks three columns (usually consecutive ones as a matter of convenience), and he arbitrarily picks a row from which to start. Then he reads off three-digit numbers from these columns moving down the page (or up, if so desired), and if necessary he continues on another page or another set of columns. For instance, if he arbitrarily chose the 11th, 12th, and 13th columns of the table on page 334 and started with row 26, he would get 910, 187, 315, 781, 173, 539, . . ., but recording only the numbers from 001 through 300, he would get 187, 173, 158, 107, 163, 252, 041, 023 from page 334, and 239, 228 continuing with the 11th, 12th, and 13th columns of page 335. Thus, the random sample consists of the income tax returns whose numbers are 187, 173, 158, 107, 163, 252, 41, 23, 239, and 228. Note that when he made this selection he ignored all numbers greater than 300, and he would also have ignored a number already chosen for the sample if it occurred again. Had the tax collector wanted to be "even more random" in this illustration, he could have used some gambling device to choose the page, the columns, and the row from which to start.

Now that we have explained what we mean by random sampling from finite populations, we should add that in actual practice this is often much easier said than done. For instance, if we wanted to take a sample in order to estimate the average weight of 10,000 live chickens, it would hardly be practical to number the live chickens 0000, 0001, 0002, 0003, . . ., 9998, and 9999, choose four-digit random numbers from a table of random numbers, and then weigh the corresponding chickens.* Similarly, it would be physically impossible (or at least highly impractical) to sample television viewers in a large city by assigning a number to each television viewer, or to sample chimpanzees living in Africa with the use of a table of random digits. In situations like this there is very little choice but to proceed according to the dictionary definition of the word "random," namely "haphazardly, without definite aim or purpose," or perhaps improve the situation somewhat by using one of the special sampling techniques described briefly in the next section. *Then we keep our fingers crossed that statistical theory intended for random samples (for instance, that of Chapters 8 through 10) can nevertheless be employed.* This is true, particularly, in situations where we have little, if any, control over the selection of the data—as in some medical research where scientists have no choice but to use whatever cases happen to be available.

Because of the difficulty of assuring the randomness of data, we must always be on the look-out for biases such as those mentioned in Chapter 1. Sometimes, a lack of randomness can be detected by looking at the data, themselves; for instance, if a stock earned $0.25, $0.32, $0.29, $0.45, $0.53, and $0.60 per share in six successive years, it would hardly seem reasonable to look upon these figures as a random sample of the earnings of this stock. Evidently, there is a general *upward trend* (things are gradually getting better), and it would be very unreasonable to analyze these data by applying theory that is ordinarily reserved for random samples. (In more advanced work in statistics, the reader will find that there exist much more objective ways of judging whether any "unusual" features of a set of data are due to chance or whether they are really indicative of a lack of randomness.)

So far we have discussed only random sampling from *finite populations*, and, as may well be expected, the situation is quite different when it comes to *infinite populations*. Since we cannot very well number the elements of an infinite population, we proceed as on page 174 and refer to the *population distribution*. Thus, we say that

> **A sample from an infinite population is random if it consists of independent observations of a random variable having the corresponding population distribution.**

As in Chapter 4, the term "independent" is meant to imply that the selection of any one sample value in no way affects the selection of any subsequent value. Thus, we refer to the lifetimes of 50 electric light bulbs, the breaking strengths of 25 pieces of nylon cord, the diameters of 100 logs shipped by a

---

* Note that we started with 0000 in this example in order to avoid having to work with five-digit numbers.

sawmill, or the manual dexterity scores of 20 job applicants as random samples, hopefully assuming that the above conditions are satisfied, namely, that in each case we are dealing with independent observations of a random variable having the corresponding population distribution. All too often, we have no choice but to justify this assumption on the grounds that there are no obvious indications that "anything is wrong," namely, that there are no obvious biases or other indications of a lack of randomness. To draw an analogy, this is like saying that *a sample is innocent of the "crime" of non-randomness unless it can be proven guilty.*

## FURTHER PROBLEMS OF SAMPLING

So far we have mentioned only random samples, and we did not even consider the possibility that under certain conditions there may be samples which are *better* (say, *cheaper, easier to obtain,* or *more reliable*) than random samples, and we did not go into any details about the question of what might be done when random sampling is impossible. Thus, let us take a brief look at the following questions:

Can we improve on simple random sampling?
What can we do when it is physically impractical or otherwise impossible to use random sampling?
What can we do when we have little or no choice about the selection of the data?
How about the cost of sampling?

The answer to the first of these questions is in the *affirmative* provided we have some knowledge about the "make-up" of the population with which we are concerned. Suppose, for instance, that age is a factor which has an important bearing on public opinion concerning an election, and that it is known that in a certain community 20 percent of all eligible voters are over 65, 50 percent are from 35 to 65, and 30 percent are under 35. In that case, a pollster who intends to interview 900 voters could *improve* on simple random sampling (that is, make his prediction of the outcome of the election more reliable) by intentionally selecting a random sample of 0.20(900) = 180 eligible voters from the over 65 group, a random sample of 0.50(900) = 450 eligible voters from the 35 to 65 group, a random sample of 0.30(900) = 270 eligible voters from the under 35 group, and combine the results. This is a form of **stratified random sampling,** namely, a kind of sampling in which definite portions of the total sample are allocated to different parts, or **strata,** of the population. Stratified sampling is widely used and it can be very effective provided one stratifies with respect to *truly relevant characteristics* of the population; unfortunately, it can be quite difficult to determine what characteristics are really relevant in a given situation.

In actual applications of stratified sampling, the selection of the individuals (or items) within the various strata is often *non-random*. Instead, interviewers are given quotas to be filled from the various strata, without too many restrictions as to *how* they are to be filled. For instance, in determining public opinion about a school board election, an interviewer may be told to interview 10 housewives who are college graduates and live in apartments, 8 professional men with advanced college degrees who own their homes, 12 male wage earners who are not college graduates and rent their homes, etc., with the actual selection being left to the interviewer's discretion. This kind of **quota sampling** is often convenient and relatively inexpensive, but as it is usually executed, the resulting samples will probably not have the essential features of random samples. As we already pointed out on page 2, there is the human element which may lead a poll taker to interview persons who all live in the same apartment house or neighborhood, who all work in the same office, factory, or store, and who may be most pleasant to talk to. Thus, quota samples are often **judgment samples** and do not lend themselves to any form of statistical evaluation.

As we already pointed out on page 180, there are situations where simple random sampling is very impractical or virtually impossible, and the examples which we used pertained among other things to the sampling of television viewers in a large city. In a situation like this we might use what is called **cluster sampling** by dividing the entire city into sections (say, of five city blocks each), randomly selecting a few of these sections, and then determining which television programs are being watched (or whatever happens to be of interest) by interviewing each viewer, or a random sample of viewers, within each of the chosen sections. Cluster sampling is used very extensively in sample surveys conducted by government agencies and private research organizations, and when the sections (subdivisions) are counties, villages, acreages, . . ., or city blocks as in our example, we refer to this kind of sampling also as **area sampling**.

In some instances, the most practical way of choosing a sample is to select, say, every 50th customer in the accounts receivable ledger, every 20th piece coming off an assembly line, or every tenth office in a very large office building. This is called **systematic sampling**, and it can be made random (at least in part) by using random numbers or some gambling device to pick the unit with which to start. The danger of this kind of sampling lies in the possible existence of *hidden periodicities*—for instance, if we selected every tenth office in a very large office building, our results would probably be biased if it so happened that every tenth office in the building is a spacious corner suite commanding a higher rental than the other offices. Similarly, our results would be very misleading if we inspected every 20th piece coming off an assembly line, and it so happened that (due to regularly recurring failure) every tenth piece is defective.

Although results based on cluster samples or systematic samples are generally not as reliable as results based on simple random samples *of the same size*, they are usually more reliable *per unit cost*. For instance, it is much cheaper to obtain data, say, concerning the eyesight of students in the same school or from several schools in the same school district than it is to obtain such data from schools throughout the whole state, and in a situation like this a cluster sample of size 1,000 may be *cheaper to obtain and more reliable* than a small random sample of size 100.

As we already indicated on page 180, we often have to keep our fingers crossed that statistical theory intended for random samples can be employed even though we had little or no control over the way in which our data were obtained. For instance, we would really have no choice but to rely on whatever records happen to be available, say, if we wanted to predict a toy manufacturer's December sales, if we wanted to estimate the frequency of a rare type of industrial accident, or if we wanted to study the patterns of customer traffic through the aisles of a department store. In situations like this we can only hope to make sure that there are no obvious indications that "anything might be wrong"—that there is no trend as in the illustration on page 180, that there are no sudden *shifts or changes* which might occur, for instance, when different measuring instruments are used for different parts of an experiment, and that there are no *regularly recurring patterns* like those which we might find in the performance of an assembly line worker in a factory, who is apt to be alert and efficient at the beginning of his shift, but tired and careless near the end.

## EXERCISES

1. What is the probability of each possible sample if a random sample of size 3 is taken from
   (a) a finite population of size 12;
   (b) a finite population of size 20?

2. What is the probability of each possible sample if a random sample of size 5 is taken from
   (a) a finite population of size 15;
   (b) a finite population of size 25?

**3.** What is the probability of each possible sample if
   (a) a random sample of size 2 is taken from a finite population of size 50;
   (b) a random sample of size 3 is taken from a finite population of size 75?

**4.** Referring to the illustration on page 177, list the 15 possible samples of size 4 which can be drawn from the finite population which consists of the elements $a$, $b$, $c$, $d$, $e$, and $f$. Also, show that if one of these samples is selected at random, the probability that it will include element $c$ is $\frac{2}{3}$. Does the same probability apply to elements $a$, $b$, $d$, $e$, and $f$?

**5.** Referring to Exercise 4, what is the probability that any specific pair of elements, say, $c$ and $e$, will be included in the sample? Does this probability apply to each possible pair?

**6.** List all possible samples of size 2 that can be drawn from the finite population which consists of the following European capitals: London, Rome, Amsterdam, Paris, Madrid, and Vienna. Show that if one such sample is chosen at random, the probability that it will include any particular one of these capitals (for example, Rome) is $\frac{1}{3}$.

**7.** On page 178 we suggested that a random sample from a finite population can also be obtained by choosing the elements *one after the other* (without replacement), making sure that in each case all of the remaining elements have the same probability of being selected. To verify this for our example, where a random sample of size 4 was taken from the finite population which consists of the elements $a$, $b$, $c$, $d$, $e$, and $f$, let us first determine the probability of getting $a$, $b$, $c$, and $d$ *in that order*. Using the generalized multiplication rule of Exercise 18 on page 112, we get $\frac{1}{6} \cdot \frac{1}{5} \cdot \frac{1}{4} \cdot \frac{1}{3} = \frac{1}{360}$, and since this probability is the same for *each possible order* in which the elements $a$, $b$, $c$, and $d$ can be obtained, of which there are 4! = 24, the desired probability of obtaining the elements in *any order* is $24 \cdot \frac{1}{360} = \frac{1}{15}$. This agrees with the probability obtained on page 177.
   (a) Use this method to show that if a random sample of size 3 is taken from a finite population of size 500, then each possible random sample has the probability $\frac{1}{20{,}708{,}500}$ (which agrees with the fact that there are 20,708,500 possible samples, as was shown on page 178).
   (b) Use this method to show that if a random sample of size $n$ is taken from a finite population of size $N$, then each possible random sample has the probability $\frac{n!}{N(N-1) \cdot \ldots \cdot (N-n+1)}$, and verify that this agrees with the formula on page 178.

CHAP. 7            FURTHER PROBLEMS OF SAMPLING    **185**

8. Suppose that the Dean of Students of a college wants to have lunch with 12 entering freshmen of a class of 427 (who are listed in the registrar's office in alphabetic order). Which ones (by numbers) will he invite, if he numbers the freshmen from 001 through 427 and choses a random sample by means of random numbers, using the last three columns of the table on page 335 beginning with the first row?

9. The accountants of a large department store regularly check samples of the invoices made out by the store's sales personnel. If a certain saleslady's invoices number 12713 through 12789 on a given day, which ones would they check if they used the 34th and 35th columns of the table on page 335 starting with the 8th row to select a random sample of five of her invoices?

10. The employees of a government agency have badges numbered serially from 1 through 907. Use the 18th, 19th, and 20th columns of the table on page 334 starting with the sixth row to select a random sample of eight of the employees to serve on a special committee.

11. **SIMULATION** Random numbers can also be used to simulate various probability distributions and games of chance.
    (a) Letting 0, 2, 4, 6, and 8 represent *heads*, and 1, 3, 5, 7, and 9 *tails*, use the sixth column of the table on page 334 to simulate 50 flips of a coin. How many heads were there and how many tails? How many times were there *three or more heads in a row* and how many times were there *three or more tails in a row?*
    (b) Describe in detail how random numbers might be used to simulate 240 rolls of a balanced die. Actually perform this simulation and indicate how many times the "die" came up 1, 2, 3, 4, 5, and 6. Do these results agree more or less with what one would expect to get with a balanced die?
    (c) To simulate the results (number of heads) we might obtain if we repeatedly flipped a pair of coins, we could proceed as in part (a) and use two random digits, one for each coin. Explain why we could also simulate this "experiment" by using two-digit random numbers and letting 00 through 24 represent *0 heads and 2 tails*, 25 through 74 represent *1 head and 1 tail*, and 75 through 99 represent *2 heads and 0 tails*. Use this alternate method to simulate 200 flips of a pair of coins, and construct a table showing how many times there were, respectively, 0, 1, and 2 heads.

12. A stratified sample of size $n = 80$ is to be taken from a group of persons in which 800 are not college graduates, 900 hold only B.A. degrees, 200 hold also M.A. degrees, and 100 others hold Ph.D. degrees. How many

should be chosen from each of these four strata so that the individual sample sizes are *proportional* to the sizes of the respective strata?

13. A stratified sample of size $n = 200$ is to be taken from among 2,000 families with incomes under $8,000, 1,500 families with incomes from $8,000 to $12,000, 500 families with incomes from $12,000 to $15,000, 800 families with incomes from $15,000 to $25,000, and 200 families with incomes over $25,000. How many should be chosen from each of the five strata so that the individual sample sizes are proportional to the respective strata?

14. The following are the number of restaurants in twenty cities: 17, 22, 11, 9, 19, 10, 6, 10, 115, 33, 8, 7, 18, 24, 8, 82, 27, 8, 12, and 14.
    (a) List the five possible systematic samples of size 4 that can be taken from this list by starting with one of the first five numbers and then taking each fifth number on the list.
    (b) Calculate the means of the five samples obtained in part (a) and verify that *their* mean equals the average (mean) number of restaurants in the given twenty cities.

15. The following are the number of commercial television stations operating in 1970 in the 50 states (listed in alphabetic order) 15, 7, 11, 7, 50, 11, 5, 0, 24, 15, 10, 7, 24, 17, 12, 12, 10, 15, 7, 5, 11, 20, 12, 9, 21, 10, 14, 7, 3, 4, 7, 27, 19, 12, 29, 9, 12, 23, 2, 11, 10, 16, 55, 3, 2, 12, 13, 9, 17, and 3. List the ten possible systematic samples of size 5 that can be taken from this list by starting with one of the first ten numbers and then taking each tenth number on the list.

## CHANCE FLUCTUATIONS OF MEANS

If *several* teachers tried a new method (say, a programmed text) on some of their students, if *several* pollsters sampled the opinions of voters in connection with a presidential election, and if *several* department store credit managers tried a new type of collection letter on some of their delinquent accounts, they should not be surprised if in each case there are differences among the respective results. As we saw earlier in this chapter, random sampling (by definition) involves an element of chance, and it stands to reason that this will affect whatever quantities we may calculate on the basis of sample data. Thus, whatever differences there may be among the respective results obtained by the different teachers, the different pollsters, and the different credit managers are what we call **chance variations** or **chance fluctuations**.

To illustrate, let us suppose that a chain of restaurants is interested in determining the average amount of time it takes their cooks to prepare the food orders of their patrons, and that (as part of a study) they obtained the following sample data, which are the times (in minutes) required by their cooks to prepare 40 food orders:

| | | | | | | | | |
|---|---|---|---|---|---|---|---|---|
| 20 | 19 | 17 | 13 | 24 | 14 | 25 | 10 | 17 | 23 |
| 17 | 28 | 15 | 31 | 27 | 24 | 22 | 19 | 24 | 15 |
| 6  | 23 | 20 | 18 | 9  | 17 | 18 | 13 | 22 | 14 |
| 24 | 19 | 29 | 23 | 26 | 20 | 19 | 20 | 16 | 24 |

The mean of these figures is $\frac{784}{40} = 19.6$ minutes, and (assuming that the sample is random) it provides an *estimate* of the quantity with which the restaurant chain is concerned, namely, the *true* average time that is required by their cooks to prepare a food order for a patron. Note that the *population mean* they are trying to estimate in this example is, in fact, the mean of the probability distribution which we tried to picture in Figure 7.1 on page 175.

Since the above estimate may, perhaps, be used in the design of the kitchen equipment of a new restaurant, in determining the size of new restaurants, and other considerations, it is only natural to ask whether a figure like this, obtained from a sample, is really **reliable**. In other words, they ought to investigate whether they are justified in expecting 19.6 minutes to be reasonably close to the true value, or whether it might not have been better, and safer, to time the preparation of 100 food orders, or perhaps even 1,000. Furthermore, they ought to check whether there might be any biases due to lack of planning or due to deficiencies in the methods which they used in getting the data, and whether they are really sampling from the right population. Surely, the data might just as well be thrown out if all the food was ordered by a bus-load of passengers which made a brief stop at the restaurant, or if all the food was ordered for an evening-long banquet of a local businessmen's club.

The question with which we are concerned mostly in this chapter is that of *reliability*, and without going into any detailed statistical theory or fancy mathematics, we can investigate it by repeating the whole sampling procedure several times. Suppose, for instance, that the management of the restaurant chain asks its statistical consultants to take ten more random samples of 40 food preparation times (that is, repeat the whole procedure ten more times), and that the means of the times which they get are, respectively,

| | | | | | | | | | |
|---|---|---|---|---|---|---|---|---|---|
| 22.0 | 20.2 | 20.0 | 21.4 | 20.6 | 20.2 | 19.1 | 20.4 | 20.2 | 18.9 |

minutes. These figures tell us quite a bit about the question with which we are concerned, namely, *the extent to which sample means vary from sample to sample.* Just by looking at the ten means we find that the smallest is 18.9 minutes, that the largest is 22.0 minutes, and that only one of them falls outside the interval from 19.0 to 22.0 minutes. Furthermore, the over-all mean of the ten samples is

$$\frac{22.0 + 20.2 + 20.0 + 21.4 + 20.6 + 20.2 + 19.1 + 20.4 + 20.2 + 18.9}{10}$$

$$= 20.3 \text{ minutes}$$

and if this were the *true mean* (which is not a bad guess as it is based on 400 observations), we could say that *six of the ten sample means differ from the true mean by less than one minute* (all those except 22.0, 21.4, 19.1, and 18.9), while *all ten of the sample means differ from the true mean by less than two minutes*, as can easily be checked.

If we now apply these results to the original sample on page 187, whose mean was 19.6 minutes, we might conclude that *6 to 4 would be fair odds* that the interval from $19.6 - 1 = 18.6$ minutes to $19.6 + 1 = 20.6$ minutes contains the true average time required by the cooks to prepare a food order for a patron, and that we can be *practically certain* that the true average time is contained in the interval from $19.6 - 2 = 17.6$ minutes to $19.6 + 2 = 21.6$ minutes. This illustrates how questions about the reliability of a sample mean *can* be answered, but we might add that it is seldom, if ever, practical to repeat experiments over and over again for this purpose. In fact, it would be wasteful, for there exists an alternate approach to the problem which is based partly on statistical theory and partly on an analysis of whatever fluctuations there are *within* a sample, namely, whatever fluctuations there are among the sample values, themselves.

Before we introduce this technique, let us observe that there are essentially two factors which affect the chance fluctuations of sample means: *the size of the sample and the variability of the population from which the sample is obtained.* So far as the sample size is concerned, it certainly stands to reason that more and more information should lead to more and more reliable results; in other words, the means of very large samples are more apt to be close to the corresponding true (population) means. So far as the variability of the population is concerned, it also stands to reason that its magnitude should be reflected in the fluctuations of the means of samples. To illustrate this point, consider the following two finite populations which consist of numbers written on individual slips of paper: *Population A* consists of the numbers 1, 2, 3, . . ., 99, and 100, each written on a separate slip of paper, and *Population B* consists of the numbers 50 and 51, each written on 50 separate slips of paper. Now, if we take a sample of size 3 from Population A, the sample mean can vary anywhere from $\bar{x} = 2$ (the mean of 1, 2, and 3) to

$\bar{x} = 99$ (the mean of 98, 99, and 100); however, if we take a sample of size 3 from Population B, the only possible samples are $\bar{x} = 50$, $\bar{x} = 50\frac{1}{3}$, $\bar{x} = 50\frac{2}{3}$, or $\bar{x} = 51$, depending on how many 50's and how many 51's we happen to draw. This shows that

**If there is very little or very much variability in a population, the same will also be true for the fluctuations of sample means.**

Returning now to the example with which we began the discussion of this section, we know the size of the sample—there are 40 observations—but what can we say about the variability of the population? The random variable with which we are concerned in this example is the time that is required by cooks of the restaurant chain to prepare food orders, and the "population" from which we are sampling is the probability distribution pictured in Figure 7.1 on page 175. In actual practice, the population distribution is, of course, unknown, for *if it were known there would be nothing to estimate*, but we can generally get some idea about the over-all shape of a population distribution by grouping the sample data and then studying their distribution. This is precisely how we obtained the histogram of Figure 7.4, which pictures the distribution of the 40 times given on page 187. Well then, the histogram of Figure 7.4 is the *image* which we get from the sample data of the population distribution of Figure 7.1, and by comparison it will enable us to *estimate* the standard deviation $\sigma$ (defined on page 137) which measures the variability of the population distribution. Since the estimation of $\sigma$ is of great importance

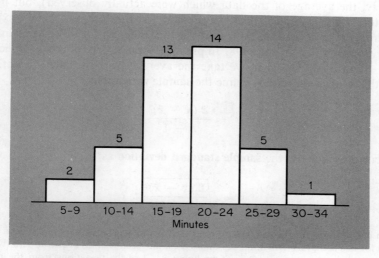

**FIGURE 7.4** Distribution of the forty food preparation times.

in most of the work that follows in this chapter and in Chapter 8, we shall discuss it separately in the section which follows before we return to the *original* problem of studying the chance fluctuations of sample means.

## THE SAMPLE STANDARD DEVIATION

In Chapter 5 we defined $\sigma^2$, the variance of a probability distribution, as the expected value (or average) of the *squared deviations from the mean*. Hence, if we actually observed several values of a random variable it would seem logical to estimate $\sigma^2$ in terms of the average of *their* squared deviations from the mean. Symbolically, if $x_1, x_2, \ldots,$ and $x_n$ are $n$ observations of a random variable with the mean $\mu$ (say, the 40 times on page 187, the ages of 25 doctors, the I.Q.'s of 120 students, ...), we would thus estimate the variance $\sigma^2$ of the corresponding probability distribution as

$$\frac{(x_1 - \mu)^2 + (x_2 - \mu)^2 + \cdots + (x_n - \mu)^2}{n}$$

which can also be written as $\frac{\Sigma (x - \mu)^2}{n}$. This would be fine, if it were not for the fact that *in most practical situations $\mu$ is unknown and will, therefore, have to be replaced with an estimate*. (For instance, in the problem of the preceding section, $\mu$ was the *unknown* true average time required to prepare the food orders.) The most obvious replacement for $\mu$ is the sample mean $\bar{x}$ (namely, the average of the data which were actually observed), but if we substitute $\bar{x}$ for $\mu$ we shall have to account for the fact that $\Sigma (x - \bar{x})^2$ *is always less than or equal to* $\Sigma (x - \mu)^2$, as the reader will be asked to demonstrate in Exercise 13 on page 198. In practice, we take care of this by dividing by $n - 1$ instead of $n$ when we take the average of the squared deviations from the mean, and we thus define the **sample variance** as

$$s^2 = \frac{\Sigma (x - \bar{x})^2}{n - 1}$$

and, correspondingly, the **sample standard deviation** as*

$$s = \sqrt{\frac{\Sigma (x - \bar{x})^2}{n - 1}}$$

* In connection with this we refer to $n - 1$ as the number of **degrees of freedom**. The reason for this is that *the sum of the deviations from the mean is always equal to zero* (which is not very difficult to prove); hence, if we know $n - 1$ of the deviations from the mean (which play such an important role in the definition of the standard deviation), the $n$th is automatically determined.

In words, the variance of a set of sample data is given by the sum of the squared deviations from their mean divided by $n - 1$, and if we calculate $s^2$ for the 40 times on page 187, we get

$$s^2 = \frac{(20 - 19.6)^2 + (19 - 19.6)^2 + \cdots + (24 - 19.6)^2}{40 - 1}$$
$$= 30.4$$

Then looking up the square root of 30.4 in the table at the end of the book, we find that the standard deviation of the 40 preparation times is approximately

$$s = \sqrt{30.4} = 5.5$$

To be honest, we did not really obtain this result by actually calculating the forty deviations from the mean $20 - 19.6$, $19 - 19.6$, ..., and $24 - 19.6$, and then dividing the sum of their squares by 39. This could have been done and the results would have been the same, but there exist short-cut formulas for $s^2$ and $s$ which can save a great deal of work. All we need for these formulas are $\Sigma x$ and $\Sigma x^2$, namely, the sum of the $x$'s and the sum of their squares, which we substitute into the expression

$$s^2 = \frac{n(\Sigma x^2) - (\Sigma x)^2}{n(n - 1)}$$

for the sample variance, and into the expression

$$s = \sqrt{\frac{n(\Sigma x^2) - (\Sigma x)^2}{n(n - 1)}}$$

for the sample standard deviation.

As we admitted earlier, we actually used these short-cut formulas in determining $s^2$ and $s$ for the 40 times. First we found that $\Sigma x = 784$ and $\Sigma x^2 = 16{,}552$, and then we obtained by substitution

$$s^2 = \frac{40(16{,}552) - (784)^2}{40 \cdot 39}$$
$$= \frac{47{,}424}{1{,}560}$$
$$= 30.4$$

and, hence, $s = \sqrt{30.4} = 5.5$. Since beginners often seem to get confused about the difference between $(\Sigma x^2)$ and $(\Sigma x)^2$, let us emphasize the point that to find $(\Sigma x^2)$ we first square each individual $x$ and then add all the squares; to find $(\Sigma x)^2$ we first add all of the $x$'s and then square their sum.

To illustrate the calculation of the sample standard deviation with an example in which the arithmetic is somewhat easier to follow, let us refer to the following data on the number of fire insurance claims submitted to a casualty insurance company on six consecutive days:

$$6 \quad 13 \quad 7 \quad 4 \quad 14 \quad 10$$

To be able to use the formula on page 190, we must first find $\bar{x}$, which in this case equals

$$\frac{6 + 13 + 7 + 4 + 14 + 10}{6} = 9$$

The remaining calculations are then performed as shown in the following table:

| $x$ | $x - \bar{x}$ | $(x - \bar{x})^2$ |
|---|---|---|
| 6  | $-3$ | 9  |
| 13 | 4    | 16 |
| 7  | $-2$ | 4  |
| 4  | $-5$ | 25 |
| 14 | 5    | 25 |
| 10 | 1    | 1  |
|    | 0    | 80 |

and using the "long method" we thus get

$$s = \sqrt{\tfrac{80}{5}} = \sqrt{16} = 4$$

Note that the sum of the entries in the middle column, namely, $\Sigma (x - \bar{x})$, is equal to 0; as the reader will be asked to demonstrate in Exercise 12 on page 198, this must always be the case, and, hence, it provides a check on the calculations.

Had we used the short-cut formula to calculate $s$ for the same data, we would first have added the $x$'s getting $\Sigma x = 6 + 13 + 7 + 4 + 14 + 10 = 54$, then we would have added their squares getting $\Sigma x^2 = 6^2 + 13^2 + 7^2 +$

$4^2 + 14^2 + 10^2 = 566$, and then we would have substituted $\Sigma x = 54$, $\Sigma x^2 = 566$, and $n = 6$ into the formula, getting

$$s = \sqrt{\frac{6(566) - (54)^2}{6 \cdot 5}} = \sqrt{\frac{480}{30}} = \sqrt{16} = 4$$

Evidently, the values of $s$ obtained by the two methods are the same. If the short-cut method seemed more involved in this last example, this is only because the mean as well as the deviations from the mean were all whole numbers. In general, the short-cut formula provides considerable simplifications, and this is true, particularly, if the work is done with a desk calculator, in which case the sum of the $x$'s and the sum of their squares can be accumulated in one operation.

If we have to calculate the standard deviation or the variance of a set of data which is already grouped, we are faced with the same problem as on page 31. However, if we proceed as in the calculation of the mean of a frequency distribution and replace each value with the *class mark* of the class to which it belongs (namely, with the *midpoint* of the corresponding class interval), we can write the formula for $s$ as

$$s = \sqrt{\frac{\Sigma (x - \bar{x})^2 \cdot f}{n - 1}}$$

where the $x$'s are the class marks, the $f$'s are the corresponding class frequencies, and $n$ is the total number of observations. Thus, for each class mark we multiply the squared deviation from the mean by the corresponding class frequency, then we divide the sum of all these products by $n - 1$, and finally we take the square root to get $s$ rather than $s^2$.

The formula which we gave in the preceding paragraph serves to *define* $s$ for grouped data, but it is seldom used in practice. To begin with, there is a short-cut formula analogous to the one on page 191, namely,

$$s = \sqrt{\frac{n(\Sigma x^2 f) - (\Sigma xf)^2}{n(n - 1)}}$$

where the $x$'s are now the class marks and the $f$'s are the corresponding class frequencies. Although this formula may look a bit involved, it makes the calculation of the standard deviation of grouped data fairly easy. Instead of having to work with the squared deviations from the mean, we have only to find $\Sigma xf$ (the sum of the products obtained by multiplying each class mark by the corresponding class frequency), $\Sigma x^2 f$ (the sum of the products obtained by multiplying the *square* of each class mark by the corresponding class frequency), and substitute into the formula.

194    CHANCE VARIATION: SAMPLING    CHAP. 7

To illustrate this technique, let us calculate the standard deviation of the 40 preparation times on page 187 on the basis of the distribution given in Figure 7.4 on page 189. Performing the necessary calculation, we get

| Time (minutes) | Class mark $x$ | $x^2$ | $f$ | $xf$ | $x^2f$ |
|---|---|---|---|---|---|
| 5–9 | 7 | 49 | 2 | 14 | 98 |
| 10–14 | 12 | 144 | 5 | 60 | 720 |
| 15–19 | 17 | 289 | 13 | 221 | 3,757 |
| 20–24 | 22 | 484 | 14 | 308 | 6,776 |
| 25–29 | 27 | 729 | 5 | 135 | 3,645 |
| 30–34 | 32 | 1,024 | 1 | 32 | 1,024 |
| | | | | 770 | 16,020 |

so that

$$s = \sqrt{\frac{40(16,020) - (770)^2}{40 \cdot 39}}$$

$$= \sqrt{\frac{47,900}{1,560}}$$

$$= \sqrt{30.7}$$

$$= 5.5$$

Note that this result is very close to the value which we obtained for the standard deviation of the ungrouped data on page 191. As things turned out, the calculation of this standard deviation was quite easy, but this was due mainly to the fact that the class marks were whole numbers. Otherwise, it is preferable by far to use the same kind of *coding* which we suggested in Exercise 11 on page 35 for the calculation of the mean; how this is done will be explained in Exercise 10 below.

In conclusion, let us point out that the various formulas which we have given for the standard deviation of a sample can serve also to *define* the standard deviation of a *finite population*—all we have to do is substitute $\mu$ for $\bar{x}$ and $N$ (the size of the population) for $n$.

## EXERCISES

1. Referring to Exercise 1 on page 34, use the basic formula which defines $s$ on page 190 to calculate the standard deviation of the ten family incomes.

2. Using the basic formula which defines $s^2$ on page 190, calculate the variance of the five percentage profits of Corporation B given in the example on page 47.

3. Find the standard deviation of the following ages of ten professional basketball players using (a) the basic formula which defines $s$ on page 190, and (b) the short-cut formula on page 191: 30, 25, 20, 24, 26, 22, 31, 29, 27, and 21 years. Time yourself on the two methods and check how much time is saved by using the short-cut formula.

4. Use the short-cut formula to calculate the variance of the 12 monthly industrial accident figures of Firm C of Exercise 1 on page 47.

5. Use the short-cut formula to calculate the standard deviation of the 1970 populations of the 16 counties of the state of Maine, given in Exercise 1 on page 43.

6. Use the short-cut formula to calculate the variance of the number of games won by the 13 teams of the National Football Conference, given in Exercise 4 on page 43.

7. Repeat Exercise 6 *twice*, once after subtracting 2 from each of the 13 figures, and once after subtracting 7 from each of the 13 figures. What simplification does this suggest for the calculation of a standard deviation?

8. Calculate the standard deviation of the following distribution of the number of hours of sleep a student got each night during the month of May:

| Hours of Sleep | Number of Days |
|---|---|
| 1– 3 | 1 |
| 4– 6 | 7 |
| 7– 9 | 10 |
| 10–12 | 9 |
| 13–15 | 2 |
| 16–18 | 2 |
| | 31 |

9. Referring to the distribution of Exercise 7 on page 25, calculate the standard deviation of the finance charges paid by the 200 customers of the furniture store.

**10. CODING** If we use the coding of Exercise 11 on page 35, the short-cut formula for the standard deviation becomes

$$s = c\sqrt{\frac{n(\Sigma u^2 f) - (\Sigma uf)^2}{n(n-1)}}$$

where all the symbols are as explained on page 36—the $u$'s are the class marks in the new scale, the $f$'s are the corresponding class frequencies, $n$ is the total number of observations, and $c$ is the class interval of the distribution. For the telephone-order distribution on page 14 we would thus get

| Class Marks $x$ | $u$ | $f$ | $uf$ | $u^2 f$ |
|---|---|---|---|---|
| 1.95 | −2 | 9 | −18 | 36 |
| 3.95 | −1 | 39 | −39 | 39 |
| 5.95 | 0 | 45 | 0 | 0 |
| 7.95 | 1 | 42 | 42 | 42 |
| 9.95 | 2 | 11 | 22 | 44 |
| 11.95 | 3 | 4 | 12 | 36 |
| | | 150 | 19 | 197 |

so that

$$s = 2\sqrt{\frac{150(197) - (19)^2}{150 \cdot 149}}$$

$$= 2\sqrt{1.3}$$

$$= 2.3 \text{ (approximately)}$$

(a) Use this method to rework the illustration on page 194.
(b) Use this method to rework Exercise 8.
(c) Use this method to rework Exercise 9.
(d) Use this method to find the standard deviation of the following distribution of the straight-time weekly earnings of a sample of 418 secretaries in manufacturing industries at New Haven, Connecticut, January, 1972 (source: *Area Wage Survey*, U.S. Department of Labor):

| Straight-Time Weekly Earnings (dollars) | Number of Secretaries |
|---|---|
| 80.00– 99.99 | 8 |
| 100.00–119.99 | 44 |
| 120.00–139.99 | 192 |
| 140.00–159.99 | 108 |
| 160.00–179.99 | 50 |
| 180.00–199.99 | 15 |
| 200.00–219.99 | 1 |
| | 418 |

**11. THE SAMPLE RANGE** When we are dealing with a very small sample, a good estimate of the population standard deviation can often be obtained on the basis of the sample range, namely, *the largest value minus the smallest*. Such estimates of the population standard deviation $\sigma$ are given by the sample range divided by the divisor $d$, which depends on the size of the sample; for random samples from populations having roughly the shape of a normal distribution its values are

| Sample size $n$ | 2 | 3 | 4 | 5 | 6 | 7 | 8 | 9 | 10 |
|---|---|---|---|---|---|---|---|---|---|
| Divisor $d$ | 1.13 | 1.69 | 2.06 | 2.33 | 2.53 | 2.70 | 2.85 | 2.97 | 3.08 |

For instance, if we take the data on the number of fire insurance claims on page 192, we find that the smallest number is 4, the largest number is 14, so that the sample range is $14 - 4 = 10$. Thus, we estimate the standard deviation of the population from which this sample of size 6 was obtained as

$$\frac{10}{2.53} = 3.95$$

and it should be observed that this figure is very close to the sample standard deviation $s = 4$ obtained for these data on page 192.

(a) Use this method to estimate $\sigma$ for the population which led to the data of Exercise 1 on page 34.

(b) Use this method to estimate $\sigma$, the true variability of the percentage profits of Corporation B in the illustration on page 47, and compare the result with that of Exercise 2.

(c) Use this method to estimate the standard deviation of the population which led to the sample data concerning the ages of professional basketball players of Exercise 3, and compare the results.

12. Show that $\Sigma(x - \bar{x}) = 0$ for any set of $x$'s whose mean is $\bar{x}$ by writing out in full what is meant by $\Sigma(x - \bar{x})$.

13. Use the binomial expansion $(a - b)^2 = a^2 - 2ab + b^2$ to show that

$$\Sigma(x - \bar{x})^2 = \Sigma[(x - \mu) - (\bar{x} - \mu)]^2 = \Sigma(x - \mu)^2 - n(\bar{x} - \mu)^2$$

which proves that $\Sigma(x - \bar{x})^2$ *is always less than or equal to* $\Sigma(x - \mu)^2$. (*Hint:* If necessary, write each of the expressions in full, that is, without summation signs.)

*The purpose of the last section has been to introduce the sample standard deviation as an estimate of the standard deviation of the population from which the sample was obtained; the exercises which follow deal with some other applications of the standard deviation of a set of data.*

14. **STANDARD UNITS** If James and George, who work in different departments of a factory, assemble 117 desk lamps and 129 electric coffee percolators, respectively, on a certain day, it is difficult to judge which one is, relatively speaking, doing a better job. However, if we know that desk lamp assembly workers average 90 desk lamps per day with a standard deviation of 18, while electric coffee percolator assembly workers average 111 electric coffee percolators per day with a standard deviation of 15, we can say that James' output is $\dfrac{117 - 90}{18} = 1.5$ standard deviations above average while George's output is only $\dfrac{129 - 111}{15} = 1.2$ standard deviations above average. Thus, even though George assembled a larger number of items than James, James' performance *relative to the workers in his department* was better than George's. What we have done here is precisely what we did on page 148 when we converted measurements into *standard units* by means of the formula $z = \dfrac{x - \mu}{\sigma}$; for observed data the formula becomes $z = \dfrac{x - \bar{x}}{s}$, and it can be of great help in the comparison of measurements or observations belonging to different sets of data.

(a) John goes fishing at a lake where the average fisherman catches nine fish in a day with a standard deviation of two fish, and Henry goes fishing at a lake where the average fisherman brings in eleven fish a day with a standard deviation of four fish. If John catches ten fish while Henry catches 12, who would seem to be the better (or luckier) fisherman?

(b) In a large city the average retail price of a head of lettuce is $0.35 (with a standard deviation of $0.03), the average retail price of a pound of tomatoes is $0.25 (with a standard deviation of $0.02), and the average retail price of a cucumber is $0.15 (with a standard deviation of $0.02). If a certain food market charges $0.39 for a head of lettuce, $0.28 for a pound of tomatoes, and $0.17 for a cucumber, which of these vegetables is *relatively* the most expensive?

**15. RELATIVE VARIATION** One disadvantage of the standard deviation as a measure of variation is that it depends on the units of measurement. Thus, if a set of measurements, say, of the height of a statue, have a standard deviation of 0.27 inches, we cannot really say whether these measurements are very precise unless we know something about the actual size of the quantity we are trying to measure. What we need in a situation like this is a measure of relative variation, such as the **coefficient of variation**

$$V = \frac{s}{\bar{x}} \cdot 100$$

which expresses the standard deviation as a percentage of the mean.

(a) Use the results of Exercise 1 on page 34 and Exercise 1 on page 194 to calculate the coefficient of variation of the ten incomes.

(b) Use the result of Exercise 4 on page 43 and the square root of the result of Exercise 6 on page 195 to calculate the coefficient of variation of the games won by the given football teams.

**16. SKEWNESS** We say that the distribution of Figure 7.5 is **symmetrical** for we can picture its histogram folded along the dashed line so that the two halves will more or less coincide. The distribution of Figure 7.6 is said to be **skewed,** on the other hand, because it has more of a "tail" on one side than it has on the other. If the "tail" is on the right, as in Figure 7.6, the mean of the distribution will generally exceed its median (and if the "tail" is on the left, the order will be reversed). Based on this difference, the so-called **Pearsonian coefficient of skewness** measures the skewness of a distribution by means of the formula

$$\frac{3(\bar{x} - \tilde{x})}{s}$$

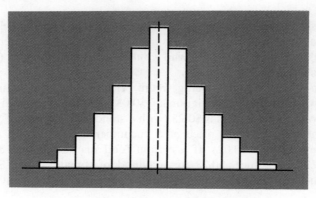

**FIGURE 7.5** A symmetrical distribution.

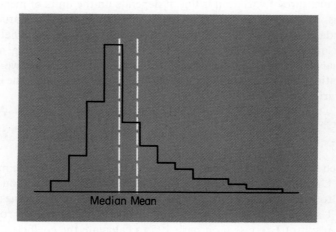

**FIGURE 7.6** A skewed distribution.

where $\tilde{x}$ denotes the median. Note that for symmetrical distributions the mean and the median coincide and the Pearsonian coefficient of skewness is equal to *zero*.

(a) Find the Pearsonian coefficient of skewness for the telephone-order distribution on page 14, using the values of the mean and the median obtained in Chapter 2 and the value of $s$ given in Exercise 10 on page 196.

(b) Find the Pearsonian coefficient of skewness for the distribution of finance charges of Exercise 7 on page 25, using the results of Exercise 7 on page 35, part (b) of Exercise 11 on page 44, and Exercise 9 on page 195.

# THE SAMPLING DISTRIBUTION OF THE MEAN

When we calculate the mean of a sample, this mean is a value of a random variable. Clearly, the values we get for $\bar{x}$ will differ from sample to sample (as they did for the ten samples on page 187), and if we knew enough about the population from which we are sampling, we could actually determine the distribution of $\bar{x}$, or at least we could make some probability statement about its variability. For instance, on page 188 we judged on the basis of very meager information that in the given example the odds were 6 to 4 that a sample mean would fall on the interval from $\mu - 1$ minutes to $\mu + 1$ minutes, namely, that the mean of a random sample (of size 40) from the given population *would differ from the true mean by less than one minute*. This argument was based on the means of ten samples (taken from one and the same population), and it stands to reason that if we took 100 samples like this or even 1,000, we should be able to get a pretty good picture about the distribution (probability function or probability density) of $\bar{x}$.

Of course, in actual practice we do not take 1,000 samples from the same population, nor do we take 100 or even ten, and we must therefore rely on other methods to obtain information about the **sampling distribution of the mean.** [This is what we call the distribution of $\bar{x}$, for it tells us how the means of random samples (from one and the same population) would vary from sample to sample.] In most practical applications *in which we have only one sample*, we have no choice but to base our argument about the sampling distribution of the mean on two theorems: *one theorem about the mean and the standard deviation of this sampling distribution, and the other, explained later on page 203, about its over-all shape.*

The mean of the sampling distribution of $\bar{x}$ will be denoted by $\mu_{\bar{x}}$ (which reads "$\mu$ sub $\bar{x}$"), and the first of the two theorems states (among other things) that $\mu_{\bar{x}}$ equals the mean of the population from which the sample is obtained. This should not come as a surprise, for it stands to reason that the means of samples from one and the same population should *on the average* equal the mean of the population, itself. The first theorem also provides a formula for the standard deviation $\sigma_{\bar{x}}$ of the sampling distribution of the mean; formally it states that for random samples of size $n$ from a population having the mean $\mu$ and the standard deviation $\sigma$,

**The sampling distribution of $\bar{x}$ has the mean**

$$\mu_{\bar{x}} = \mu$$

### and the standard deviation*

$$\sigma_{\bar{x}} = \frac{\sigma}{\sqrt{n}}$$

It is customary to refer to $\sigma_{\bar{x}}$, the standard deviation of the sampling distribution of the mean, as the **standard error of the mean**. Its role in statistics is fundamental, for *it measures the extent to which sample means vary, or fluctuate, due to chance*, and this is *The Problem* with which we have been concerned in this chapter.

The quotient $\sigma/\sqrt{n}$ demonstrates two things: *If $\sigma$ is large and there is considerable variation in the population from which we are sampling, we can expect a proportionally large variation in the sampling distribution of the mean. So far as n is concerned, it is apparent from the formula that the larger the size of the sample, the smaller will be the chance fluctuations of the mean and the closer we can expect the mean of a sample to be to $\mu$, the mean of the population.* Both of these properties make sense, intuitively, as we already indicated on page 188.

Let us now take another look at the example of the preceding sections, where we tried to estimate the true average time required by the cooks of a restaurant chain to prepare food orders for the patrons. On page 191 we showed that the standard deviation of the 40 sample values is $s = 5.5$, and if we now use this figure as an estimate of $\sigma$, the standard deviation of the population from which the sample was obtained, we can argue that for $n = 40$ the standard error of the mean is approximately

$$\frac{\sigma}{\sqrt{n}} = \frac{5.5}{\sqrt{40}} = 0.87$$

Having obtained this result by means of the formula for $\sigma_{\bar{x}}$, it would be interesting to see how close we will come if we actually calculate the standard deviation of the ten sample means on page 187; after all, their fluctuations are also indicative of the variability of the sampling distribution of the mean, and their standard deviation constitutes an estimate of $\sigma_{\bar{x}}$ (for random samples of size 40 from the given population). If we add the ten means on

---

\* When the population from which we are sampling is *finite*, the formula for $\sigma_{\bar{x}}$ must be modified by multiplying $\sigma/\sqrt{n}$ by the so-called **finite population correction factor** $\sqrt{\frac{N-n}{N-1}}$, where $N$ is the size of the population. In actual practice, this modification is used only when the sample constitutes a substantial portion of the population (say, 5 percent or more), for otherwise the effect of the correction factor is negligible (see Exercise 2 on page 206).

page 187 and also their squares, we get 203.0 and 4,128.62, and if we substitute these figures together with $n = 10$ into the short-cut formula for $s$, we obtain

$$s = \sqrt{\frac{10(4{,}128.62) - (203.0)^2}{10 \cdot 9}} = \sqrt{0.86} = 0.93$$

This is fairly close, about as close as we can expect in an example like this, to 0.87, our first estimate of $\sigma_{\bar{x}}$.

To demonstrate how the formula for the standard error of the mean is used in actual practice, we really need the second theorem referred to on page 201, but even without it we can get some idea by referring to *Chebyshev's Theorem* (see page 139) according to which we can now say that

**The probability of getting a sample mean which differs from the population mean $\mu$ by at least $k$ times $\sigma_{\bar{x}}$ is at most $1/k^2$.**

For instance, if we take a random sample of size 49 from an infinite (or very large) population with $\sigma = 21$, then the probability that we will get a sample mean which differs from the *true* (population) mean by at least $k = 2$ standard errors, namely, by at least

$$k \cdot \frac{\sigma}{\sqrt{n}} = 2 \cdot \frac{21}{\sqrt{49}} = 6$$

is *at most* $1/2^2 = 0.25$. Similarly, if we take a random sample of size 64 from an infinite (or very large) population with $\sigma = 4$, then the probability that we will get a sample mean which differs from the *true* (population) mean by at least $k = 3$ standard errors, namely, by at least

$$3 \cdot \frac{4}{\sqrt{64}} = 1.5$$

is *at most* $1/3^2 = 0.11$. In the first example Chebyshev's theorem tells us that the probability of a sample mean being "off" by at least 6 is *at most* 0.25, and in the second example it tells us that the probability of a sample mean being "off" by at least 1.5 is *at most* 0.11.

If we want to be more specific about probabilities like these, we shall have to refer to the second theorem mentioned on page 201, namely, the **Central Limit Theorem,** which states that

**If $n$ is large, the sampling distribution of the mean can be approximated closely with a normal distribution.**

It is difficult to say precisely how large $n$ must be before the theorem applies, but unless the population has a very unusual shape, the approximation will be good even when $n$ is relatively small—certainly when $n$ is 30 or more. (Incidentally, when the population, itself, has the shape of a normal distribution, the theorem applies regardless of the size of the sample.)

In case the reader still has some difficulties picturing the significance of a sampling distribution and the Central Limit Theorem, suppose that the owner of a very large orange grove packages his oranges into plastic bags, each containing 36 oranges. If we are interested in the *true* average weight of all these oranges, the weights of all the oranges picked in this grove constitute the *population* from which the workers who pack the bags are "sampling," and it will have a certain mean $\mu$ and a certain standard deviation $\sigma$. If we actually calculated the mean weight for each bag of 36 oranges, we would have hundreds, or perhaps even thousands, of sample means which could be grouped into a distribution, say, the one pictured in Figure 7.7. It is this kind of distribution, namely, a distribution of sample means, which according to the Central Limit Theorem should have roughly the shape of a normal distribution (provided the sample size $n$ is sufficiently large).

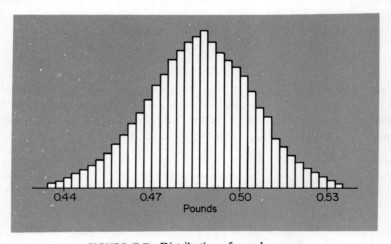

**FIGURE 7.7** Distribution of sample means.

If the mean and the standard deviation of *all* the oranges picked in the grove happened to be $\mu = 0.48$ pounds and $\sigma = 0.09$ pounds, we could use Table I to determine all sorts of probabilities about the average weight of the oranges in any given bag (assuming that the workers do not throw away any oranges which they consider too small). For instance, the probability that the average weight of the oranges in any one bag will exceed 0.50 pounds is given by the area of the shaded region of Figure 7.8, namely, the area to the right of

$$z = \frac{0.50 - \mu_{\bar{x}}}{\sigma_{\bar{x}}} = \frac{0.50 - 0.48}{0.09/\sqrt{36}} = 1.33$$

Looking up the entry corresponding to $z = 1.33$ in Table I, we find that the desired probability is $0.5000 - 0.4082 = 0.0918$, and in Exercise 4 on page 207 the reader will be asked to use the same kind of argument to find the probabilities that the oranges contained in one of the plastic bags will average less than 0.44 pounds in weight, and that they will average anywhere from 0.45 to 0.51 pounds in weight.

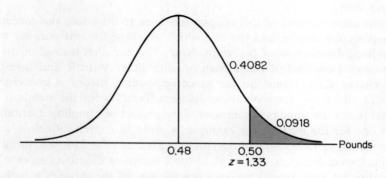

**FIGURE 7.8** Sampling distribution of the mean.

Returning to the first of the two examples in which we use Chebyshev's theorem on page 203, we can now say that if a random sample of size 49 is taken from an infinite population with $\sigma = 21$, then the probability of getting a sample mean which differs from $\mu$ (the *true* mean of the population) by at least 6 is given by the total area of the shaded regions of Figure 7.9. Since this is the normal curve area to the left of

$$z = \frac{-6}{21/\sqrt{49}} = -2$$

and to the right of $z = 2$, we find that the answer is *twice* $0.5000 - 0.4772$, namely, *twice* 0.0228 or roughly 0.05. Thus, whereas we were able to say according to Chebyshev's theorem that the probability of getting an $\bar{x}$ which

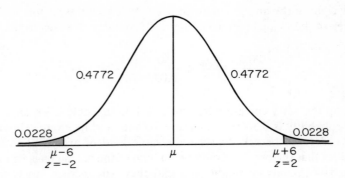

**FIGURE 7.9** Sampling distribution of the mean.

is off by at least 6 is *at most 0.25*, we can now say that this probability is just about 0.05.

The main purpose of this chapter has been to introduce the concept of a *sampling distribution*, and the one which we chose for this purpose was the sampling distribution of the mean. Note, however, that instead of the mean we could have studied the median or some other statistic and investigated its chance fluctuations. So far as corresponding theory is concerned, we would, of course, have obtained different formulas for the standard errors (that is, for the standard deviations of the respective sampling distributions); the one for the median, for example, is given in Exercise 9 below, and the one for the sample standard deviation is given on page 214. One thing which helps tremendously is that most of these sampling distributions can be approximated with normal curves when the size of the samples is sufficiently large.

## EXERCISES

1. What happens to the standard error of the mean for random samples from an infinite population when
   (a) the sample size is increased from 50 to 200;
   (b) the sample size is increased from 23 to 575;
   (c) the sample size is decreased from 640 to 40?

2. What is the value of the *finite population correction factor* when
   (a) $n = 500$ and $N = 5,000$;
   (b) $n = 400$ and $N = 40,000$?

3. On page 205 we showed that if a random sample of size 49 is taken from an infinite population with $\sigma = 21$, the probability of getting a sample

mean which differs from the population mean $\mu$ by at least 6 is approximately 0.05. What would this probability be if the population were finite and of size $N = 250$? *Explain why it stands to reason that this probability should be less than 0.05.*

4. With reference to the illustration on page 205, use similar methods to find
   (a) the probability that the oranges contained in one of the plastic bags will average less than 0.44 pounds;
   (b) the probability that the oranges contained in one of the plastic bags will average anywhere from 0.45 to 0.51 pounds.

5. On page 203 we used Chebyshev's theorem to show that if a random sample of size 64 is taken from an infinite population with $\sigma = 4$, the probability that the sample mean will be off either way by at least 1.5 is *at most* 0.11. What would we get for this probability if we used the Central Limit Theorem?

6. The mean of a random sample of size $n = 36$ is used to estimate the mean of a very large population (consisting of the lifetimes of a certain kind of television tube) which has a standard deviation of $\sigma = 120$ hours. What can we assert about the probability that the value of $\bar{x}$ will be off either way by less than 45 hours if
   (a) we use Chebyshev's theorem;
   (b) we use the Central Limit Theorem?

7. The mean of a random sample of size $n = 49$ is used to estimate the mean of a very large population (consisting of the number of hot dogs sold at the student center each day) which has a standard deviation of $\sigma = 21$ hot dogs. Using the Central Limit Theorem, what can we say about the probability that the value of $\bar{x}$ which we get will be off either way by less than (a) 1.5 hot dogs, (b) 2.5 hot dogs, and (c) 4.5 hot dogs?

8. If the packages which a firm ships by truck from New Orleans to Miami have a mean weight of 195 pounds and a standard deviation of 22 pounds, what is the probability that the combined gross weight of 100 of these packages will overload a truck designed to carry 20,000 pounds?

9. As we indicated in Chapter 2 (see, for example, Exercise 7 on page 43), the sample mean is generally *more reliable* than the sample median, that is, it is subject to smaller chance fluctuations. Theoretically, this is expressed by the fact that for random samples from very large populations having roughly the shape of a normal distribution, the **standard error of the median** (namely, the standard deviation of the sampling distribution of the median) is approximately $1.25 \cdot \dfrac{\sigma}{\sqrt{n}}$. Verify that the

mean of a random sample of size 80 is just about as reliable an estimate of a population mean as the median of a random sample of size 125.

10. Show that if the mean of a random sample of size $n$ is used to estimate the mean of an infinite population with the standard deviation $\sigma$, there is a *fifty-fifty chance* that the error is less than $0.67 \cdot \dfrac{\sigma}{\sqrt{n}}$. It has been the custom to refer to this quantity, or more precisely $0.6745 \cdot \dfrac{\sigma}{\sqrt{n}}$, as the **probable error of the mean;** nowadays, this term is used mainly in military applications.

11. If a random sample of size 64 is taken from a very large population (consisting of the amounts of the personal loans made by a large bank during a year) which has the standard deviation $\sigma = \$46.00$, determine the probable error (see Exercise 10) and explain its significance.

# 8

# THE ANALYSIS OF MEASUREMENTS

## INTRODUCTION

In the beginning of Chapter 1 we explained that statistical inference is the science of basing decisions on numerical data; now we might say more specifically that statistical inference is *the science of making generalizations about populations on the basis of samples*. Traditionally, statistical inference has been divided into **problems of estimation** and **tests of hypotheses**—in problems of estimation we try to determine the parameters (statistical descriptions) of populations, and in tests of hypotheses we face the task of having to accept or reject specific assertions about populations.

Problems of estimation can be found everywhere: in business, in science, as well as in everyday life. *In business*, a Chamber of Commerce may want to know the average income of the families in its community, and a real-estate developer may want to know how many cars can be expected to drive by a certain location per day; *in science*, a mineralogist may wish to determine the average iron content of a given ore, and a biologist may want to know how many mutations will be produced in mice by a certain radiation; finally, in *everyday life*, a commuter may want to know how long on the average it will take him to drive to work, and a serious gardener may want to know what proportion of his tulip bulbs he can expect to bloom.

In each of the examples of the preceding paragraph somebody was interested in determining the "true" value of some quantity, so that they were all *problems of estimation*. They would have been *tests of hypotheses*, however, if the Chamber of Commerce had wanted to decide on the basis of sample data whether the average family income of its community is really $10,600, if the commuter had wanted to determine whether he can really expect (in the sense of an average) that it will take him 12.4 minutes to drive to work, . . ., or if the gardener had wanted to check whether it is true that 80 percent of the tulip bulbs can be expected to bloom. Now it must be decided in each case whether to accept or reject a hypothesis (namely, an assertion or claim) about the parameter of a population.

Note that in each pair of the examples (those from business, those from science, and those from everyday life) the first concerned **measurements** while

210   THE ANALYSIS OF MEASUREMENTS                                CHAP. 8

the other concerned **count data.** The incomes, amounts of iron, and driving times are all quantities one has to measure in some way, while the number of cars that drive by a location, the number of mice that show mutations, and the number of tulip bulbs that bloom are all quantities one has to count. Since the statistical treatment of measurements differs in most instances from that of count data, we shall study the two separately—this chapter will be devoted to methods which apply mostly to measurements, while Chapter 9 will be devoted to methods which apply mostly to counts.

## THE ESTIMATION OF MEANS

In the analysis of measurements we are most often concerned with the *mean* of the population (or the *means* of the populations) from which our data are obtained. To illustrate some of the problems we face in the estimation of means, let us refer to a study in which the owner of a laundry wants to estimate *the true average number of pounds of detergent used each day in his washing machines.* His information consists of the following data on the number of pounds of detergent used on 50 days (randomly selected from the records of his laundry):

| | | | | | | | | | |
|---|---|---|---|---|---|---|---|---|---|
| 152 | 156 | 145 | 116 | 122 | 131 | 181 | 130 | 167 | 172 |
| 142 | 143 | 129 | 127 | 154 | 95  | 136 | 111 | 117 | 139 |
| 82  | 121 | 126 | 135 | 138 | 140 | 151 | 146 | 124 | 141 |
| 188 | 136 | 137 | 100 | 143 | 133 | 175 | 160 | 145 | 118 |
| 139 | 196 | 148 | 163 | 107 | 168 | 158 | 123 | 113 | 154 |

The mean of this sample is $\bar{x} = 139.46$ pounds, and in the absence of any other information this figure may well have to serve as an estimate of $\mu$, the *true* average number of pounds of detergent used per day in the washing machines.

An estimate like this is called a **point estimate,** as it consists of a single number, namely, a single point on the real number scale. Although this is the most common way of expressing an estimate, point estimates have the bad feature that they do not tell us how good they are—*they do not tell us how close we can expect them to be to the quantities they are supposed to estimate.* For instance, if a television advertisement claimed "on the basis of scientific evidence" that 75 percent of all doctors prescribe a certain brand of pain reliever, this would not mean very much if their claim were based on interviews of four doctors among whom three happen to prefer the given brand of pain reliever.

On the other hand, it might be meaningful and significant if the claim were based on interviews of 100 doctors, or perhaps 400 or more.

This demonstrates why *point estimates should always be accompanied by some information which makes it possible to judge their merits*. So far as the mean is concerned, we already saw in Chapter 7 that its chance fluctuations (and, hence, its reliability and its merit) depend on two things—the size of the sample and the variability (standard deviation) of the population. Thus, we might supplement the above point estimate of the average number of pounds of detergent used in the washing machines (namely $\bar{x} = 139.46$ pounds) with the information that it is the mean of a sample of size $n = 50$, whose standard deviation is $s = 23.24$ pounds (as can easily be verified by calculating $s$ in accordance with the formula on page 193). Although this does not tell us the *exact* value of the standard deviation $\sigma$ of the population, 23.24 pounds is at least an estimate.

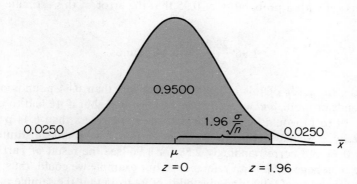

**FIGURE 8.1** Sampling distribution of the mean.

Scientific reports often present sample means together with the corresponding values of $n$ and $s$, but this does not supply the reader with a coherent picture unless he has had some formal training in statistics. To make the supplementary information meaningful even to the layman, let us refer briefly to the two theorems of Chapter 7 concerning the sampling distribution of the mean (namely, the ones on pages 201 and 203), and to the result of part (b) of Exercise 6 on page 156, according to which 95 percent of the area under the standard normal curve falls between $z = -1.96$ and $z = 1.96$. Thus, making use of the fact that the sampling distribution of the mean can be approximated closely with a normal distribution having the mean $\mu$ and the standard deviation $\sigma_{\bar{x}} = \dfrac{\sigma}{\sqrt{n}}$, we find from Figure 8.1 that a sample mean will differ from the population mean $\mu$ by less than $1.96 \cdot \dfrac{\sigma}{\sqrt{n}}$ just about 95 percent of the time. In other words,

**The probability that a sample mean will be "off" either way by less than** $1.96 \cdot \dfrac{\sigma}{\sqrt{n}}$ **is 0.95.**

The application of this result (which, incidentally, applies only to samples from infinite, or very large, populations) involves one difficulty—in order to judge the size of the error we might make when we use the mean of a sample to estimate the mean of a population, *we must know* $\sigma$, *the population standard deviation*. In actual practice this is seldom the case, but we can get around this difficulty by replacing $\sigma$ with an estimate, usually the sample standard deviation $s$. Generally, this is reasonable provided the sample size is not too small (that is, when $n$ is 30 or more).

Returning now to our numerical example and using 139.46 pounds (the sample mean from page 210) as an estimate of the true average number of pounds of detergent used each day in the washing machines of the laundry, we can assert with a probability of 0.95 that the error of this estimate is less than

$$1.96 \cdot \frac{23.24}{\sqrt{50}} = 6.44 \text{ pounds}$$

Of course, the error of this estimate is either less than 6.44 pounds or it is 6.44 pounds or more, *and we really don't know which*, but if we had to bet, 95 to 5 (or 19 to 1) would be *fair odds* that the error is less than 6.44 pounds. Incidentally, if we had wanted to be "more certain," we could substitute, say, 0.99 for 0.95 and correspondingly 2.58 for 1.96 [see the result of part (f) of Exercise 6 on page 156]. With reference to our example we could, thus, assert with a probability of 0.99 (and give odds of 99 to 1) that the sample mean is "off" by less than

$$2.58 \cdot \frac{23.24}{\sqrt{50}} = 8.48 \text{ pounds}$$

The error which we make when we use a sample mean to estimate the mean of a population is given by the difference $\bar{x} - \mu$, and the fact that the *magnitude* of this error is less than $1.96 \cdot \dfrac{\sigma}{\sqrt{n}}$ can be expressed in the following form:

$$-1.96 \cdot \frac{\sigma}{\sqrt{n}} < \bar{x} - \mu < 1.96 \cdot \frac{\sigma}{\sqrt{n}}$$

(In case the reader is not familiar with **inequality signs,** $a < b$ means "$a$ is less than $b$," while $a > b$ means "$a$ is greater than $b$"; also, $a \leqslant b$ means "$a$ is less than or equal to $b$," and $a \geqslant b$ means "$a$ is greater than or equal to $b$.")

If we now apply some relatively simple algebra, we can rewrite the above inequality as

$$\bar{x} - 1.96 \cdot \frac{\sigma}{\sqrt{n}} < \mu < \bar{x} + 1.96 \cdot \frac{\sigma}{\sqrt{n}}$$

and we can assert with a probability of 0.95 that it will hold for any given sample. In other words, if $\bar{x}$ is the mean of a random sample of size $n$ from an infinite (or very large) population with the standard deviation $\sigma$,

The probability is 0.95 that the interval from $\bar{x} - 1.96 \cdot \frac{\sigma}{\sqrt{n}}$ to $\bar{x} + 1.96 \cdot \frac{\sigma}{\sqrt{n}}$ will contain the true mean of the population.

An interval like this is called a **confidence interval,** its endpoints are called **confidence limits,** and the probability that such an interval will "do its job," namely, that it will contain the quantity we are trying to estimate, is called the **degree of confidence.** Note that if we had wanted the degree of confidence to be 0.99 instead of 0.95, we would have substituted 2.58 for 1.96, as on page 212.

When $\sigma$ is unknown (as is the case in most practical applications) and $n$ is 30 or more, we simply proceed as before and substitute for $\sigma$ the sample standard deviation $s$. For instance, if we refer again to the example on page 210 where we had $n = 50$, $\bar{x} = 139.46$ pounds, and $s = 23.24$ pounds, we obtain the following **0.95 large-sample confidence interval** for the true average number of pounds of detergent used each day in the washing machines:

$$139.46 - 1.96 \cdot \frac{23.24}{\sqrt{50}} < \mu < 139.46 + 1.96 \cdot \frac{23.24}{\sqrt{50}}$$

$$133.02 \text{ pounds} < \mu < 145.90 \text{ pounds}$$

Had we wanted to calculate a 0.99 confidence interval for this example, we would have obtained

$$130.98 \text{ pounds} < \mu < 147.94 \text{ pounds}$$

(see Exercise 12 on page 217), and this emphasizes the important fact that *if we increase the degree of certainty (namely, the degree of confidence), the confidence interval will become wider and, hence, tell us less about the quantity we are trying to estimate.*

When we estimate the mean of a population with the use of a confidence interval, we refer to this kind of estimate as an **interval estimate**. In contrast to point estimates, interval estimates require no further elaboration about their reliability—this is taken care of indirectly by their width and the degree of confidence.

## THE ESTIMATION OF $\sigma$

So far we have concerned ourselves only with the estimation of means, but *in principle* the methods that are used to estimate other population parameters are very much the same. By studying the sampling distributions of appropriate statistics, mathematicians have been able to develop formulas for confidence intervals for population standard deviations, population medians, population quartiles, ..., and, as can well be imagined, the corresponding theory is quite involved. However, as we already indicated on page 206, the problem is greatly simplified by the fact that *for large samples most of these sampling distributions can be approximated closely with normal curves.*

For instance, if we are dealing with large samples, the sampling distribution of the standard deviation $s$ can be approximated closely with a normal distribution having the mean $\sigma$ and the standard deviation

$$\sigma_s = \frac{\sigma}{\sqrt{2n}}$$

called the **standard error of s** in accordance with the terminology introduced on page 202. Now, if we reason as on page 213, we arrive at the result that 95 percent of the time a sample value of $s$ will fall on the interval from $\sigma - 1.96 \cdot \frac{\sigma}{\sqrt{2n}}$ to $\sigma + 1.96 \cdot \frac{\sigma}{\sqrt{2n}}$, and fairly straightforward algebra leads to the following **0.95 large-sample confidence interval for** $\sigma$

$$\frac{s}{1 + \frac{1.96}{\sqrt{2n}}} < \sigma < \frac{s}{1 - \frac{1.96}{\sqrt{2n}}}$$

(To obtain a corresponding 0.99 large sample confidence interval for $\sigma$ we have only to substitute 2.58 for 1.96.)

Referring again to the example on page 210, where we were interested in the average number of pounds of detergent used each day in the washing machines of a laundry, and substituting $n = 50$ and $s = 23.24$ into the above formula, we get

$$\frac{23.24}{1 + \frac{1.96}{\sqrt{100}}} < \sigma < \frac{23.24}{1 - \frac{1.96}{\sqrt{100}}}$$

$$19.43 \text{ pounds} < \sigma < 28.91 \text{ pounds}$$

Thus, we can assert with a probability of 0.95 that the interval from 19.43 pounds to 28.91 pounds contains $\sigma$, the *true* value of the standard deviation which measures the variability in the number of pounds of detergent used per day in the washing machines of the laundry. This may well make us think twice before we substitute $s = 23.24$ for $\sigma$ (as on page 212); all we can really say is that 19 to 1 would be fair odds if we had to bet that the interval from 19.43 pounds to 28.91 pounds contains the true value of $\sigma$. Note also that the method we have described here applies only to *large samples;* there exist small-sample techniques for the estimation of $\sigma$, but they will not be discussed in this text.

## EXERCISES

1. The training department of a large company would like to know how many hours on the average it will take a trainee to complete a self-study course in office procedures. If 100 randomly selected trainees took on the average 8.2 hours with a standard deviation of 0.8 hours, what can we say with a probability of 0.95 about the amount of time by which the training department might be "off" if they estimated the true average time it takes a trainee to complete the course as 8.2 hours?

2. Suppose that the department store mentioned on page 14 uses the mean of $\bar{x} = 6.20$ minutes to estimate the true mean time required to take telephone orders. Using the fact that the standard deviation of this sample of size 150 is 2.3 minutes (see Exercise 10 on page 196), what can we assert with a probability of 0.95 about the possible size of their error?

3. Suppose that the management of the restaurant chain on page 187 uses the mean of $\bar{x} = 19.6$ minutes as an estimate of the true average time it takes one of their cooks to prepare a food order. Making use of the fact that the standard deviation of this sample of size $n = 40$ is 5.5 minutes, what can we assert with a probability of 0.99 about the possible size of their error?

**4.** To assist in the design of new elevators, the engineering department of an elevator company wants to know the average number of stops one of their elevators makes per day. Finding that 81 of their elevators, each observed for one day, averaged 620 stops with a standard deviation of 31.6 stops, what can they say with a probability of 0.99 about the amount by which they might be "off" if they used 620 as an estimate of the true average number of stops one of their elevators will make per day?

**5. DETERMINATION OF SAMPLE SIZE** An interesting by-product of the theory discussed on page 211 is that it sometimes enables us to determine the sample size which is required to attain a desired degree of reliability. If we want to use the mean of a random sample to estimate the mean of a population and we want to be able to assert with a probability of 0.95 that this estimate will be "off" either way by less than some quantity $E$, we can write

$$E = 1.96 \cdot \frac{\sigma}{\sqrt{n}}$$

and, upon solving this equation for $n$, we get

$$n = \left(\frac{1.96 \cdot \sigma}{E}\right)^2$$

To illustrate this technique, suppose that a college dean wants to determine the average time it takes a student to get from one class to the next. He wants to be able to assert with a probability of 0.95 that his estimate, the mean of a suitable random sample, will not be "off" by more than 0.25 minutes. Suppose also that preliminary studies have shown that it is reasonable to let $\sigma = 1.5$ minutes. Substituting $E = 0.25$ and $\sigma = 1.5$ into the above formula for $n$, he gets

$$n = \left(\frac{1.96 \cdot 1.50}{0.25}\right)^2 = 139$$

rounded *up* to the nearest whole number, and he concludes that a random sample of size 139 will be enough for the job. *Note that this method can be used only when we know (at least approximately) the value of the standard deviation of the population whose mean we are trying to estimate.*

(a) An efficiency expert would like to determine the average time it takes a housewife to mop and wax the family-room floor. If preliminary studies have shown that the standard deviation is 14 minutes, how large a sample will he need to be able to assert with a

probability of 0.95 that his estimate, the sample mean, will be "off" by at most 3 minutes?
(b) A park ranger wants to know the average size fish in a certain lake. How large a sample will he have to take to be able to assert with a probability of 0.95 that the sample mean will not be "off" by more than 0.5 inches? Assume that it is known from previous studies that it is reasonable to let $\sigma = 4.1$ inches.
(c) Suppose we want to estimate the average I.Q. of adults living in a large rural area, and we want to be able to assert with a probability of 0.99 that our estimate (the mean of a suitable random sample) will be "off" by at most 2.5. How large a sample will we need if it is known that the standard deviation of all their I.Q.'s is about $\sigma = 12$? (*Hint:* Substitute 2.58 for 1.96 into the formula for $n$.)
(d) Using this hint, recalculate the required sample size of part (a), so that the efficiency expert can assert with a probability of 0.99 that his estimate will be "off" by at most 3 minutes.

**6.** Use the information of Exercise 1 to construct a 0.95 confidence interval for the true average time it takes a trainee to complete the given self-study course in office procedures.

**7.** Using the data of Exercise 2, construct (a) a 0.95 confidence interval, and (b) a 0.99 confidence interval for the true average time required to take the telephone orders.

**8.** Referring to Exercise 3, construct a 0.95 confidence interval for the true average time it takes one of the cooks to prepare a food order.

**9.** Referring to Exercise 4, construct a 0.95 confidence interval for the true average number of stops made by one of the manufacturer's elevators.

**10.** A police department would like to determine the true average speed of cars traveling on a highway where the maximum speed limit is 60 miles per hour. If 900 cars are timed at an average speed of 61.2 miles per hour with a standard deviation of 12.4 miles per hour, construct a 0.95 confidence interval for the speed at which a car can be *expected* (in the sense of an average) to travel on this highway.

**11.** Over a period of 36 weeks, a company's weekly maintenance cost of its office machinery averages $42.15 with a standard deviation of $12.25. Construct a 0.99 confidence interval for the true average weekly maintenance cost of the company's office machinery.

**12.** Verify the 0.99 confidence interval given on page 213 for the true average number of pounds of detergent used each day in the washing machines.

**13. SMALL-SAMPLE CONFIDENCE INTERVALS FOR $\mu$** To be able to construct confidence intervals for $\mu$ on the basis of small samples (namely, when $n$ is less than 30), we shall have to assume that the population from which we are sampling has roughly the shape of a normal distribution. Then, we can base confidence intervals for $\mu$ on the $t$ **distribution,** a continuous distribution which is in many respects very similar to the normal distribution (see Figure 8.2). It is also symmetrical and has a

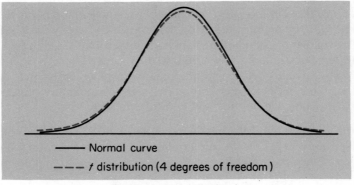

**FIGURE 8.2** $t$ distribution.

zero mean, but its shape depends on a parameter called the **number of degrees of freedom** or simply the **degrees of freedom,** which, as the distribution will be used here, is given by *the sample size minus one.* (In other applications of the $t$ distribution, for instance in Exercise 5 on page 238, the number of degrees of freedom will be given by a different expression.) Based on the $t$ distribution and reasoning similar to that on page 213, we obtain the following **0.95 small-sample confidence interval for $\mu$:**

$$\bar{x} - t_{.025} \cdot \frac{s}{\sqrt{n}} < \mu < \bar{x} + t_{.025} \cdot \frac{s}{\sqrt{n}}$$

which differs from the one on page 213 only insofar as $\sigma$ is replaced by $s$ (which we did anyhow) and that 1.96 is replaced by the quantity $t_{.025}$, which is such that 95 percent of the area under the curve of the $t$ distribution falls between $-t_{.025}$ and $t_{.025}$ (see Figure 8.3). Since the values of $t_{.025}$ depend on the number of degrees of freedom, they will have to be looked up in Table II (which, incidentally, also contains the values of $t_{.005}$ which must be used instead of $t_{.025}$ when the degree of confidence is 0.99). To illustrate the calculation of a small-sample confidence interval

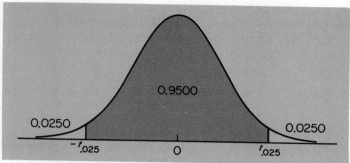

**FIGURE 8.3** $t$ distribution.

for $\mu$, suppose that 14 tractor-trailer trucks traveling between Montreal and Toronto have a mean gross weight of $\bar{x} = 49,000$ pounds with a standard deviation of $s = 3,500$ pounds. Since $t_{.025}$ for $14 - 1 = 13$ degrees of freedom equals 2.160 according to Table II, substitution into the formula yields

$$49,000 - 2.160 \cdot \frac{3,500}{\sqrt{14}} < \mu < 49,000 + 2.160 \cdot \frac{3,500}{\sqrt{14}}$$

$$46,980 \text{ pounds} < \mu < 51,020 \text{ pounds}$$

Information like this may be important in designing tractors, trailers, highways, establishing weight limits on highways, and so forth.

(a) In order to determine how often to refill their automatic coffee vending machines, a vending machine company found that in eight randomly selected locations its machines dispensed on the average 352 cups of coffee per day with a standard deviation of 47 cups of coffee. Construct a 0.95 confidence interval for the number of cups of coffee such a vending machine can be expected (in the sense of an average) to dispense per day.

(b) In 11 tests of a particular brand of ping-pong ball, it was found that the balls lasted on the average 12,641 hits before losing their shape with a standard deviation of 1,012 hits. Construct a 0.95 confidence interval for this brand of ping-pong ball's true average ability to retain its shape.

(c) In establishing the authenticity of an ancient coin, its weight is often of critical importance. If four experts independently weighed a Phoenician tetradrachm and obtained a mean of 14.32 grams and a standard deviation of 0.04 grams, construct a 0.99 confidence interval for the true weight of this coin.

(d) If five pieces of steel (randomly selected from a day's output of a certain mill) had a mean compressive strength of 62,473 pounds per square inch and a standard deviation of 8,592 pounds per square inch, construct a 0.99 confidence interval for the true average compressive strength of this steel.

14. With reference to Exercise 1, construct a 0.95 confidence interval for the true standard deviation of the time it takes the company's trainees to complete the self-study course in office procedures.

15. Using the information of Exercise 2, construct a 0.99 confidence interval for the true standard deviation of the time that is required to take the store's telephone orders.

16. Using the data of Exercise 3, construct a 0.99 confidence interval for the true standard deviation of the time it takes one of the restaurant chain's cooks to prepare a food order.

## TESTS CONCERNING MEANS

In this section we shall study methods which enable us to decide whether to accept or reject hypotheses (namely, assumptions or claims) about the means of populations. As we already pointed out on page 209, this includes such problems as deciding whether the average family income in a given community is really $10,600, or whether it really takes 12.4 minutes, on the average, for a certain person to drive to work.

In practice, we base most decisions like this on the *difference* between the mean of a sample and the assumed value of the population from which the sample was obtained—*if the difference is small, the sample mean supports the assumption about the mean of the population, and if the difference is large, the sample mean tends to refute it*. For instance, if the Chamber of Commerce of the above-mentioned community takes a random sample of 55 family incomes and obtains a mean of $10,580, it may well consider the small difference of $10,600 − $10,580 = $20 as *supporting evidence* for the contention that the true average family income in the community is $\mu$ = $10,600. On the other hand, if it obtained a mean of $9,740 or $12,500 (namely, a value which is much less or much greater than $10,600), it may well feel that the assumption $\mu$ = $10,600 *cannot be correct*.

If the difference between a sample mean and the assumed value of the mean of the population is *very small or very large*, we really do not need high-powered statistical techniques to decide whether the sample evidence supports or refutes the hypothesis about $\mu$. However, what do we do when the difference between a sample mean and the assumed mean of the population is *neither very small nor very large?* What if the Chamber of Commerce obtained a mean of $10,740 which differs from the assumed value by $10,740 − $10,600 = $140, or what if it obtained a mean of $10,360 which differs from the assumed value by $10,360 − $10,600 = −$240? All this really boils down to the question "Where do we draw the line?"

To answer this question we shall again have to study the chance fluctuations of means of random samples from the given population, or in other words, we shall have to refer to the sampling distribution of the mean. If the hypothesis $\mu = \$10{,}600$ is correct and if it can be assumed (perhaps, on the basis of previous studies) that the incomes of all the families in the community have the standard deviation $\sigma = \$780$, we can picture this sampling distribution as in Figures 8.4 and 8.5. According to the theorem on page 201, the mean of this sampling distribution equals the mean of the population,

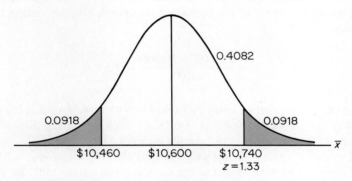

**FIGURE 8.4**  Sampling distribution of the mean.

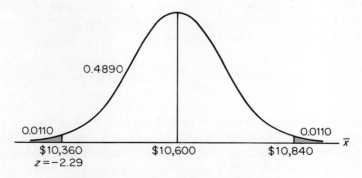

**FIGURE 8.5**  Sampling distribution of the mean.

namely, $10,600 (if the assumption is correct), and its standard deviation is given by

$$\sigma_{\bar{x}} = \frac{\sigma}{\sqrt{n}} = \frac{780}{\sqrt{55}} = 105$$

provided the population is so large that it can be treated as if it were infinite. Also, since the sample size is $n = 55$, we can use the Central Limit Theorem to approximate the sampling distribution of the mean with a normal curve.

Suppose now that the Chamber of Commerce actually obtains a mean of $10,740 for a sample of size $n = 55$, which is "off" by $140, and *it wants to know the probability of being "off" either way by $140 or more*, namely, the probability of getting a sample mean which is *at least* $10,740 or *at most* $10,600 − $140 = $10,460. The answer to this question is given by the total area of the shaded regions of Figure 8.4, which (due to the symmetry of the normal distribution) is *twice* the area under the curve to the right of $10,740. Making use of the result of the preceding paragraph and converting to standard units, we get

$$z = \frac{10,740 - 10,600}{105} = 1.33$$

and since the corresponding entry in Table I is 0.4082, we find that the desired probability is $2(0.5000 - 0.4082) = 0.1836$. Thus, the Chamber of Commerce may well conclude that the sample mean of $10,740 does *not* constitute any real evidence against the assumption that the true mean is $\mu = \$10,600$; after all, they could have expected the sample mean to be "off" by $140 or more *better than 18 percent of the time.*

If we use the same method to see how the Chamber of Commerce might react if they obtained a sample mean of $10,360, we find that *the probability of their being "off" by $240 or more* is given by the total area of the shaded regions of Figure 8.5, namely, *twice* the area under the curve to the left of $10,360. Converting to standard units, we get

$$z = \frac{10,360 - 10,600}{105} = -2.29$$

and since the corresponding entry in Table I is 0.4890, we find that the desired probability is $2(0.5000 - 0.4890) = 0.0220$. In this case the Chamber of Commerce may well conclude that the true average family income in the community is *not* $10,600; after all, the odds *against* their being "off" by $240 or more are better than 44 to 1 (0.9780 to 0.0220 to be more exact).

We can now answer the question posed on page 221, namely, "Where do we draw the line?" In practice, we proceed as is illustrated in Figure 8.6: *We specify the probability with which we are willing to risk rejecting the hypothesis about the population mean even though it is true, usually, 0.05 or 0.01. Then we convert the sample mean into standard units (as in our two illustrations) and reject the hypothesis about the population mean if z is less than − 1.96 or greater than 1.96, or if z is less than −2.58 or greater than 2.58.* This procedure is called a **test of significance,** for it enables us to decide whether a difference between a sample mean and the assumed value of a population mean can be attributed to chance, or whether it is **statistically significant,** namely, too large to be "reasonably" attributed to chance. Correspondingly, the probability with which we are willing to risk rejecting the hypothesis about the population mean even though it is true is called the **level of significance,** and it is usually denoted by the Greek letter $\alpha$ *(alpha).* The

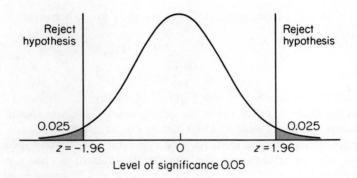

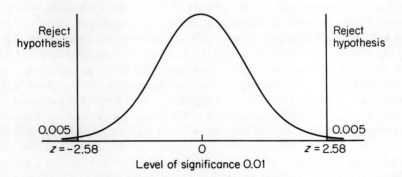

**FIGURE 8.6** Two-tail test criteria.

choice of $\alpha$, usually 0.05 or 0.01, depends in each case on the consequences (risks or penalties) of erroneously rejecting a hypothesis which is really true.

To give another example, suppose that a production process for making nails is considered to be *under control* if the nails produced have a mean length of 3.00 inches with a standard deviation of 0.09 inches. Whether or not the process is to be regarded as "under control" is decided each morning by a quality control engineer who bases his decision on the mean of a random sample of 36 nails and uses the level of significance $\alpha = 0.05$. Thus, let us see whether he should ask for an adjustment of the process on a day on which he obtains a mean of $\bar{x} = 3.03$ inches (which is a bit higher than the required 3.00 inches). Given $n = 36$ and $\sigma = 0.09$ inches, we find that the standard error of the mean is

$$\sigma_{\bar{x}} = \frac{0.09}{\sqrt{36}} = 0.015$$

so that $\bar{x} = 3.03$ inches, converted into standard units, becomes

$$z = \frac{3.03 - 3.00}{0.015} = 2.00$$

Since this value exceeds 1.96 (see Figure 8.6), the hypothesis $\mu = 3.00$ inches must be *rejected*—in other words, the quality control engineer should ask that the process (perhaps, some machine setting) be adjusted.

In both of our illustrations, the one concerning the family incomes and the one concerning the nails, the population standard deviation $\sigma$ was assumed to be known. If $\sigma$ is unknown but $n$ is *large*, 30 or more, we simply substitute for it the sample standard deviation $s$; however, if $\sigma$ is unknown and $n$ is less than 30, we will have to use an alternate small-sample technique which will be explained in Exercise 9 on page 231.

It is customary to refer to the criteria of Figure 8.6 as **two-sided tests** or as **two-tail tests,** because the hypothesis about the population mean is rejected for values of $\bar{x}$ falling into either "tail" of its sampling distribution. To give an example of a **one-tail test,** let us refer to the real-estate developer mentioned on page 209, and let us suppose that he is interested, in particular, in the hypothesis that on the average *at least* 9,000 cars drive by the given location each day.* Thus, he will reject the hypothesis only when $\bar{x}$ is *too small*, since large values of $\bar{x}$, values greater than 9,000, would support rather than refute the contention that on the average at least 9,000 cars drive by the given

---

* On page 209 we gave this as an example of *count data*, to be studied separately in Chapter 9, but the total which we count in a given situation can often be regarded as a measurement and, hence, be treated by the methods of this chapter.

location each day. As we cannot perform a significance test without assigning a *specific value* to the population mean $\mu$, we handle problems like this by testing the hypothesis $\mu = 9,000$ against the **alternative hypothesis** that $\mu$ is less than 9,000. This means that we will either accept the hypothesis $\mu = 9,000$ or we will accept the **one-sided alternative** that $\mu$ is less than 9,000. Returning for a moment to the example on page 222, where we were concerned with the family incomes, note that we were, in fact, testing the hypothesis $\mu = \$10,600$ against the **two-sided alternative** that the true average family income in the community is either less than or greater than $10,600, namely, the alternative that $\mu \neq \$10,600$.

The general procedure which we use to test the hypothesis that a population mean $\mu$ *equals a given value* against the one-sided alternative that it is *less than that value* is illustrated in Figure 8.7. If the level of significance is $\alpha = 0.05$, we reject the hypothesis (and accept the one-sided alternative) when $z$, the sample mean converted into standard units, is less than $-1.64$; if the level of significance is $\alpha = 0.01$, we reject the hypothesis (and accept the one-sided alternative) when $z$ is less than $-2.33$. These dividing lines, or

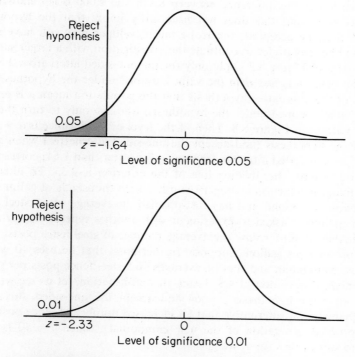

**FIGURE 8.7** One-tail test criteria (alternative hypothesis $\mu$ less than given value).

critical values, of the criterion were obtained by approximating the sampling distribution of the mean with a normal curve; in fact, they are the results of parts (a) and (e) of Exercise 6 on page 156.

To illustrate how this kind of test is performed in actual practice, suppose that the real-estate developer has somebody count the number of cars that drive by the given location on 60 randomly selected days and that he obtains $\bar{x} = 8{,}826$ and $s = 963$. Since $n = 60$ is large enough to substitute $s$ for $\sigma$, we get

$$\frac{\sigma}{\sqrt{n}} = \frac{963}{\sqrt{60}} = 124.32$$

for the standard error of the mean, and, hence,

$$z = \frac{8{,}826 - 9{,}000}{124.32} = -1.40$$

Since this value is *not* less than $-1.64$, the hypothesis $\mu = 9{,}000$ *cannot be rejected* at the level of significance $\alpha = 0.05$. Even though the sample mean *is* less than 9,000, the difference between 8,826 and 9,000 is **not statistically significant**. Note that this does not necessarily imply that the hypothesis $\mu = 9{,}000$ *must* be accepted; the real-estate developer may still have some doubts, and he may prefer to continue the investigation with a larger sample.

The criteria of Figure 8.7 apply only for the one-sided alternative that the population mean $\mu$ is *less than* the value assumed under the hypothesis. To take care of the alternative hypothesis that the population mean $\mu$ is *greater than* the value assumed under the hypothesis, we have only to turn the diagrams around as in Figure 8.8. Then, if the level of significance is $\alpha = 0.05$, we reject the hypothesis (and accept the one-sided alternative) when $z$, the sample mean converted into standard units, is greater than 1.64; correspondingly, for $\alpha = 0.01$, the dividing line of the criterion is 2.33. To illustrate, suppose that a production manager (who has used thousands of gallons of a rust-proofing compound and has rust-proofed an average of 64 steel fence posts per gallon) is asked to experiment with another type of rust-proofing to see whether it will *increase* the average number of steel fence posts which are rust-proofed per gallon. Suppose, furthermore, that he uses 40 gallons of the new compound, and that he averages 66.5 steel fence posts per gallon with a standard deviation of 4.8. Using this information, let us see whether we can reject the hypothesis $\mu = 64$ and accept the one-sided alternative $\mu > 64$ (namely, the alternative that an increased number of fence posts can be rust-proofed per gallon of the new compound). Since $n = 40$ is large enough to substitute $s$ for $\sigma$, we get

$$\frac{\sigma}{\sqrt{n}} = \frac{4.8}{\sqrt{40}} = 0.76$$

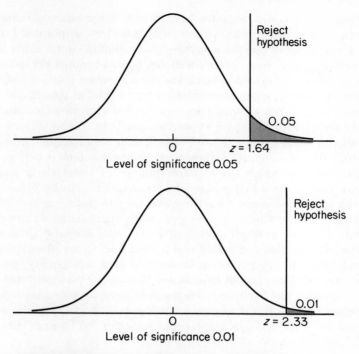

**FIGURE 8.8** One-tail test criteria (alternative hypothesis $\mu$ greater than given value).

for the standard error of the mean, and, hence,

$$z = \frac{66.5 - 64.0}{0.76} = 3.29$$

Since this value exceeds 2.33, the "critical value" for a level of significance of 0.01, we conclude that the new rust-proofing compound has *quite definitely* increased the number of steel fence posts that can be rust-proofed per gallon. It is generally a sound practice to *specify the level of significance beforehand;* this will spare us the temptation of choosing a level of significance which happens to suit our purpose *after we have seen the data or performed the calculations* (see Exercise 6 on page 230).

Since the general problem of testing hypotheses and constructing statistical decision criteria often tends to confuse the beginner, it helps to proceed systematically as outlined in the following steps:

1. **We formulate the hypothesis to be tested in such a way that the probability of erroneously rejecting it can be calculated.**

To follow this first step, we often have to assume the exact opposite of what we may want to prove. For instance, if it had been our purpose (in the last illustration) to *prove* that the new rust-proofing compound actually increased the number of steel fence posts which can be rust-proofed per gallon of the new compound, we would nevertheless have assumed that $\mu = 64$, namely, that the new rust-proofing compound is only as good as the old one. Had we assumed that $\mu > 64$, we would not have known what value to substitute for $\mu$ in the formula for $z$, and we would not have been able to perform the test. Similarly, to prove that a new machine will reduce spoilage of materials more effectively than another, we assume that the new machine is only as good as the old one (for which the average spoilage of materials is presumably known); and to show that a new sales technique is actually better than the one which has been in use, we assume that the new technique is only as good as the old one (for which we know how much merchandise we can expect to be sold). Since we assumed in each case that there is *no difference* (between the new rust-proofing compound and the old one, in the effectiveness of the two machines, and in the merits of the two sales techniques), we refer to hypotheses like these as **null hypotheses.** Nowadays, the term "null hypothesis" is used for any hypothesis *which is set up primarily to see whether it can be rejected;* in fact, this is why the symbol $\mu_0$ is used to denote the value of $\mu$ assumed in a test of significance—the subscript "0" stands for "zero," "null," or "nought."

Actually, the idea of setting up a null hypothesis is not uncommon even in non-statistical thinking. It is precisely what is done in criminal court procedures, where the accused is assumed to be innocent unless his guilt can be established beyond any reasonable doubt. The assumption that the accused is *not guilty* is a null hypothesis; if it cannot be rejected the accused will go free, but this does not necessarily imply that he is really innocent. Note also that the question of what we mean here by "reasonable doubt" is precisely the problem of assigning a level of significance.

2. **We formulate the alternative hypothesis which is to be accepted when the null hypothesis must be rejected.**

In the first example of this section, the one dealing with the family incomes, the null hypothesis was $\mu = \$10{,}600$ and the alternative hypothesis was $\mu \neq \$10{,}600$ (namely, the hypothesis that the true average family income is not equal to $\$10{,}600$). *Thus, we use the two-sided alternative $\mu \neq \mu_0$ if we want to reject the null hypothesis regardless of whether $\mu_0$ happens to be too small or too large.*

The choice of an appropriate one-sided alternative depends mostly on what we hope to be able to show, or better, perhaps, *where we want to put the burden of proof.* Suppose, for instance, that in the rust-proofing compound example the new compound is *much more expensive,* so that it would not be worthwhile to use the new compound unless it were really better than the old compound. Thus, we would test the null hypothesis $\mu = 64$ against the one-sided alternative $\mu > 64$, and we would conclude that the new compound is *better* only if the null hypothesis can be rejected. On the other hand, if the new rust-proofing compound is *much cheaper,* it would be worthwhile to use it unless the old compound were really better. Thus, we would test the null hypothesis $\mu = 64$ against the one-sided alternative $\mu < 64$, and we would conclude that the new compound is *not worse* than the old one so long as the null hypothesis *cannot be rejected.* In the first case we would thus be putting the burden of proof on the more expensive *new* compound, and in the second case we would be putting the burden of proof on the more expensive *old* compound.

Having formulated a null hypothesis and a suitable alternative, we then specify the *level of significance* (namely, the probability of erroneously rejecting the null hypothesis) and proceed with the following steps:

3. Using the sample data, we calculate the value of the statistic

$$z = \frac{\bar{x} - \mu_0}{\sigma/\sqrt{n}}$$

that is, we convert the sample mean into standard units, replacing $\sigma$ (if necessary) with the sample standard deviation $s$.

4. Finally, we base our decision on the appropriate criterion of Figure 8.6, 8.7, or 8.8, depending on the choice of the alternative hypothesis and the level of significance.

As we already pointed out on page 224, the choice of the level of significance depends on the consequences (risks or penalties) of erroneously rejecting a null hypothesis which is actually true. Generally, we use $\alpha = 0.05$ or $\alpha = 0.01$, and preferably the smaller of the two values when the consequences of this kind of error are really serious.

If we *cannot reject* the null hypothesis, we have the option of either accepting it or reserving judgment. In the latter case we say that "the difference between $\bar{x}$ and $\mu_0$ is *not statistically significant.*" In other words, we admit that we are unable to *disprove* the null hypothesis, but by the same token we do not have to accept it either. This is the possibility which we suggested in the example on page 226.

## EXERCISES

1. A construction company is considering the purchase of a new rivet gun. If their old rivet gun averages 230 rivets an hour and the company samples the performance of the new gun for several one-hour periods, against what alternative should they test the null hypothesis $\mu = 230$ if (a) they do not want to spend the money for the new rivet gun unless it can definitely be proven superior, and (b) they want to buy the new gun which is cheaper to operate unless it is actually inferior to the old one?

2. A large company is considering changing its filing system from the standard file cabinet type in which a clerk takes on the average 2.3 minutes to locate a file, to a new micro-film system. If one department of the company is now trying out the micro-film system, against what alternative should they test the null hypothesis $\mu = 2.3$ minutes if (a) the new system is to be recommended so long as it is not slower than the old one, and (b) the new system is to be recommended only if it is actually faster than the old one?

3. The dean of students of a university wants to test the hypothesis that on the average a student spends $62.00 per month for personal expenses. What can the dean conclude at the level of significance $\alpha = 0.05$ if 225 students (selected at random from the registrar's files) averaged $64.75 in a given month with a standard deviation of $13.60?

4. A manufacturer of woolen cloth has been producing a grade of woolen cloth with a water absorption capacity of 4.02 ounces per yard, and a standard deviation of 0.44 ounces. (Water absorption capacity is one factor contributing to the "warmth" of woolen cloth.) If tests of 50 samples of the same grade cloth produced by an experimental new process average 4.15 ounces, what can the manufacturer conclude at the level of significance $\alpha = 0.01$?

5. According to specifications, the mean time required to inflate a rubber life raft is to be 8.5 seconds. What can one conclude (at the level of significance $\alpha = 0.05$) about a particular shipment of such life rafts, if a random sample of 64 of the rafts have a mean inflation time of 8.6 seconds with a standard deviation of 0.7 seconds?

6. A rental agent who is anxious to rent a location for a restaurant in a shopping mall assures a restaurant owner that during an average business day at least 3,500 persons pass through the mall. Being a cautious business man, the restaurant owner conducts his own study (to make sure that on the average at least 3,500 persons pass through the mall per day) and he obtains a mean of 3,350 persons with a standard deviation of 436

persons for 32 randomly selected days. What can he conclude (a) at the level of significance $\alpha = 0.05$, and (b) at the level of significance $\alpha = 0.01$?

7. A Marine sergeant claims that his average recruit can get through a particularly difficult obstacle course in 15.0 minutes. Is this figure substantiated by the performance of 150 recruits who averaged 15.4 minutes with a standard deviation of 2.4 minutes? Use (a) the level of significance $\alpha = 0.05$, and (b) the level of significance $\alpha = 0.01$.

8. While investigating a soda-pop manufacturer, an investigator finds that a random sample of 100 "12-ounce size" cans of root beer average only 11.86 ounces per can with a standard deviation of 0.23 ounces. Using an appropriate one-sided alternative and the level of significance $\alpha = 0.05$, test whether this constitutes evidence on which the investigator can base a finding that the manufacturer is producing cans of root beer which, on the average, contain less than 12 ounces?

9. **SMALL-SAMPLE TESTS** For small samples (that is, when $n$ is less than 30), we proceed as in Exercise 13 on page 218, assume that the population from which we are sampling has roughly the shape of a normal distribution, and base our argument on the $t$ distribution. (If the assumption about the shape of the population is unreasonable, we have to substitute one of the so-called **nonparametric tests,** which may be found in more advanced texts.) So far as the outline on pages 227 through 229 is concerned, the first two steps remain the same, but in the third step we calculate the statistic

$$t = \frac{\bar{x} - \mu_0}{s/\sqrt{n}}$$

instead of $z$ (which is really not much of a change, for even in the large-sample tests we replaced $\sigma$ with $s$ when the population standard deviation was unknown). So far as the fourth step is concerned, we base our decision on the value of $t$ and we still refer to the criteria of Figures 8.6, 8.7, and 8.8, but in Figure 8.6 we substitute $t_{.025}$ for 1.96 and $t_{.005}$ for 2.58, while in Figures 8.7 and 8.8 we substitute $t_{.05}$ for 1.64 and $t_{.01}$ for 2.33. As in the small-sample confidence interval for $\mu$, the number of *degrees of freedom* is $n - 1$ and the required values of $t_{.05}$, $t_{.025}$, $t_{.01}$, and $t_{.005}$ are given in Table II on page 326. To illustrate this technique, suppose that the specifications for a certain kind of fishing line call for a mean breaking strength of 22.0 pounds. If seven pieces (randomly selected from different reels of a large shipment) have a mean breaking strength of 21.3 pounds and a standard deviation of 2.1 pounds, let us use the small-sample technique to test at the level of significance $\alpha = 0.05$ whether the shipment meets specifications. Substituting into the formula for $t$, we get

$$t = \frac{\bar{x} - \mu_0}{s/\sqrt{n}} = \frac{21.3 - 22.0}{2.1/\sqrt{7}} = -0.88$$

and since this is *not* less than $-1.943$, where $1.943$ is the value of $t_{.05}$ for $7 - 1 = 6$ degrees of freedom, we find that the null hypothesis $\mu_0 = 22.0$ pounds cannot be rejected against the alternative that the line is too weak.

(a) In ten test jumps, a newly-designed track shoe enables a high jumper to jump on the average 6.97 feet with a standard deviation of 0.22 feet. What does this tell us about the high jumper's claim that with these new shoes he can average 7.0 feet?

(b) In fourteen test drives over a marked racing course, a racing car with a newly-designed steam engine averaged 4.22 minutes with a standard deviation of 0.14 minutes. What does this tell us about the claim that on the average the car can complete the course in less than 4.30 minutes? (Using the level of significance $\alpha = 0.05$, test the null hypothesis $\mu = 4.30$ against the alternative which puts the burden of proof on the car with the new engine.)

(c) A new weight-reducing drug given to eight overweight women helped them to lose 14, 9, 17, 11, 10, 18, 12, and 13 pounds within a period of one month. Use a level of significance of 0.05 to test the claim that on the average the drug will help an overweight woman to lose at least 10 pounds within a month. (Test the hypothesis $\mu = 10$ against a suitable one-sided alternative so that the burden of proof will be on the new drug.)

(d) A random sample of 15 graduates of a secretarial school averaged 73.6 words per minute with a standard deviation of 8.1 words per minute on a typing test. Use the level of significance $\alpha = 0.01$ to test the school's claim that their graduates average better than 75 words per minute.

**10. TYPE I AND TYPE II ERRORS**  The error which we make when we *reject a true null hypothesis* is often called a **Type I error,** and the error which we make when we *accept a false null hypothesis* is then called a **Type II error.** For instance, in the example on page 222 we would have committed a Type I error if our sample data had led us to *reject* the null hypothesis $\mu = \$10,600$ even though the average family income is $\$10,600$, and we would have committed a Type II error if our sample data had led us to *accept* the null hypothesis $\mu = \$10,600$ even though the true average family income is, say, $\$11,000$.

(a) With reference to the illustration on page 224 where the quality control engineer tests the null hypothesis that the nails have a mean length of 3.00 inches, explain under what conditions he would be committing a Type I error and under what conditions he would be committing a Type II error.
(b) With reference to the illustration on page 226 where the real-estate developer tests the null hypothesis that the average number of cars that drive by a certain location is 9,000 (against the alternative that their average number is less), explain under what conditions he would be committing a Type I error and under what conditions he would be committing a Type II error.
(c) Suppose that a psychological testing service is asked to check whether an executive is emotionally fit to assume the presidency of a large corporation. What type of error would they be committing if they erroneously rejected the null hypothesis that he is fit for the job? What type of error would they be committing if they erroneously accepted the null hypothesis that he is fit for the job?
(d) A life insurance salesman wants to test the null hypothesis that 70 percent of his sales are made to married persons (against the alternative that the percentage is actually less). Explain under what conditions he would be committing a Type I error and under what conditions he would be committing a Type II error.

**11. THE PROBABILITY OF A TYPE II ERROR** The probability of a Type I error is what we have been referring to as the level of significance, and it has always been *specified* in the test criteria which we have discussed. So far as the probability of a Type II error is concerned, the situation is much more complicated, as the reader will discover by working this example. Referring to the problem on page 226 where the real-estate developer was concerned with the average number of cars that pass a certain location, let us first determine the value of $\bar{x}$ which corresponds to the dividing line of the $\alpha = 0.05$ criterion of Figure 8.7, namely, $z = -1.64$. Substituting $n = 60$, $\sigma = 963$, $\mu_0 = 9{,}000$, and $z = -1.64$ into the formula for $z$ on page 229, we get

$$-1.64 = \frac{\bar{x} - 9{,}000}{963/\sqrt{60}}$$

and, hence,

$$\bar{x} = 9{,}000 - 1.64 \cdot \frac{963}{\sqrt{60}} = 9{,}000 - 204 = 8{,}796$$

rounded to the nearest whole number. Now we can calculate the probability of a Type II error, say, when the true average number of cars that pass the given location is 8,750. We would be committing this kind of error if we got a sample mean greater than 8,796 when $\mu = 8{,}750$, and the answer is therefore given by the area of the shaded region of Figure 8.9. Converting 8,796 into standard units, we get

$$z = \frac{8{,}796 - 8{,}750}{963/\sqrt{60}} = 0.37$$

and since the corresponding entry in Table I is 0.1443, we find that the probability of a Type II error is $0.5000 - 0.1443 = 0.3557$, or approximately 0.36, when the true average number of cars that pass each day by the given location is 8,750.

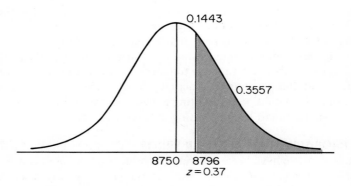

FIGURE 8.9  Probability of Type II error.

(a) Duplicating the last part of these calculations, show that if $\mu = 8{,}700$ the probability of a Type II error is about 0.22, if $\mu = 8{,}600$ the probability of a Type II error is about 0.06, if $\mu = 8{,}500$ the probability of a Type II error is about 0.009, and if $\mu = 8{,}900$ the probability of a Type II error is about 0.80.

(b) Draw a system of perpendicular axes on which $\mu$ is measured along the horizontal axis and the probability of a Type II error is measured along the vertical axis; plot the points which correspond to $\mu = 8{,}900$, $\mu = 8{,}750$, $\mu = 8{,}700$, $\mu = 8{,}600$, and $\mu = 8{,}500$; and join them by means of a smooth curve. This kind of curve is called an **operating characteristic curve**, and it provides us with an over-all picture of the probabilities of Type II errors to which we are exposed when the true average number of cars that pass the given location each day is not 9,000. [Of course, all this work applies only to this particular example, and in most practical situations we are satisfied specifying the probability of a Type I error (namely, the level of significance), keeping our fingers crossed that the probabilities of *serious* Type II errors are "within reason."]

12. **PROBABILITIES OF TYPE I AND TYPE II ERRORS** Unfortunately, the relationship between the probabilities of Type I errors and Type II errors is such that *if we reduce one this will automatically increase the other* (provided everything else remains unchanged). To illustrate, let us refer to the example on page 226 and reduce the probability of a Type I error from 0.05 to 0.01. If we proceed as in Exercise 11, we obtain $\bar{x}$, the dividing line of the criterion, by solving the equation

$$-2.33 = \frac{\bar{x} - 9{,}000}{963/\sqrt{60}}$$

which yields $\bar{x} = 8{,}710$. Then, if $\mu = 8{,}750$ (as in Exercise 11), the probability of a Type II error is given by the normal curve area to the right of

$$z = \frac{8{,}710 - 8{,}750}{963/\sqrt{60}} = -0.32$$

and since the entry in Table I corresponding to $z = 0.32$ is 0.1255, we find that the probability of a Type II error has increased from 0.36 to $0.5000 + 0.1255 = 0.6255$, or approximately 0.63. Use the same kind of argument to recalculate the probabilities of Type II errors for $\alpha = 0.01$ and the alternative hypotheses $\mu = 8{,}500$, $\mu = 8{,}600$, $\mu = 8{,}700$, and $\mu = 8{,}900$, and compare the results with those obtained in part (a) of Exercise 11 corresponding to $\alpha = 0.05$.

13. **PROBABILITIES OF TYPE I AND TYPE II ERRORS, CONTINUED** To illustrate the effect that changes in the sample size have on the probabilities of Type I and Type II errors, suppose that in the example on page 226 the sample size had been $n = 100$ instead of $n = 60$. Duplicating the work of Exercise 11, show that if the probability of a Type I error is kept at $\alpha = 0.05$, the probability of a Type II error is reduced from 0.36 to 0.17.

## DIFFERENCES BETWEEN MEANS

There are many situations in which we must decide whether an observed difference between two sample means can be attributed to chance, or whether it is indicative of the fact that the two samples come from populations with unequal means. For instance, if a sample of one kind of vacuum tube lasted on the average 2,815 hours while a sample of another kind of vacuum tube lasted on the average 1,963 hours, we may want to decide (or have to decide) whether there really is a difference in the average performance of the two kinds of vacuum tubes, or whether the difference between the two sample means can be attributed to chance. In other words, *we have to decide whether the difference between the two means is statistically significant.* Similarly, we may want to decide on the basis of sample data whether the average mileage yield of one kind of gasoline is higher than that of another, whether the average number of mistakes made by clerks who have taken the company's training program is smaller than the average number of mistakes made by those who have not taken the program, whether the average wrinkle resistance of one kind of fabric is greater than that of another kind of fabric, and so on.

The test which we shall use to decide whether an observed difference between two sample means can be attributed to chance is based on the following theory: *If $\bar{x}_1$ and $\bar{x}_2$ are the means of independent random samples from two populations with the means $\mu_1$ and $\mu_2$ and the standard deviations $\sigma_1$ and $\sigma_2$, then the sampling distribution of the statistic $\bar{x}_1 - \bar{x}_2$ has the mean $\mu_1 - \mu_2$ and the standard deviation*

$$\sigma_{\bar{x}_1 - \bar{x}_2} = \sqrt{\frac{\sigma_1^2}{n_1} + \frac{\sigma_2^2}{n_2}}$$

*appropriately called the* **standard error of the difference between two means.** * Also, for large samples (namely, when neither $n_1$ nor $n_2$ is less than 30), the sampling distribution of $\bar{x}_1 - \bar{x}_2$ can be approximated closely with a normal distribution, and, if necessary, the sample standard deviations $s_1$ and $s_2$ can be substituted for $\sigma_1$ and $\sigma_2$. Thus, we can base the test of the null hypothesis $\mu_1 = \mu_2$ on the statistic

$$z = \frac{\bar{x}_1 - \bar{x}_2}{\sqrt{\frac{s_1^2}{n_1} + \frac{s_2^2}{n_2}}}$$

which has approximately the standard normal distribution as it was obtained by *subtracting* from $\bar{x}_1 - \bar{x}_2$ the mean of its sampling distribution (which

---

* By "independent" samples we mean that the selection of one sample should in no way affect the selection of the other. Thus, the theory does *not* apply to "before and after" kinds of comparisons, nor does it apply, for example, if we want to compare the ages of husbands and wives. A special method for handling situations of this kind is given in Exercise 6 on page 239.

under the null hypothesis equals $\mu_1 - \mu_2 = 0$), and then *dividing* by the expression for the standard error of the difference between two means with the sample standard deviations $s_1$ and $s_2$ substituted for $\sigma_1$ and $\sigma_2$. Depending on whether the *alternative hypothesis* is $\mu_1 - \mu_2 \neq 0$, $\mu_1 - \mu_2 < 0$, or $\mu_1 - \mu_2 > 0$, we can thus base the test of the null hypothesis on the criteria of Figures 8.6, 8.7, and 8.8.

To illustrate this *large-sample* technique, suppose that we want to investigate whether there *is* a difference between the salaries received by male and female Class A accounting clerks in Cincinnati, Ohio. Sample data show that $n_1 = 97$ male Class A accounting clerks received average weekly earnings of $147.50 with a standard deviation of $20.54, while $n_2 = 199$ female Class A accounting clerks received average weekly earnings of $130.00 with a standard deviation of $21.23. Following the steps outlined on pages 227 through 229, we begin by formulating the null hypothesis $\mu_1 = \mu_2$ and the alternative hypothesis $\mu_1 - \mu_2 > 0$, so that the rejection of the null hypothesis will support the contention that in Cincinnati male Class A accounting clerks receive higher average weekly earnings than female Class A accounting clerks. Also, we shall specify the level of significance as $\alpha = 0.01$. Then, substituting $n_1 = 97$, $n_2 = 199$, $\bar{x}_1 = 147.50$, $\bar{x}_2 = 130.00$, $s_1 = 20.54$, and $s_2 = 21.23$ into the formula for $z$, we get

$$z = \frac{147.50 - 130.00}{\sqrt{\frac{(20.54)^2}{97} + \frac{(21.23)^2}{199}}} = 6.81$$

and since this greatly *exceeds* 2.33, the dividing line of the $\alpha = 0.01$ criterion of Figure 8.8, the null hypothesis will have to be *rejected*. In other words, these sample data confirm the claim that male Class A accounting clerks receive higher average weekly earnings than female Class A accounting clerks (at least, in the city to which the data pertain).

## EXERCISES

1. A consumer testing organization wants to find out whether either of two brands of work shoes has longer lasting soles than the other. If 40 randomly selected workers wear out the soles of a Brand $X$ shoe on the average in 80 days with a standard deviation of 5.2, while 40 randomly selected workers wear out the soles of a Brand $Y$ shoe on the average in 84 days with a standard deviation of 3.1, test at the level of significance $\alpha = 0.05$ whether the difference between the two average wearing lives is significant.

**238** THE ANALYSIS OF MEASUREMENTS CHAP. 8

2. A manufacturer of rubber bands claims that its rubber bands are superior to those of a competitor on the basis of a study which showed that a sample of 60 of its rubber bands could stretch on the average 7.6 inches (before breaking) with a standard deviation of 0.65 inches, while a sample of 40 rubber bands made by the competitor had an average stretch of only 7.3 inches with a standard deviation of 1.33 inches. Use a suitable one-sided alternative and the level of significance $\alpha = 0.05$ to check whether the claim is justified.

3. A sample study of the number of pieces of chalk the average professor uses was conducted at two universities. If 80 professors at one university averaged 13.7 pieces of chalk per month with a standard deviation of 2.4, while 60 professors at another university averaged 12.2 pieces of chalk with a standard deviation of 3.3, test at the level of significance $\alpha = 0.01$ whether the difference between these two sample means is significant.

4. A sample survey conducted in two suburbs showed that in one suburb a sample of 60 houses averaged 17.4 windows with a standard deviation of 4.3 windows, while in the other suburb a sample of 70 houses averaged 15.3 windows with a standard deviation of 3.6 windows. If a window washer wants to decide in which of the two suburbs he should open his business, what decision would he reach in checking the claim that the average house in the first suburb has more windows than the average house in the second suburb, if he used a level of significance of $\alpha = 0.05$?

5. **SMALL-SAMPLE TESTS** For small samples (that is, when either sample size is less than 30), we can base our argument on the $t$ distribution as in Exercise 9 on page 231, provided the two populations from which we are sampling have roughly the shape of normal distributions and *equal standard deviations*. So far as the null hypothesis and the alternative hypothesis are concerned, they are the same as in the large-sample test discussed in the text, and the decision criteria are identical with those of Exercise 9 on page 231, but the formula for the statistic on which we base our decision is now

$$t = \frac{\bar{x}_1 - \bar{x}_2}{\sqrt{\dfrac{(n_1 - 1)s_1^2 + (n_2 - 1)s_2^2}{n_1 + n_2 - 2}} \sqrt{\dfrac{1}{n_1} + \dfrac{1}{n_2}}}$$

Also, the number of *degrees of freedom* (needed to look up the appropriate value of $t_{.05}$, $t_{.025}$, $t_{.01}$, or $t_{.005}$ in Table II) is now $n_1 + n_2 - 2$. To illustrate, suppose that an office manager wants to compare two brands of desk calculators that are used in his office, and that six randomly selected workers solve an average of 67.2 problems in a day with a machine of

Brand A and that six randomly selected workers solve an average of 62.4 problems in a day with a machine of Brand B; the corresponding standard deviations are, respectively, 4.3 problems and 3.1 problems. Calculating $t$ to test the null hypothesis $\mu_1 = \mu_2$ against the alternative $\mu_1 \neq \mu_2$, the office manager gets

$$t = \frac{67.2 - 62.4}{\sqrt{\frac{(6-1)(4.3)^2 + (6-1)(3.1)^2}{6+6-2}} \sqrt{\frac{1}{6} + \frac{1}{6}}} = 2.22$$

which is *not* significant at the level of significance $\alpha = 0.05$, since $t_{.025}$ for $6 + 6 - 2 = 10$ degrees of freedom equals 2.228. Even though the difference between the two sample means is *fairly large*, the samples are so small that the results are not conclusive.

(a) To find out whether the students in an experimental mathematics program have learned faster than the students (of equal ability) in the standard program, a teacher gave an examination to ten students in each program. If the ten students in the standard program averaged 84.27 with a standard deviation of 3.6, while those in the experimental program averaged 88.10 with a standard deviation of 2.3, test at the level of significance $\alpha = 0.01$ whether the difference between these two sample means is significant.

(b) If 18 randomly selected trees in a city park have a mean height of 38 feet with a standard deviation of 10 feet, while ten randomly selected trees in the same city's playground have a mean height of 54 feet with a standard deviation of 15 feet, check on the claim that the trees in the playground are taller *on the average* than those in the park. Use the alternative hypothesis that the claim is true and the level of significance $\alpha = 0.05$.

(c) With reference to the illustration on page 47, use the level of significance $\alpha = 0.05$ to check whether the difference between the average percentage profits of Corporations A and C is significant.

**6. DEPENDENT SAMPLES** As we pointed out on page 236, the methods of this section do not apply unless the two samples are *independent*. Suppose, then, that we want to study the effectiveness of a new speed reading program and that we are given the following data on the "before and after" speeds at 80 percent comprehension of ten students in the program:

325 and 360,  392 and 441,  374 and 391,  273 and 264,  505 and 528
243 and 250,  425 and 423,  302 and 327,  512 and 522,  609 and 651

**240**  THE ANALYSIS OF MEASUREMENTS  CHAP. 8

The average "before" speed is 396 words per minute and the average "after" speed is 415.7, and in order to test whether the difference between these two means is significant, we proceed as follows: we calculate the *change* in speed for each student, getting +35, +49, +17, −9, +23, +7, −2, +25, +10, and +42, and *then we test the null hypothesis that these differences constitute a random sample from a population with the mean* $\mu = 0$. Since $n = 10$ in this example, use the small-sample test of Exercise 9 on page 231 to test this null hypothesis against the alternative hypothesis $\mu > 0$ (which would imply that the speed reading program is, indeed, effective) at the level of significance $\alpha = 0.05$.

**7. DEPENDENT SAMPLES (CONTINUED)**  To determine the effectiveness of an industrial safety program, the following data were collected over a period of one year on the average weekly loss of man hours due to accidents in 12 plants "before and after" the program was put into operation:

| | | | |
|---|---|---|---|
| 38 and 29, | 75 and 63, | 25 and 23, | 129 and 117 |
| 47 and 48, | 53 and 41, | 12 and 14, | 76 and 73 |
| 13 and 17, | 24 and 19, | 38 and 36, | 23 and 25 |

Use the method of Exercise 6 and the level of significance $\alpha = 0.01$ to check whether we can say that the safety program *is* effective.

## DIFFERENCES AMONG k MEANS
## (ANALYSIS OF VARIANCE)

Let us now generalize the work of the preceding section and consider the problem of deciding whether observed differences among *more than two* sample means can be attributed to chance, or whether they are indicative of actual differences among the means of the corresponding populations. For instance, we may want to decide on the basis of sample data whether there really is a difference in the useful life of three types of color television tubes, we may want to compare the average salaries received by samples of cotton textile workers in several areas of the United States, we may want to see whether there really is a difference in the average amount of life insurance sold by the salesmen of different companies, we may want to judge whether there really is a difference in the average amount of air pollution in several cities, and so on. To illustrate, suppose that a major automobile rental agency wants to know whether there really are differences in the mileages obtained with Super Grade, Premium Grade, and Standard Grade tires, all produced by the same manufacturer. Now suppose that random samples of four tires of each kind (taken from the performance records maintained by the automobile rental agency) yield the following data, all rounded to the nearest thousand miles:

| Super Grade | Premium Grade | Standard Grade |
|---|---|---|
| 48 | 38 | 29 |
| 38 | 42 | 35 |
| 44 | 41 | 25 |
| 42 | 35 | 27 |

The means of these three samples are, respectively, 43, 39, and 29 thousand miles, and what we would like to know is whether the differences among these means are actually significant; after all, the size of the samples is *very small*.

If we let $\mu_1$, $\mu_2$, and $\mu_3$ denote the true average mileages for the three grades of tires, we shall want to decide on the basis of the given data whether or not it is reasonable to say that these $\mu$'s are all equal; in other words, we shall want to test the null hypothesis $\mu_1 = \mu_2 = \mu_3$ against the alternative hypothesis that $\mu_1$, $\mu_2$, and $\mu_3$ are *not all equal*. Evidently, this null hypothesis would be supported if the sample means were all nearly the same size, and the alternative hypothesis would be supported if the differences among the sample means were large. Hence, we need a precise measure of the discrepancies among the $\bar{x}$'s, and the most obvious, perhaps, is their standard deviation or their variance, which we can calculate according to the formula on page 193. Since $\dfrac{43 + 39 + 29}{3} = 37$, we get

$$s_{\bar{x}}^2 = \frac{(43 - 37)^2 + (39 - 37)^2 + (29 - 37)^2}{3 - 1} = 52$$

where we used the subscript $\bar{x}$ to indicate that this variance measures the variability of the sample means.

Let us now make an assumption which is critical to the method of analysis we shall employ: *It will be assumed that the populations from which we are sampling can be approximated closely with normal distributions having the same standard deviation* $\sigma$. If the null hypothesis $\mu_1 = \mu_2 = \mu_3$ is true, we can then look upon our three samples as samples from *one and the same population*, and, hence, upon $s_{\bar{x}}^2 = 52$ as an estimate of $\sigma_{\bar{x}}^2$, the *square* of the standard error of the mean. Now, if we make use of the theorem on page 202 according to which $\sigma_{\bar{x}} = \dfrac{\sigma}{\sqrt{n}}$ for samples from infinite populations, we can look upon

$s_{\bar{x}}^2 = 52$ as an estimate of $\sigma_{\bar{x}}^2 = \left(\dfrac{\sigma}{\sqrt{n}}\right)^2 = \dfrac{\sigma^2}{n}$, and, therefore, upon $n \cdot s_{\bar{x}}^2 = 4 \cdot 52 = 208$ as an estimate of $\sigma^2$, the common variance of the three populations.

If $\sigma^2$ were known, we could compare $n \cdot s_{\bar{x}}^2 = 208$ with $\sigma^2$ and *reject* the null hypothesis if 208 were much larger than $\sigma^2$. However, in our example (as in most practical problems) $\sigma^2$ is unknown and we have no choice but to estimate it on the basis of the given data. Having assumed, in fact, that the three samples came from identical populations, we could use any one of their variances (namely, $s_1^2$, $s_2^2$, or $s_3^2$) as an estimate of $\sigma^2$ and, hence, we can also use their *mean*

$$\begin{aligned}\dfrac{s_1^2 + s_2^2 + s_3^2}{3} = \dfrac{1}{3}\Bigg[ &\dfrac{(48-43)^2 + (38-43)^2 + (44-43)^2 + (42-43)^2}{4-1} \\ &+ \dfrac{(38-39)^2 + (42-39)^2 + (41-39)^2 + (35-39)^2}{4-1} \\ &+ \dfrac{(29-29)^2 + (35-29)^2 + (25-29)^2 + (27-29)^2}{4-1} \Bigg] \\ = &\dfrac{138}{9} \text{ (or } 15\tfrac{1}{3})\end{aligned}$$

which utilizes *all* the information we have about the population variance $\sigma^2$. We now have two estimates of $\sigma^2$,

$$n \cdot s_{\bar{x}}^2 = 208 \quad \text{and} \quad \dfrac{s_1^2 + s_2^2 + s_3^2}{3} = 15\tfrac{1}{3}$$

and if the *first estimate* (which is based on the variation among the sample means) is much larger than the *second estimate* (which is based on the variation within the samples and, hence, measures variation that is due to chance), it stands to reason that the null hypothesis should be rejected. After all, *in that case the variation among the sample means would be greater than it should be if it were due only to chance.*

To put the comparison of these two estimates of $\sigma^2$ on a rigorous basis, we use the statistic

$$F = \dfrac{\text{estimate of } \sigma^2 \text{ based on the variation among the } \bar{x}\text{'s}}{\text{estimate of } \sigma^2 \text{ based on the variation within the samples}}$$

which is appropriately called a **variance ratio.** If the null hypothesis is true, the sampling distribution of this statistic is the so-called $F$ **distribution,** an example of which is shown in Figure 8.10. Since the null hypothesis will be rejected only when $F$ is *large* (namely, when the variability of the $\bar{x}$'s is too great to be attributed to chance), we ultimately base our decision on the criterion of Figure 8.10. Here $F_{.05}$ is such that the area under the curve to its right is equal to 0.05, and it provides the dividing line of the criterion when the level of significance is $\alpha = 0.05$; correspondingly, $F_{.01}$ provides the divid-

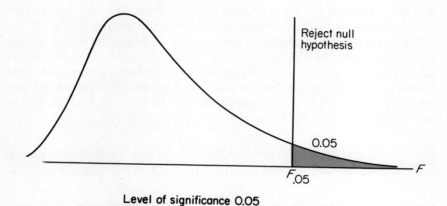

**Level of significance 0.05**

**FIGURE 8.10** $F$ distribution.

ing line of the criterion when the level of significance is $\alpha = 0.01$. These quantities, which depend on two parameters called, respectively, the **numerator** and **denominator degrees of freedom,** are given in Tables IVa and IVb at the end of the book. When we compare the means of $k$ samples of size $n$, the number of degrees of freedom for the numerator of $F$ is $k - 1$, and the number of degrees of freedom for the denominator of $F$ is $k(n - 1)$.*

Returning now to our numerical example, we find that $F_{.05} = 4.26$ for $k - 1 = 3 - 1 = 2$ and $k(n - 1) = 3(4 - 1) = 9$ degrees of freedom, and since this value is exceeded by

$$F = \frac{208}{\frac{138}{9}} = 13.57$$

the value of the $F$ statistic which we obtained in our example, the null hypothesis will have to be *rejected*. In other words, we conclude that there *is* a differ-

---

* So far as the formula for the numerator degrees of freedom is concerned, note that the numerator is $n \cdot s_{\bar{x}}^2$ where $s_{\bar{x}}^2$ is the variance of $k$ means and, hence, has $k - 1$ degrees of freedom in accordance with the terminology introduced in the footnote to page 190. So far as the formula for the denominator degrees of freedom is concerned, note that the denominator is the mean of $k$ sample variances with each having $n - 1$ degrees of freedom.

ence in the true average mileage obtained with the three grades of tires produced by the same manufacturer. (Incidentally, had we used the level of significance $\alpha = 0.01$, we would have found that $F_{.01} = 8.02$ for two and nine degrees of freedom, and the result would have been the same.) Actually, the result should not come as a surprise since the differences in the mileage yields were quite apparent from the sample data; but, as we already indicated on page 241, the samples were very small and we just wanted to make sure that the differences among the means *are* significant. This will not be quite so obvious in some of the exercises at the end of this section.

If we have $k$ random samples of size $n$ and let $x_{ij}$ denote the $i$th observation of the $j$th sample (for instance, $x_{31}$ is the *third* observation of the *first* sample and for the data on page 241 it equals 44, $x_{42}$ is the *fourth* observation of the *second* sample and for the data on page 241 it equals 35, and $x_{33}$ is the *third* observation of the *third* sample and for the data on page 241 it equals 25), we can write the formula for $F$ directly as

$$F = \frac{k(n-1)[n \cdot \Sigma\,(\bar{x}_j - \bar{x})^2]}{(k-1)[\Sigma\,\Sigma\,(x_{ij} - \bar{x}_j)^2]}$$

where $\bar{x}_j$ is the mean of the $j$th sample and $\bar{x}$ is the **grand mean** of all the data. The expression bracketed in the *numerator* is $n$ times the sum of the squares of the differences between the sample means and the grand mean, and in actual practice it is usually calculated by means of the short-cut formula

$$n \cdot \Sigma\,(\bar{x}_j - \bar{x})^2 = \frac{1}{n} \cdot \Sigma\,T_j^2 - \frac{T^2}{kn}$$

where $T_j$ is the total of the observations in the $j$th sample and $T$ is the grand total of all the data. The expression bracketed in the *denominator* is the sum of the squares of the differences between the individual observations and the means of the respective samples; in fact, it is a **double summation** as we indicated by means of the $\Sigma\,\Sigma$, for we are summing on $i$ as well as $j$ (in other words, it applies to *all* of the data). In actual practice, this quantity is usually calculated by means of the short-cut formula

$$\Sigma\,\Sigma\,(x_{ij} - \bar{x}_j)^2 = \Sigma\,\Sigma\,x_{ij}^2 - \frac{1}{n} \cdot \Sigma\,T_j^2$$

where $\Sigma\,\Sigma\,x_{ij}^2$ is the sum of the squares of all of the given observations, and the quantity which we subtract was already determined in connection with the short-cut formula for the expression bracketed in the numerator of $F$.

Had we used these formulas in our numerical example, we would have obtained $T_1 = 172$, $T_2 = 156$, $T_3 = 116$, $T = 444$, and $\Sigma \Sigma x_{ij}^2 = 16{,}982$, so that

$$n \cdot \Sigma (\bar{x}_j - \bar{x})^2 = \frac{1}{4}(172^2 + 156^2 + 116^2) - \frac{444^2}{3 \cdot 4}$$

$$= 416$$

$$\Sigma \Sigma (x_{ij} - \bar{x}_j)^2 = 16{,}982 - \frac{1}{4}(172^2 + 156^2 + 116^2)$$

$$= 138$$

and, hence,

$$F = \frac{3(4-1)416}{(3-1)138} = 13.57$$

which is identical with the result obtained before. The advantages of the short-cut technique may not be apparent from this example, but this is due to the fact that all of the sample means happened to be whole numbers.

The method which we have studied in this section belongs to a very important branch of statistics called **analysis of variance**. We, so to speak, analyze what part of the total variation of our data might be attributed to specific "causes," or *sources of variation* (for instance, the three grades of tires of our example), and we then compare it with that part of the variation of the data which can be attributed to chance.

## EXERCISES

1. The following are the number of pieces produced by three factory workers (employed by a firm to do the same job) on four consecutive days:

   | | | | | |
   |---|---|---|---|---|
   | Mr. Green: | 63, | 75, | 57, | 81 |
   | Mr. Black: | 71, | 78, | 68, | 79 |
   | Mr. Brown: | 80, | 69, | 84, | 83 |

   Use the level of significance $\alpha = 0.05$ to test the null hypothesis that there is no difference in the average daily productivity of these three workers; first calculate the value of $F$ *without* the short-cut formulas on page 244, and then use these formulas to check the result.

**2.** The following are the number of words per minute which a secretary typed on four occasions on each of four different typewriters:

| Typewriter I | Typewriter II | Typewriter III | Typewriter IV |
|---|---|---|---|
| 71 | 65 | 75 | 62 |
| 69 | 63 | 79 | 59 |
| 75 | 69 | 81 | 68 |
| 77 | 67 | 77 | 63 |

Use the level of significance $\alpha = 0.05$ to test the null hypothesis that there is no difference in the true average number of words per minute the secretary can type on the four typewriters; first calculate the value of $F$ without the short-cut formulas, and then use them to check the result.

**3.** To study the effectiveness of four different kinds of packaging, a processor of breakfast foods puts a breakfast food packaged in the four different ways into five different supermarkets, obtaining the following sales (between 9 A.M. and noon on a given day):

*Packaging A:* 24, 34, 36, 24, 27
*Packaging B:* 25, 27, 38, 36, 25
*Packaging C:* 41, 35, 30, 29, 32
*Packaging D:* 35, 22, 26, 22, 30

Use the short-cut formulas on page 244 and the level of significance $\alpha = 0.05$ to test the null hypothesis that the differences in packaging have no effect on the breakfast food's sales.

**4.** The following are the grades obtained by ten students randomly selected from three different colleges in a test designed to measure their knowledge of current events:

*College 1:* 73, 86, 60, 95, 60, 79, 91, 74, 51, 97
*College 2:* 78, 80, 62, 43, 65, 82, 53, 37, 58, 55
*College 3:* 62, 88, 56, 62, 80, 69, 57, 64, 85, 61

Use the short-cut formulas on page 244 and the level of significance $\alpha = 0.01$ to check whether the differences among the means of these three samples are significant.

5. With reference to the illustration on page 47, check (at the level of significance $\alpha = 0.05$) whether the differences among the average percentage profits of the three corporations are significant. Use the short-cut formulas on page 244 to calculate $F$.

# 9

# THE ANALYSIS OF COUNT DATA

## INTRODUCTION

Many problems in business, science, and everyday life deal with *count data* that are used to estimate, or test hypotheses about, proportions, percentages, and probabilities. In principle, the work of this chapter will be very similar to that of Chapter 8: In *problems of estimation* we shall again construct confidence intervals and worry about the possible size of our error, and in the *testing of hypotheses* we shall again have to be careful in formulating null hypotheses and their alternatives, in deciding between one-tail tests and two-tail tests, in choosing the level of significance, and so on. The main difference is that the parameters with which we will now be concerned are "true" proportions (percentages, or probabilities) instead of population means. Hence, we shall base our methods on *counts* instead of measurements.

## THE ESTIMATION OF PROPORTIONS

The kind of information that is usually available for the estimation of a "true" proportion is the **relative frequency** with which an event has occurred, namely, a **sample proportion.** If an event has occurred $x$ times out of $n$, the relative frequency of its occurrence is $x/n$, and we generally use this sample proportion to estimate the corresponding "true" proportion, which we shall denote $p$. For example, if in a sample of 150 dwelling units in the state of Indiana, 108 were occupied by their owners, then $\frac{x}{n} = \frac{108}{150} = 0.72$, and (if the sample is random) we can use this figure as an estimate of the true proportion of dwelling units in the state of Indiana that are occupied by their owners. Note that since a *percentage* is simply a proportion multiplied by 100 and a *probability* can be interpreted as a proportion "in the long run," we could also say that we are thus estimating that 72 percent of all dwelling units in the state of Indiana are occupied by their owners, or that we are estimating as 0.72 the probability that a randomly selected dwelling unit in the state of

Indiana will be occupied by its owner. We have made this point to emphasize that *the problem of estimating a true percentage or a true probability is the same as that of estimating a true proportion.*

Throughout this section it will be assumed that the situations with which we are dealing satisfy (at least approximately) the conditions underlying the *binomial distribution*. Our information will always tell us how many successes there are in a given number of independent trials, and that in each trial the probability of a success has the same value $p$. Thus, the sampling distribution of the *number of successes* (namely, the sampling distribution of the *count* on which our methods will be based) is the binomial distribution for which we indicated on pages 134 and 138 that the mean and the standard deviation are given by the formulas $\mu = np$ and $\sigma = \sqrt{np(1-p)}$.

The fact that this formula for $\sigma$ involves the quantity $p$ (which we want to estimate) leads to some difficulties, but we can avoid them, at least for the moment, by constructing confidence intervals for $p$ with the use of Tables Va and Vb at the end of the book. (So far as the construction of these special tables is concerned, let us merely point out that the statisticians who made them used methods which are very similar to what we did on page 213, except that they referred to very detailed tables of binomial probabilities instead of areas under the normal curve.) Tables Va and Vb are very easy to use and require no calculations. If a sample proportion $x/n$ is less than or equal to 0.50, we begin by marking its value on the *bottom scale;* then we go up vertically until we reach the two contour lines (curves) which correspond to the size of the sample, and finally we read the confidence limits for $p$ off the *left-hand scale*, as is indicated in Figure 9.1. If the sample proportion exceeds 0.50, we mark its value on the scale which is *at the top*, go down vertically until we reach the two contour lines (curves) which correspond to the size of the sample, and then we read the confidence limits for $p$ off the *right-hand scale*, as is indicated in Figure 9.2.

To illustrate the use of these tables in a situation where the sample proportion is less than or equal to 0.50, suppose that in a random sample of 200 motorists in a certain community there were 64 who had not fastened their seat belts. Marking $\frac{64}{200} = 0.32$ on the bottom scale of Table Va and proceeding as in Figure 9.1, we find that

$$0.26 < p < 0.39$$

is a 0.95 confidence interval for the true proportion of motorists (in that community) who do not use their seat belts. Had we wanted to use a degree of confidence of 0.99, Table Vb would have yielded the interval

$$0.24 < p < 0.41$$

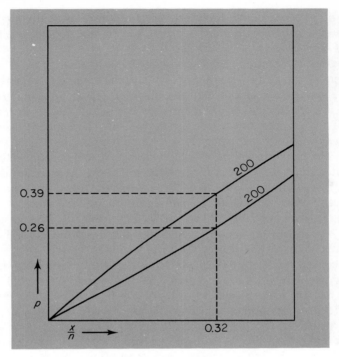

**FIGURE 9.1** Confidence limits for $p$.

and this illustrates again that *an increase in the degree of confidence will lead to a wider interval and, hence, to less specific information about the quantity we are trying to estimate.*

To give an example where the sample proportion exceeds 0.50, let us refer again to the problem on page 248 which concerned the proportion of dwelling units in the state of Indiana that were occupied by their owners. The sample proportion of owner-occupied dwelling units was $\frac{108}{150} = 0.72$, and the first thing we discover is that there are no curves corresponding to $n = 150$ in Tables Va and Vb. Reading between the lines, however, as in Figure 9.2, we find that a 0.95 confidence interval for the *true* proportion of dwelling units in the state of Indiana that were occupied by their owners (at the time of the survey) is given by

$$0.63 < p < 0.79$$

and that a corresponding 0.99 confidence interval, obtained from Table Vb instead of Table Va, is given by

$$0.61 < p < 0.81$$

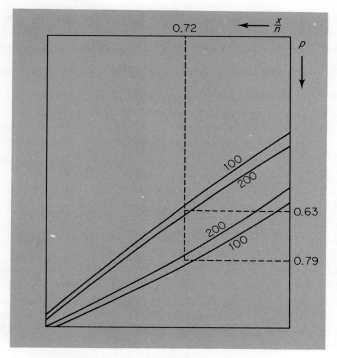

**FIGURE 9.2** Confidence limits for $p$.

Had we not wanted to "read between the lines" in this example, we could have constructed an approximate confidence interval for $p$ on the basis of the *normal curve approximation* to the binomial distribution which we discussed in Chapter 6. This approximate large-sample technique can also be used when tables like Tables Va and Vb are not available, when it is desired to use a different degree of confidence, or when the sample size exceeds 1,000. Using the formulas for $\mu$ and $\sigma$ on pages 134 and 138, the normal curve approximation to the binomial distribution enables us to assert with the probability 0.95 that $x$, the number of successes, will fall within 1.96 standard deviations of the mean, namely, on the interval from $np - 1.96\sqrt{np(1-p)}$ to $np + 1.96\sqrt{np(1-p)}$. This can be written as

$$np - 1.96\sqrt{np(1-p)} < x < np + 1.96\sqrt{np(1-p)}$$

which becomes

$$\frac{x}{n} - 1.96\sqrt{\frac{p(1-p)}{n}} < p < \frac{x}{n} + 1.96\sqrt{\frac{p(1-p)}{n}}$$

after considerable algebraic manipulation (which the reader will be asked to perform in Exercise 14 on page 256). This double inequality *could* serve as a confidence interval for $p$ if it were not for the fact that the quantity $p$, itself, appears on both sides in the expression $\sqrt{\dfrac{p(1-p)}{n}}$. This expression is referred to as the **standard error of a proportion,** for it is, in fact, the standard deviation of the sampling distribution of $x/n$ (see Exercise 15 on page 257).

There are several ways in which we can get around this difficulty. Easiest, perhaps, is to substitute the sample proportion $x/n$ for $p$ in $\sqrt{\dfrac{p(1-p)}{n}}$, which leads to the following 0.95 *large-sample confidence interval for p:*

$$\frac{x}{n} - 1.96\sqrt{\frac{\frac{x}{n}\left(1 - \frac{x}{n}\right)}{n}} < p < \frac{x}{n} + 1.96\sqrt{\frac{\frac{x}{n}\left(1 - \frac{x}{n}\right)}{n}}$$

If we want to change the degree of confidence to 0.99, we have only to substitute 2.58 into this formula for 1.96.

To illustrate this large-sample technique, let us refer again to the example on page 249, where 64 of 200 motorists failed to fasten their seat belts. Substituting $x = 64$ and $n = 200$ into the above 0.95 confidence interval formula for $p$, we obtain

$$\frac{64}{200} - 1.96 \cdot \sqrt{\frac{\frac{64}{200}\left(1 - \frac{64}{200}\right)}{200}} < p < \frac{64}{200} + 1.96 \cdot \sqrt{\frac{\frac{64}{200}\left(1 - \frac{64}{200}\right)}{200}}$$

or

$$0.255 < p < 0.385$$

which is very close to the interval obtained previously with the use of Table Va.

The large-sample theory which we have presented here can also be used to judge the possible size of the error we make when we use a sample proportion as an estimate of the corresponding "true" proportion $p$. Corresponding to what we said on page 211, the normal curve approximation to the binomial distribution permits us to assert that

> The probability is 0.95 that the numerical difference between $x/n$ and $p$, namely, the size of our error either way, is less than

$$1.96\sqrt{\frac{\frac{x}{n}\left(1 - \frac{x}{n}\right)}{n}}$$

where we again substituted $x/n$ for $p$ in the formula for the standard error of a proportion. Of course, if we want the probability (with which we make this assertion about the error) to be 0.99, we have only to substitute 2.58 for 1.96.

To illustrate this technique, let us refer again to the example on page 248, where 108 of 150 dwelling units were occupied by their owners. Now if we use $\frac{108}{150} = 0.72$ as a *point estimate* of the true proportion of dwelling units in the state of Indiana that were occupied by their owners (at the time of the survey), we can assert with a probability of 0.95 that this estimate is "off" either way by less than

$$1.96 \sqrt{\frac{0.72(1 - 0.72)}{150}} = 0.07$$

rounded to two decimals. In Exercise 12 on page 254 the reader will be asked to show that 99 to 1 would be *fair odds* in this example that the estimate is not "off" by more than 0.09 (rounded to two decimals).

## EXERCISES

1. In a sample survey, 360 of 1,000 persons interviewed in a large city said that they believed that parking facilities were inadequate. Use Table Va to construct a 0.95 confidence interval for the true proportion of persons in this city who believe that parking facilities are inadequate.

2. In a random sample of 400 claims filed against the major medical policies of a group of casualty insurance companies, 192 exceeded $2,500. Use Table Vb to construct a 0.99 confidence interval for the true proportion of major medical claims filed against this group of insurance companies which exceed $2,500.

3. In a random sample of 40 purchases made at a large department store, 22 were made with credit cards. Use Table Va to construct a 0.95 confidence interval for the true *percentage* of purchases made at this store with credit cards.

4. In a random sample of 100 vacationists visiting a large resort city, 78 arrived by automobile. Use Table Vb to construct a 0.99 confidence interval for the *probability* that a vacationist visiting this resort city will arrive by automobile.

5. In a random sample of 500 residents of a large Western city, 420 stated that they are protected by some form of health insurance.
   (a) Use Table Va to construct a 0.95 confidence interval for the true

proportion of the residents of this city who are protected by some form of health insurance.

(b) Repeat part (a), using the large-sample confidence interval formula on page 252 and compare the results.

6. In a random sample of 700 patrons of a restaurant, only 126 had not been there before.

(a) Use Table Vb to construct a 0.99 confidence interval for the probability that any one patron (approached by an interviewer) will be there for the first time.

(b) Repeat part (a), using the large-sample confidence interval formula on page 252 and compare the results.

7. Repeat Exercise 1 using the large-sample confidence interval formula on page 252 and compare the results.

8. Repeat Exercise 2 using the large-sample confidence interval formula on page 252 and compare the results.

9. In a random sample of 600 television viewers contacted by phone in a certain suburban area, 210 were watching the movie on Channel 12.

(a) Use Table Va to construct a 0.95 confidence interval for the true proportion of television viewers in that area who were watching that movie.

(b) If we use $\frac{210}{600} = 0.35$ as an estimate of the true proportion of television viewers in the given area who were watching that movie, what can we say with a probability of 0.95 about the possible size of our error?

10. In a random sample of 400 persons applying to a finance company for a loan, 64 were refused and did not obtain the loan.

(a) Use Table Vb to construct a 0.99 confidence interval for the true *percentage* of the loan applications which this finance company refuses.

(b) If we estimate the true percentage of applicants who do not obtain a loan from this finance company as $\frac{64}{400} \cdot 100 = 16$, what can we say with a probability of 0.99 about the possible size of our error?

11. With reference to Exercise 4, what can we say with a probability of 0.95 about the possible size of our error if we estimate as 0.78 the *probability* that any one vacationist visiting the given resort will arrive by automobile?

12. Verify the statement concerning the occupacy of dwelling units in Indiana which we made on page 253, namely, that 99 to 1 would be *fair odds* in the given example that the estimate is not "off" by more than 0.09 (rounded to two decimals).

**13. DETERMINATION OF SAMPLE SIZE** As in the estimation of means (see Exercise 5 on page 216), the formula which enables us to judge the possible size of our error can also be used to determine *how large a sample is needed to attain a desired degree of reliability*. If we want to be able to assert with a probability of 0.95 that a sample proportion will be "off" either way by less than some quantity $E$, we can write

$$E = 1.96\sqrt{\frac{p(1-p)}{n}}$$

and, upon solving this equation for $n$, we get

$$n = p(1-p)\left[\frac{1.96}{E}\right]^2$$

Unfortunately, this formula requires knowledge of $p$, the true proportion we are trying to estimate, which means that it cannot be used unless we have some prior (perhaps, indirect) information about the possible size of $p$. To get around this difficulty, we shall make use of the fact which the reader will be asked to verify in part (d) of this exercise, namely, that $p(1-p) = \frac{1}{4} - (p - \frac{1}{2})^2$ and, hence, that the *maximum* value of $p(1-p)$ is $\frac{1}{4}$. Thus, *when we have no idea about the possible size of $p$, we use the formula*

$$n = \frac{1}{4}\left[\frac{1.96}{E}\right]^2$$

and since this may make the sample size *unnecessarily large*, we can assert with a probability of *at least* 0.95 that the sample proportion we get will be "off" by less than $E$. To illustrate this technique, let us suppose that an airconditioning company wants to estimate what proportion of the residents in a given area have airconditioning units, and that it wants to be "95 percent sure" that the error of its estimate will be less than $E = 0.06$. If they have no idea about the true value of $p$, they substitute the given value of $E$, namely, 0.06, into the second of the two formulas for $n$, getting

$$n = \frac{1}{4}\left[\frac{1.96}{0.06}\right]^2 = 267$$

rounded *up* to the nearest whole number. Thus, if their estimate is a sample proportion based on a random sample of size $n = 267$, they can assert with a probability of *at least* 0.95 that it will be "off" by less than

0.06. Had they known from other studies that the proportion they are trying to estimate is in the neighborhood of, say, 0.25, the first formula for $n$ would have yielded

$$n = (0.25)(0.75)\left[\frac{1.96}{0.06}\right]^2 = 201$$

rounded up to the nearest whole number. This illustrates how prior information about the possible size of $p$ can lead to an appreciable reduction in the required size of a sample.

(a) Suppose we want to estimate what proportion of the dresses sold in a certain department store are returned for a refund (or a charge credit). How large a sample will we need to be able to assert with a probability of at least 0.95 that the error of our estimate (the sample proportion) will be less than 0.05? What would be the answer to this question if we guessed (or assumed) that the proportion we are trying to estimate is in the neighborhood of 0.15?

(b) The leader of a national labor union wants to estimate what percentage of the members of his union favor a new pension plan. How large a sample will he have to take to be able to assert with a probability of at least 0.95 that his estimate, the sample *percentage*, will be within 3 percent of the correct value? How large a sample would he have to take to be able to make this assertion with a probability of at least 0.99? (*Hint:* Substitute 2.58 for 1.96 into the formula for $n$.)

(c) If we want to estimate what proportion of the teenagers in the metropolitan Phoenix area prefer a vacation at the beach to a vacation in the mountains, how large a random sample will we need to be able to assert with a probability of 0.95 that our sample proportion will differ from the true proportion by less than 0.02? What would be the answer to this question if we suspected that the proportion we are trying to estimate should be pretty close to 0.90?

(d) Making use of the *binomial expansion* $(a + b)^2 = a^2 + 2ab + b^2$, verify that $p(1 - p) = \frac{1}{4} - (p - \frac{1}{2})^2$. Explain why the maximum value of $p(1 - p)$ is $\frac{1}{4}$, which occurs at $p = \frac{1}{2}$.

14. Show that if we add $1.96\sqrt{np(1 - p)}$ to the expressions on both sides of the inequality $np - 1.96\sqrt{np(1 - p)} < x$ and then divide by $n$, we get the inequality $p < \frac{x}{n} + \frac{1.96}{n}\sqrt{np(1 - p)}$ and, hence,

$$p < \frac{x}{n} + 1.96\sqrt{\frac{p(1 - p)}{n}}.$$

Use a similar argument to change the other part of the double inequality on page 252.

15. Since the *proportion* of successes is simply the *number* of successes divided by $n$, the mean and the standard deviation of the sampling distribution of the *proportion* of successes may be obtained by *dividing by $n$* the mean and the standard deviation of the sampling distribution of the *number* of successes. Use this argument to obtain formulas for the mean and the standard deviation of the sampling distribution of the proportion of successes in $n$ "binomial" trials.

## TESTS CONCERNING PROPORTIONS

Having learned in Chapter 8 what is involved in the test of a (statistical) hypothesis, the reader should find it easy to understand the various other tests which we shall study in this book. In this section we will be concerned with tests which enable us to decide, on the basis of sample data, *whether the "true" value of a proportion (percentage, or probability) equals a given constant*. These tests will enable one to check, say, whether the true proportion of shoppers who can identify the trade-mark of a highly advertised beverage is 0.45, whether it is true that less than 40 percent of all sales made by a certain furniture store are for cash (the other sales being made on credit), or whether the probability that a new automatic washing machine will require repairs within the one-year warranty period is greater than 0.30.

Questions of this kind are generally decided on the basis of the observed number, or proportion, of "successes" in $n$ trials, and it will be assumed throughout this section that these trials are independent and that the probability of a success is the same for each trial. In other words, we shall assume that we can use the binomial distribution. When $n$ is *small*, tests concerning the "true" proportion of "successes" (namely, tests concerning the parameter $p$ of the binomial distribution) are usually based directly on tables of binomial probabilities such as Table XI. To illustrate, let us refer to the second of the three examples of the preceding paragraph, and let us suppose that in a random sample of 15 furniture sales *only two* were made for cash. To test the null hypothesis $p = 0.40$ against the alternative hypothesis $p < 0.40$ at the level of significance $\alpha = 0.05$, we shall have to determine the probability that *at most two* of 15 such furniture sales are made for cash when $p = 0.40$, and see whether it is less than 0.05. Since Table XI shows that this probability is $0.005 + 0.022 = 0.027$, which is *less than* 0.05, we conclude that the null hypothesis will have to be rejected, namely, that the true proportion of the store's furniture sales that are made for cash is *less than* 0.40. To picture this

258   THE ANALYSIS OF COUNT DATA   CHAP. 9

as in Figure 8.7 on page 225, we find that our sample value $x = 2$ falls into the left-hand "tail" of the distribution. Note that if the level of significance had been $\alpha = 0.01$, the null hypothesis $p = 0.40$ could not have been rejected.

When $n$ is *large*, tests concerning "true" proportions are usually based on the normal curve approximation to the binomial distribution, and (as we indicated on page 167) this approximation is satisfactory so long as $np$ and $n(1 - p)$ both exceed 5. Making use of the formulas on pages 134 and 138, according to which the mean and the standard deviation of the binomial distribution are given by $\mu = np$ and $\sigma = \sqrt{np(1 - p)}$, we base all large-sample tests of the null hypothesis $p = p_0$ on the statistic

$$z = \frac{x - np_0}{\sqrt{np_0(1 - p_0)}}$$

whose sampling distribution is approximately the *standard normal distribution*. Note that we are thus converting $x$, the observed number of "successes," into *standard units;* first we subtract from it the mean of its sampling distribution and then we divide by the standard deviation. The actual test criteria which we shall use are again those of Figures 8.6, 8.7, and 8.8. For the two-sided alternative $p \neq p_0$ we use the criteria of Figure 8.6 on page 223, for the one-sided alternative $p < p_0$ we use the criteria of Figure 8.7 on page 225, and for the one-sided alternative $p > p_0$ we use the criteria of Figure 8.8 on page 227.

To illustrate this kind of test, suppose that in connection with the third example on page 257 a sample check revealed that 110 of 300 new washing machines required repairs within the one-year warranty period. Assuming that we are interested primarily in knowing whether the true proportion requiring repairs within the first year might be greater than 0.30, we shall test the null hypothesis $p = 0.30$ against the alternative $p > 0.30$, say, at the level of significance $\alpha = 0.01$. Substituting $x = 110$, $n = 300$, and $p_0 = 0.30$ into the above formula for $z$, we get

$$z = \frac{110 - 300(0.30)}{\sqrt{300(0.30)(0.70)}} = 2.52$$

and since this exceeds 2.33 (the dividing line of the $\alpha = 0.01$ criterion of Figure 8.8), we conclude that the null hypothesis will have to be rejected. In other words, we conclude that more than 30 percent of the washing machines require repairs within the one-year warranty period.

To give an example where it would be appropriate to use a two-tail test, let us return to the first of the three examples on page 257, and let us suppose that 236 of 500 (randomly selected) shoppers can identify the trade-mark of the highly advertised beverage. What we shall want to know is whether 0.45 is the correct proportion (of shoppers who can identify the trade-mark), or whether this figure is *too high or too low*. Substituting $x = 236$, $n = 500$, and $p = 0.45$ into the formula for $z$, we get

$$z = \frac{236 - 500(0.45)}{\sqrt{500(0.45)(0.55)}} = 0.99$$

and since this value falls *between* $-1.96$ and $1.96$ (the two dividing lines of the $\alpha = 0.05$ criterion of Figure 8.6), we conclude that the null hypothesis *cannot be rejected*. Depending on whether or not we have to commit ourselves one way or the other, we can accept 0.45 as the correct proportion (of shoppers who can identify the trade-mark of the beverage), or else reserve judgment. In the latter case we would simply say that the difference between $500(0.45) = 225$ and 236, namely, the difference between what we expected and what we got, is *not statistically significant*.

## EXERCISES

1. The advertising manager of a canned soup company claimed that 80 percent of the population recognize the soup company's jingle and associate it with the brand of soup. To check this claim, a market research organization interviewed 300 persons and found that only 232 of them recognized the jingle and associated it with the correct brand of soup. What can they conclude at the level of significance $\alpha = 0.05$, if they use these data to test the null hypothesis $p = 0.80$ (corresponding to the manager's claim) against the two-sided alternative $p \neq 0.80$?

2. A manufacturer of a spot remover claims that his product fails to remove at most 10 percent of all spots. To check the null hypothesis $p = 0.10$ against the alternative hypothesis $p > 0.10$ at the level of significance $\alpha = 0.05$, a consumer protection agency tries the product on a random sample of 150 spots and finds that it fails to remove 23. What can the consumer protection agency conclude? What could they conclude if they used the level of significance $\alpha = 0.01$?

3. Suppose we want to investigate the claim that 50 percent of the workers in a factory are opposed to the firm's new wage incentive plan. If we want to base our decision on the opinions of 15 workers (selected at random), how few or how many of them would have to be opposed to the firm's plan so that we can reject the null hypothesis $p = 0.50$ at the level of significance $\alpha = 0.05$? (*Hint:* Refer to Table XI.)

**4.** A promoter wants to check a sports writer's claim that the probability is only 0.10 that he will have a full house at one of his boxing matches. Testing the null hypothesis $p = 0.10$ against the alternative hypothesis $p > 0.10$, what can he conclude at the level of significance $\alpha = 0.01$ if he had a full house in 32 of 200 previous boxing matches, randomly chosen from among his many boxing promotions?

**5.** A professor claims that 70 percent of his students are willing to write an extra book report in lieu of taking a final examination. What can we conclude about this claim, namely, the hypothesis $p = 0.70$, if in a random sample of 90 students there were only 43 who said that they would be willing to write an extra book report in lieu of taking a final examination. Use the two-sided hypothesis $p \neq 0.70$ and the level of significance $\alpha = 0.05$.

**6.** Suppose that a producer of television commercials claims that at most 5 percent of all viewers will object to the noise level of a certain commercial. If nine persons are shown this commercial, at least how many will have to object about the noise level before we can reject the null hypothesis $p = 0.05$ at the level of significance $\alpha = 0.01$? (*Hint:* Use Table XI.)

**7.** In a sample survey of 200 college students it was found that 125 of them use Brand Q tooth paste. Test the null hypothesis that the true proportion of college students who brush with Brand Q is 0.70 against the alternative hypothesis that this figure is incorrect one way or the other, using (a) the level of significance $\alpha = 0.05$, and (b) the level of significance $\alpha = 0.01$.

**8.** A self-service laundry in a college town claims that at least 40 percent of its users are students. Test the null hypothesis $p = 0.40$ against the alternative $p > 0.40$ if 150 randomly selected users of the laundry included 78 students, using (a) the level of significance $\alpha = 0.05$, and (b) the level of significance $\alpha = 0.01$.

**9.** The proprietor of a shoe repair shop claims that 30 percent of his customers' shoes have leather soles. What can we conclude about this claim at the level of significance $\alpha = 0.05$ if a sample of 80 pairs of his customers' shoes included 29 with leather soles?

## DIFFERENCES AMONG PROPORTIONS

If we were to follow the pattern of Chapter 8, we would continue our study of the analysis of count data by looking at methods which enable us to decide whether observed differences between *two* sample proportions are significant or whether they can be attributed to chance. This kind of problem would arise, for example, if we found that 180 of 300 business executives interviewed in

Montreal and 208 of 400 business executives interviewed in New York felt that business conditions would improve, and we wanted to judge whether the difference between the corresponding proportions, namely, $\frac{180}{300} = 0.60$ and $\frac{208}{400} = 0.52$ can be attributed to chance. Similarly, if 184 of 500 television viewers who watch a certain news program in Detroit, Michigan, feel that local news coverage is adequate, and the same sentiment is expressed by 148 of 500 viewers who watch the *same* program in Windsor, Ontario, it would be of interest to know whether the difference between $\frac{184}{500} = 0.368$ and $\frac{148}{500} = 0.296$ is significant, namely, whether viewers in Detroit really have a better opinion of the local news coverage on that station than viewers in Windsor.

The method which we shall study in this section is actually *more general*—it applies to problems in which we have to decide whether differences among *two or more* sample proportions are significant, or whether they can be attributed to chance. Thus, we will be able to judge, for example, whether there really is a difference in the effectiveness of four kinds of weight-reducing diets if 26 of 200 persons on Diet A failed to lose 10 pounds, while the corresponding figures for 200 persons on Diets B, C, and D, are, respectively, 22, 17, and 31. Similarly, the method will also apply if we add to the first example of the preceding paragraph that among 300 business executives interviewed in Toronto there were 138 who felt that business conditions would improve. In fact, let us use this example to illustrate the method of analysis which we shall employ, and let us begin by presenting all the information together in the following table

|  | Montreal | New York | Toronto |
|---|---|---|---|
| Predict Business Improvement | 180 | 208 | 138 |
| Undecided or Predict Business Decline | 120 | 192 | 162 |
| Totals | 300 | 400 | 300 |

where the entries of the second row were obtained simply by subtraction from the corresponding totals. Using this table, we shall now investigate whether the differences among the *three* proportions of business executives who feel that business conditions will improve, namely, $\frac{180}{300} = 0.60$, $\frac{208}{400} = 0.52$, and $\frac{138}{300} = 0.46$, can be attributed to chance.

If we let $p_1$, $p_2$, and $p_3$ denote the *true* proportions of business executives in the three given cities who feel that business conditions will improve, the null hypothesis we shall want to test is

$$p_1 = p_2 = p_3$$

and the alternative hypothesis is that $p_1$, $p_2$, and $p_3$ are *not all equal*. If the null hypothesis is true and the differences *are* due to chance, we can combine all of the data and look upon the three samples as *one sample from one and the same population*, and then estimate the true proportion of all business executives (in the three cities) who feel that business conditions will improve as

$$\frac{180 + 208 + 138}{300 + 400 + 300} = 0.526$$

Then, using this estimate, we can say that we could have *expected* 300(0.526) = 157.8 of the business executives interviewed in Montreal to say that business conditions will improve, that we could have *expected* 400(0.526) = 210.4 of the business executives interviewed in New York to say that business conditions will improve, and that we could have *expected* 300(0.526) = 157.8 of the business executives interviewed in Toronto to say that business conditions will improve. To find the corresponding *expectations* for the number of business executives interviewed who are undecided or feel that business will decline, we have only to subtract the figures we have just calculated from the respective totals, getting $300 - 157.8 = 142.2$, $400 - 210.4 = 189.6$, and $300 - 157.8 = 142.2$. These results are summarized in the following table, where the expected frequencies are shown in parentheses below the ones that were actually observed:

|  | *Montreal* | *New York* | *Toronto* |
|---|---|---|---|
| Predict Business Improvement | 180 (157.8) | 208 (210.4) | 138 (157.8) |
| Undecided or Predict Business Decline | 120 (142.2) | 192 (189.6) | 162 (142.2) |

To test the null hypothesis $p_1 = p_2 = p_3$, we can now compare the frequencies which were actually observed with those which we could have *expected* if the null hypothesis were true. If the two sets of frequencies are very much alike, it stands to reason that the null hypothesis should be *accepted*—after all, we would then have obtained almost exactly what we could have expected if the null hypothesis were true. On the other hand, if the discrepancies between the two sets of frequencies are large, the null hypothesis will probably have to be *rejected*, for in that case we did not get what we could have expected if the null hypothesis were true.

To judge *how large* the discrepancies between the two sets of frequencies must be before the null hypothesis can be rejected (note the word "probably" in the preceding sentence), we use the following statistic, where the observed frequencies are denoted by the letter $f$ and the expected frequencies by the letter $e$:

$$\chi^2 = \Sigma \frac{(f - e)^2}{e}$$

This statistic is called **chi-square** ($\chi$ is the Greek letter *chi*), and it should be observed that the summation extends over all of the "cells" of the table. In other words, $\chi^2$ is the *sum* of the quantities which we obtain by dividing the squared difference $(f - e)^2$ by $e$ *separately* for each cell of the table. Returning to our numerical example, we thus obtain

$$\chi^2 = \frac{(180 - 157.8)^2}{157.8} + \frac{(208 - 210.4)^2}{210.4} + \frac{(138 - 157.8)^2}{157.8}$$
$$+ \frac{(120 - 142.2)^2}{142.2} + \frac{(192 - 189.6)^2}{189.6} + \frac{(162 - 142.2)^2}{142.2}$$
$$= 11.89$$

As is apparent from its formula, $\chi^2$ will be *small* when there is a close agreement between the $f$'s and the $e$'s, and it will be *large* when there are sizable discrepancies between the respective frequencies. If the null hypothesis is true, the sampling distribution of the $\chi^2$ statistic is approximately the **chi-square distribution,** an example of which is shown in Figure 9.3. Since the null hypothesis will be rejected only when $\chi^2$ is too large (namely, when the

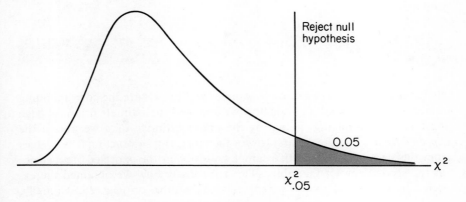

Level of significance 0.05

**FIGURE 9.3** Chi-square distribution.

discrepancies between the $f$'s and the $e$'s are too great to be attributed to chance), we ultimately base our decision on the criterion of Figure 9.3. Here $\chi^2_{.05}$ is such that the area under the curve to its right equals 0.05, and it provided the dividing line of the criterion when the level of significance is $\alpha = 0.05$; correspondingly, $\chi^2_{.01}$ provides the dividing line of the criterion when the level of significance is $\alpha = 0.01$. These quantities, which depend on a parameter which (again) is called the **number of degrees of freedom,** or simply the **degrees of freedom,** are given in Table III. It equals 2 in our example, and *in general it equals $k - 1$ when we compare $k$ sample proportions.* Intuitively, we can justify this formula with the argument that once we have calculated any $k - 1$ of the expected frequencies in either row of the table, all of the other expected frequencies can be obtained by subtraction from the fixed totals of the rows and columns (see Exercise 8 on page 267 and also the discussion on page 271).

Returning now to our numerical example, we find that $\chi^2_{.05} = 5.991$ for $3 - 1 = 2$ degrees of freedom. Since this is exceeded by 11.89, the value of the $\chi^2$ statistic which we obtained in our example, we find that the null hypothesis will have to be rejected. In other words, we conclude that there is some difference of opinion among business executives in Montreal, New York, and Toronto, as to whether business conditions will improve. It should also be noted that the value of $\chi^2_{.01}$ for $3 - 1 = 2$ degrees of freedom is 9.210 (according to Table III), so that the null hypothesis would also have been rejected at this level of significance.

In general, if we have $k$ sample proportions and want to test the null hypothesis $p_1 = p_2 = \ldots = p_k \; (=p)$ against the alternative that these "true" proportions are *not all equal,* we first calculate the expected frequencies as on page 262. Combining the data, we estimate the true value of the common proportion $p$ as

$$\frac{x_1 + x_2 + \cdots + x_k}{n_1 + n_2 + \cdots + n_k}$$

where the $n$'s are the sizes of the $k$ samples and the $x$'s are the corresponding number of "successes." (This is how we obtained the estimate 0.526 on page 262.) Then, multiplying the $n$'s by the above estimate of $p$ we obtain the *expected frequencies* for the first row of the table, and subtracting these values from the sizes of the respective samples we obtain the expected frequencies for the second row of the table. (This is precisely how we obtained the expected frequencies shown in parentheses in the table on page 262.) Finally, we calculate $\chi^2$ according to the formula on page 263, with $\dfrac{(f-e)^2}{e}$ determined separately for each of the $2k$ cells of the table, and we *reject* the null hypothesis that the differences among the sample proportions are due to chance if the value which we obtain for $\chi^2$ *exceeds* $\chi^2_{.05}$ or $\chi^2_{.01}$ (depending on the level of significance) for $k-1$ degrees of freedom.

When we calculate the expected frequencies, it is customary to round (if necessary) to the nearest integer or to one decimal. The entries of Table III are given to three decimals, but there is seldom any need to carry more than two decimals in calculating the value of the $\chi^2$ statistic, itself. Let us also remind the reader that the test which we have been discussing is only *approximate*, and it is for this reason that it is best not to use the test when one of the *expected frequencies* is less than 5. (If this is the case, we can sometimes "salvage" the situation by combining some of the samples.)

## EXERCISES

1. A razor blade company wants to know how well their slogan is recognized in the neighboring cities of Kansas City, Kansas, and Kansas City, Missouri. If 235 of 500 persons questioned in Kansas City, Kansas, and 297 of 700 persons questioned in Kansas City, Missouri, associated the slogan with the correct brand of razor blades, can they conclude that there is a difference in recognition of the slogan in these two cities? Use the level of significance $\alpha = 0.05$.

2. In a consumer study it was found that among 400 randomly selected women in Seattle, 128 preferred mint-flavored to regular-flavored tooth paste, while among 400 randomly selected women in New Orleans, 165 preferred the mint-flavored to the regular-flavored tooth paste. Use the level of significance $\alpha = 0.01$ to test the null hypothesis that there is no difference between the corresponding true proportions.

3. On page 261 we referred to a study which showed that 184 of 500 persons who watched a certain news program in Detroit and 148 of 500 persons who watched the same news program in Windsor, Ontario, thought that local news coverage was adequate. Use the level of significance $\alpha = 0.05$ to test the null hypothesis that there is no difference between the corresponding true proportions.

4. On page 261 we referred to a study which showed that 26 of 200 persons on Diet A failed to lose 10 pounds, while the corresponding figures for 200 persons on Diets B, C, and D were, respectively, 22, 17, and 31. Use the level of significance $\alpha = 0.01$ to test the null hypothesis that there is no difference in the effectiveness of the four diets, namely, that there is no difference between the corresponding true proportions.

266    THE ANALYSIS OF COUNT DATA    CHAP. 9

**5.** The dean of students of a large university found that 210 of 250 students opposed late Friday afternoon classes, 52 of 80 faculty members opposed late Friday afternoon classes, and 12 of 30 staff personnel opposed late Friday afternoon classes. Assuming that these are random sample data from large populations of students, faculty members, and staff personnel, test the null hypothesis that there is no difference between the corresponding true proportions of persons opposed to late Friday afternoon classes. Use the level of significance $\alpha = 0.05$.

**6.** The management of a large firm wants to know whether there is a relationship between the quality of a salesman's work and his score on an introvert–extrovert test. If there are 28 introverts and 32 extroverts in a sample of 60 salesmen whose work is rated below average, 64 introverts and 56 extroverts in a sample of 120 salesmen whose work is rated average, and 37 introverts and 43 extroverts in a sample of 80 salesmen whose work is rated above average, what can we conclude from these data (using a level of significance of 0.05)?

**7. TWO SAMPLE PROPORTIONS (ONE-SIDED ALTERNATIVE)** When we compare two sample proportions, we are sometimes interested in testing the null hypothesis $p_1 = p_2$ against the one-sided alternative $p_1 < p_2$ or $p_1 > p_2$. This would have been the case, for example, in Exercise 3 if it had been our objective to prove that television viewers in Detroit are more satisfied with the local news coverage of the television station than viewers in Windsor, Ontario, and it would be the case if we wanted to see whether an item in a sales aptitude test *really discriminates between good salesmen and poor salesmen*, say, if 82 of 100 good salesmen and 51 of 100 poor salesmen answered it correctly. In situations like this we base our decision on the following statistic *provided that both samples are large*:

$$z = \frac{\dfrac{x_1}{n_1} - \dfrac{x_2}{n_2}}{\sqrt{p(1-p)\left(\dfrac{1}{n_1} + \dfrac{1}{n_2}\right)}} \quad \text{with } p = \frac{x_1 + x_2}{n_1 + n_2}$$

The sampling distribution of this statistic is approximately the *standard normal distribution* when the null hypothesis $p_1 = p_2$ is true, and the criteria on which we base the tests are again those of Figure 8.7 on page 225 when the alternative hypothesis is $p_1 < p_2$, and those of Figure 8.8 on page 227 when the alternative hypothesis is $p_1 > p_2$. For instance, in the *item analysis* referred to above, $p_1$ is the true proportion of good salesmen who will answer the question correctly, $p_2$ is the true proportion of poor salesmen who will answer the question correctly, and we shall want to test the null hypothesis $p_1 = p_2$ against the one-sided alternative $p_1 > p_2$, say, at the level of significance $\alpha = 0.01$. Substituting the given data, namely, $x_1 = 82$, $n_1 = 100$, $x_2 = 51$, and $n_2 = 100$, into the formula for $p$ and then into the formula for $z$, we get

$$p = \frac{82 + 51}{100 + 100} = \frac{133}{200} = 0.665$$

and, hence,

$$z = \frac{\frac{82}{100} - \frac{51}{100}}{\sqrt{(0.665)(0.335)(\frac{1}{100} + \frac{1}{100})}} = 4.64$$

Since this exceeds 2.33, the $\alpha = 0.01$ criterion of Figure 8.8, we conclude that the test item definitely discriminates between good salesmen and poor salesmen.

(a) Repeat Exercise 3 using the alternative hypothesis $p_1 < p_2$, where $p_1$ is the true proportion for television viewers in Windsor, Ontario, while $p_2$ is the true proportion for television viewers in Detroit.

(b) In trying to decide which of two snowblowers has a better starter, we test each kind 120 times. If the first kind of snowblower fails to start on the first try 15 times and the second kind of snowblower fails to start on the first try 10 times, can we conclude that the second kind is really better (at starting)? Use the level of significance $\alpha = 0.05$.

8. Making use of the fact that the expected number of successes for the $k$ samples are obtained by multiplying $\dfrac{x_1 + x_2 + \cdots + x_k}{n_1 + n_2 + \cdots + n_k}$, respectively, by $n_1, n_2, \ldots,$ and $n_k$ show that *the sum of the expected number of successes for the k samples equals the sum of the observed number of successes.*

## CONTINGENCY TABLES

The chi-square criterion which we introduced in the preceding section plays an important role in many different kinds of problems involving count data. In this section, we shall apply it to *two* kinds of problems which differ conceptually, but are analyzed in the same way. In the first kind of problem we deal with trials having *more than two* possible outcomes—for instance, the dividends paid on the common stock of a corporation can decrease, remain the same, or increase; a college undergraduate can be a freshman, a sophomore, a junior, or a senior; and in the example on page 261 each of the business executives in Montreal and New York might have been asked whether he feels that business will improve, whether he feels that business will decline, or whether he is undecided. Thus, we might have obtained the data shown in the following table:

|  | Montreal | New York |
|---|---|---|
| Predict Business Improvement | 180 | 208 |
| Undecided | 19 | 24 |
| Predict Business Decline | 101 | 168 |
| Totals | 300 | 400 |

As in the examples of the preceding section, *the column totals are fixed,* they represent the sizes of the respective samples, while the *row totals* (180 + 208 = 388, 19 + 24 = 43, and 101 + 168 = 269) *depend on the answers of the business executives interviewed, and, hence, on chance.*

In the second kind of problem we shall study in this section, *the column totals also depend on chance.* Suppose, for instance, that a large corporation wants to know whether there really is a relationship between the amount of formal education received by their office employees and the salaries which they are paid. Suppose, furthermore, that a sample of 400 cases taken from the very extensive files of the corporation's office employees yielded the results shown in the following table:

| Highest Education Completed | High | Salary Medium | Low | |
|---|---|---|---|---|
| College | 47 | 29 | 12 | 88 |
| High School | 70 | 95 | 52 | 217 |
| Grammar School | 31 | 80 | 84 | 195 |
| | 148 | 204 | 148 | 500 |

Tables like this are called **two-way tables** or **contingency tables,** and it is common practice to describe them by giving their number of rows and columns. Thus, the above table is a 3 × 3 table, the first one on this page is a 3 × 2 table, and in general a table with $r$ rows and $k$ columns is referred to as an **r × k table** (where $r \times k$ reads "$r$ by $k$").

CONTINGENCY TABLES  269

Before we demonstrate how $r \times k$ tables are analyzed, let us examine briefly what null hypotheses we shall actually want to test. In the first kind of problem, the one on page 267 where the column totals are fixed, we shall want to test the null hypothesis that the probabilities of choosing a business executive who predicts that business will improve, choosing a business executive who is undecided, or choosing a business executive who predicts that business will decline are the same for both cities. In other words, we shall want to test the null hypothesis that business executives' predictions about the outlook for business are *independent* of the location where we conduct the poll (at least, so far as Montreal and New York are concerned). In the second kind of problem we are also concerned with a null hypothesis of *independence*, namely, the hypothesis that the salaries received by office employees of the given corporation do not depend on the extent of the employees' education.

To illustrate how an $r \times k$ table is analyzed, let us refer to the second of these examples, and let us begin by calculating the *expected frequencies* as on page 268. If the null hypothesis of independence is true, the probability of randomly selecting an office employee (from the corporation's files) who has completed college *and* who is earning a high salary is the *product* of the probability that he will have completed college and the probability that he is earning a high salary. (This is simply an application of the Special Multiplication Rule on page 108.) Since $47 + 29 + 12 = 88$ of the 500 office employees in the sample had completed college while $47 + 70 + 31 = 148$ are earning high salaries, we *estimate* the probability of choosing an office employee who has completed college as $\frac{88}{500}$, and the probability of choosing an office employee who is earning a high salary as $\frac{148}{500}$. Hence, we *estimate* the probability of choosing an office employee who has completed college *and* is earning a high salary as the product $\frac{88}{500} \cdot \frac{148}{500}$, and in a sample of size 500 we would expect to find

$$500 \left( \frac{88}{500} \cdot \frac{148}{500} \right) = \frac{88 \cdot 148}{500} = 26.0$$

office employees who fit this description. Note that this result was obtained by evaluating $\frac{88 \cdot 148}{500}$, namely, by multiplying the total of the first row by the total of the first column and then dividing by the grand total for the entire table. Using exactly the same sort of reasoning, we can show that this rule applies to any cell of an $r \times k$ table, namely, that

> The expected frequency of any cell of an $r \times k$ table is the product of the totals of the row and the column to which it belongs, divided by the grand total for the entire table.

Thus, we obtain an expected frequency of $\frac{88 \cdot 204}{500} = 35.9$ for the second cell of the first row, and expected frequencies of $\frac{217 \cdot 148}{500} = 64.2$ and $\frac{217 \cdot 204}{500} = 88.5$ for the first two cells of the second row.

Using an argument similar to that of Exercise 8 on page 267 it can be shown that *the sum of the expected frequencies for any row or column must equal the sum of the corresponding observed frequencies*. Thus, we find by subtraction that the expected frequency for the third cell of the first row is*

$$88 - 26.0 - 35.9 = 26.1$$

that the expected frequency for the third cell of the second row is

$$217 - 64.2 - 88.5 = 64.3$$

and that the expected frequencies for the three cells of the third row are, respectively,

$$148 - 26.0 - 64.2 = 57.8$$
$$204 - 35.9 - 88.5 = 79.6$$

and

$$148 - 26.1 - 64.3 = 57.6$$

All these expected frequencies were rounded to one decimal, and they are summarized in the following table, where they are shown in parentheses below the corresponding frequencies which were actually observed:

| Highest Education Completed | High | Salary Medium | Low |
|---|---|---|---|
| College | 47 (26.0) | 29 (35.9) | 12 (26.1) |
| High School | 70 (64.2) | 95 (88.5) | 52 (64.3) |
| Grammar School | 31 (57.8) | 80 (79.6) | 84 (57.6) |

* Since the first and third columns have the same total of 148, the corresponding expected frequencies in these two columns should actually be equal. The differences we got (between 26.0 and 26.1, for example) are due to rounding.

From here on, the work is like that of the preceding section: Writing the observed and expected frequencies again as $f$ and $e$, we calculate $\chi^2$ according to the formula on page 263, namely,

$$\chi^2 = \Sigma \frac{(f-e)^2}{e}$$

where the quantity $\frac{(f-e)^2}{e}$ is again determined separately for each individual cell. Then we *reject* the null hypothesis of *independence* if the value which we obtain for $\chi^2$ exceeds $\chi^2_{.05}$ or $\chi^2_{.01}$ (depending on the level of significance) for $(r-1)(k-1)$ degrees of freedom. As before, $r$ is the number of rows and $k$ is the number of columns, and to justify the formula for the degrees of freedom, intuitively at least, let us point out that after $(r-1)(k-1)$ of the expected frequencies have been calculated according to the rule on page 269, all of the others can be obtained by subtraction from the totals of appropriate rows or columns. In our numerical example we had $r = 3$ and $k = 3$, and it should be observed that after we had determined $(r-1)(k-1) = (3-1)(3-1) = 4$ of the expected frequencies according to the rule on page 269, all of the others were obtained by subtraction from the totals of rows or columns.

Returning now to our numerical example, we find that

$$\chi^2 = \frac{(47-26.0)^2}{26.0} + \frac{(29-35.9)^2}{35.9} + \frac{(12-26.1)^2}{26.1}$$
$$+ \frac{(70-64.2)^2}{64.2} + \frac{(95-88.5)^2}{88.5} + \frac{(52-64.3)^2}{64.3}$$
$$+ \frac{(31-57.8)^2}{57.8} + \frac{(80-79.6)^2}{79.6} + \frac{(84-57.6)^2}{57.6}$$
$$= 53.79$$

and since $\chi^2_{.01} = 13.277$ for $(3-1)(3-1) = 4$ degrees of freedom (see Table III at the end of the book), we find that the null hypothesis will have to be *rejected*. This means that there *is* a relationship between the extent of the office employees' education and the salaries which they receive.

The method by which we analyzed this $3 \times 3$ table applies also in the first kind of problem described on page 268, namely, when the column totals are fixed sample sizes and do not depend on chance. The expected frequencies are determined in the same way, but the formula according to which we multiply the row total by the column total and then divide by the grand total has to be justified differently (see Exercise 6 on page 273). Let us also point out that, as in the preceding section, the chi-square criterion is only approximate and, hence, should be used only when each expected cell frequency is at least 5. If one or more of the expected cell frequencies is less than 5, we can sometimes "salvage" the situation by combining several of the cells and subtracting one degree of freedom for each cell which is thus eliminated.

## EXERCISES

1. Analyze the 3 × 2 table on page 268 and decide, at the level of significance $\alpha = 0.05$, whether the differences between the opinions expressed by the business executives in Montreal and New York are significant.

2. A sample survey, designed to show how students attending a large university get from their residences to their classes, yielded the following results:

|  | Freshman | Sophomore | Junior | Senior |
|---|---|---|---|---|
| Walk | 104 | 87 | 89 | 72 |
| Automobile | 22 | 29 | 35 | 43 |
| Bicycle | 46 | 34 | 37 | 32 |
| Other | 28 | 50 | 39 | 53 |
| Totals | 200 | 200 | 200 | 200 |

Test the null hypothesis that, so far as getting from their residences to their classes is concerned, the methods of transportation used by freshmen, sophomores, juniors, and seniors are the same. Use the level of significance $\alpha = 0.05$.

3. The following sample data are from a study of dress lengths (measured in inches above the knee) worn by adult women in the United States in 1973:

| Length (inches) | Housewives | Working Women | Students |
|---|---|---|---|
| 0 | 18 | 12 | 5 |
| 2 | 80 | 29 | 61 |
| 4 | 74 | 47 | 97 |
| More than 4 | 16 | 12 | 49 |
| Sample sizes | 188 | 100 | 212 |

Use the level of significance $\alpha = 0.01$ to test the null hypothesis that the true proportions of adult women in these categories of dress length are the same for housewives, working women, and students.

**4.** The following data pertain to the performance and the standard of clothing worn by a sample of 700 security salesmen:

|  | Amount of Security Sales | | |
|---|---|---|---|
|  | Low | Average | High |
| Very Well Dressed | 50 | 237 | 226 |
| Well Dressed | 54 | 69 | 33 |
| Poorly Dressed | 10 | 12 | 9 |

Use the level of significance $\alpha = 0.01$ to test the null hypothesis that there is no relationship between the level of sales of security salesmen and the standard of clothing which they wear.

**5.** A sanitation inspector for a national chain of gasoline stations obtained the following results in a sample survey in which he inspected 600 of the chain's gasoline stations in three parts of the country:

|  | Standard of Sanitation | | |
|---|---|---|---|
|  | Below Average | Average | Above Average |
| East | 30 | 96 | 55 |
| Midwest | 39 | 112 | 51 |
| West | 46 | 103 | 68 |

Use the level of significance $\alpha = 0.05$ to test the null hypothesis that there is no difference in the standard of sanitation of the chain's gasoline stations in the three parts of the country.

**6.** Using an argument similar to that on page 264, show that the rule on page 269 for calculating the expected cell frequencies applies also when the column totals are fixed sample sizes and do not depend on chance.

## GOODNESS OF FIT

The chi-square criterion can also be used to compare observed frequency distributions with distributions which we might *expect* according to mathematical theory or assumptions. To illustrate, let us refer to the binomial distribution on page 121 which pertained to the number of heads we might obtain in four flips of a balanced coin. As we showed on that page, the probabilities of getting 0, 1, 2, 3, or 4 heads are, respectively, $\frac{1}{16}$, $\frac{4}{16}$, $\frac{6}{16}$, $\frac{4}{16}$, and $\frac{1}{16}$, and it should be noted that these probabilities apply also when we flip four balanced coins together (instead of flipping one balanced coin four times).

Suppose now that we actually flip four coins together 480 times, and that the results we obtain are shown in the "observed frequency" column of the following table:

| Number of Heads | Probability | Observed Frequency $f$ | Expected Frequency $e$ |
|---|---|---|---|
| 0 | $\frac{1}{16}$ | 26 | 30 |
| 1 | $\frac{4}{16}$ | 104 | 120 |
| 2 | $\frac{6}{16}$ | 171 | 180 |
| 3 | $\frac{4}{16}$ | 142 | 120 |
| 4 | $\frac{1}{16}$ | 37 | 30 |
| | | 480 | 480 |

The expected frequencies were obtained by multiplying the corresponding probabilities by 480, and it should be observed that there are quite some discrepancies between the $f$'s and the $e$'s. In fact, there are altogether

$$0(26) + 1(104) + 2(171) + 3(142) + 4(37) = 1{,}020 \text{ heads}$$

which is quite a bit more than the 960 heads we would expect in $4 \cdot 480 = 1{,}920$ flips of balanced coins.

To check whether the differences between the observed and expected frequencies can be attributed to chance, we use the same chi-square statistics as in the two preceding sections, namely,

$$\chi^2 = \Sigma \frac{(f-e)^2}{e}$$

where $\frac{(f-e)^2}{e}$ is calculated separately for each class of the distribution.

Then, if the value of $\chi^2$ is *too large* we *reject* the null hypothesis that the coins are all properly balanced and randomly flipped, namely, that the binomial distribution with $p = \frac{1}{2}$ and $n = 4$ provides a *good fit* to the observed distribution. More specifically, we reject the null hypothesis at the level of significance $\alpha = 0.05$ if the value we obtain for $\chi^2$ *exceeds* $\chi^2_{.05}$ for the appropriate number of degrees of freedom. When we fit a binomial distribution as in our example, the number of degrees of freedom is $k - 1$, where $k$ is the number of classes in the distribution. More generally, the number of degrees of freedom in a chi-square test of *goodness of fit* is $k - m$, where $k$ is the number of terms $\frac{(f - e)^2}{e}$ which we have to add, and $m$ is the number of quantities *which we have to obtain from the observed distribution* in order to calculate the expected frequencies. In our example we had to know only the total number of times the four coins had been flipped (namely, 480), so that $m = 1$ and the number of degrees of freedom is $k - 1$.

Returning now to our numerical example and substituting into the formula for $\chi^2$ the $f$'s and the $e$'s of the table on page 274, we get

$$\chi^2 = \frac{(26 - 30)^2}{30} + \frac{(104 - 120)^2}{120} + \frac{(171 - 180)^2}{180}$$

$$+ \frac{(142 - 120)^2}{120} + \frac{(37 - 30)^2}{30}$$

$$= 8.78$$

Since this is less than 9.488, the value of $\chi^2_{.05}$ for $5 - 1 = 4$ degrees of freedom, we find that the null hypothesis (that the coins are all properly balanced and randomly flipped) *cannot be rejected*. (In fact, if the reader drew superimposed histograms, he would find that the expected distribution provides a *fairly good fit* to the distribution which was originally observed.)

The method which we have illustrated in this section can be used quite generally to see how well a distribution which we *expect* on the basis of theory or assumptions fits observed data. It is not limited to the binomial distribution referred to in our example, and (although we shall not discuss it in this book) it can be used even to show how well a *continuous distribution curve* fits a distribution of observed data.

## EXERCISES

1. The following is the distribution of the number of green Christmas tree ornaments found in a gross of clear plastic boxes, which were randomly packed with three red or green ornaments:

   | Number of Green Ornaments | Number of Plastic Boxes |
   |---|---|
   | 0 | 15 |
   | 1 | 66 |
   | 2 | 41 |
   | 3 | 22 |
   | Total | 144 |

   (a) Assuming that there is a fifty-fifty chance for each ornament to be red or green, look up the corresponding probabilities for 0, 1, 2, and 3 green ornaments in Table XI, and find the corresponding expected frequencies for 144 clear plastic boxes, each containing three ornaments.
   (b) Use the $\chi^2$ criterion and the level of significance $\alpha = 0.05$ to decide whether the binomial distribution with $n = 3$ and $p = \frac{1}{2}$ provides a good fit to the observed distribution.

2. To determine whether the switchboard of an office can be expected to receive an equal number of telephone calls during each hour of the afternoon, a record was kept of the time at which telephone calls were received during the six hours from noon to 6 P.M. The results were as shown in the following table:

   | | Number of Telephone Calls |
   |---|---|
   | First hour | 31 |
   | Second hour | 52 |
   | Third hour | 46 |
   | Fourth hour | 39 |
   | Fifth hour | 50 |
   | Sixth hour | 22 |
   | | 240 |

Using the fact that the expected frequencies are 40 telephone calls for each hour ($\frac{240}{6} = 40$), calculate $\chi^2$ and test the null hypothesis that in general an equal number of telephone calls can be expected during the six hours from noon to 6 P.M. Let the level of significance be $\alpha = 0.01$.

3. Using any four columns of Table IX (that is, a total of 200 random digits), construct a table showing how many times each of the digits 0, 1, 2, ..., and 9 occurred. Compare these observed frequencies with the corresponding expected frequencies (which are all equal to 20) by means of the $\chi^2$ statistic, and decide at the level of significance $\alpha = 0.05$ whether this **uniform distribution** provides a good fit to the observed distribution.

# 10

# THE ANALYSIS OF PAIRED DATA

**INTRODUCTION**

There are many statistical studies in which we deal with *paired* measurements or observations—the ages and the prices of used cars, the supply and the demand for wheat, the grades and the I.Q.'s of students in a course in finance, family expenditures on food and clothing, a country's imports and exports, the amount of fertilizer applied and the size of the crop, and so forth. In studies like these we are usually interested in one aspect or another of the following question:

> Is there a relationship, or dependence, between the two variables under consideration, and if so, what is its strength and what is its "nature" or its form?

To answer questions like this we usually try to express the relationship between the variables in terms of a mathematical equation. Among the many equations that can be used for this purpose, the simplest (and also the most widely used) is the **linear equation** in two variables

$$y = a + bx$$

where $x$ and $y$ are the two variables and $a$ and $b$ are numerical constants. The name "linear" is accounted for by the fact that, when plotted on ordinary graph paper, all points whose $x$- and $y$-coordinates satisfy an equation of the form $y = a + bx$ will fall on a straight line. To illustrate, let us consider the equation

$$y = 11.0 + 1.7x$$

where $x$ stands for the number of catalogs (in thousands) which a mail order business distributes in a city and $y$ represents the corresponding number of mail orders (also in thousands) they can expect to receive. The graph of this equation is shown in Figure 10.1, and any pair of values of $x$ and $y$ which are such that $y = 11.0 + 1.7x$ form a point $(x, y)$ that falls on the line. In

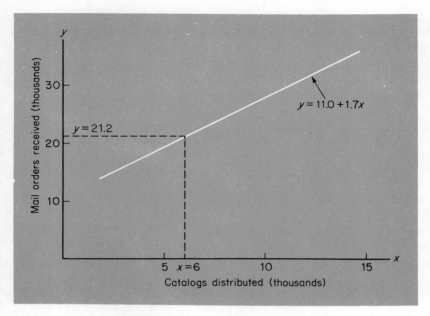

**FIGURE 10.1** Graph of linear equation.

most practical applications we use lines like this to make *predictions or forecasts*, and in our example we might predict, say, that if the mail order business distributed $x = 6$ thousand catalogs in a city, they can expect to receive $y = 11.0 + 1.7(6) = 21.2$ thousand mail orders (or 21,000 rounded to the nearest thousand).

Whenever we use paired measurements or observations to arrive at an equation which represents, or expresses, the relationship between the corresponding two variables, we face *three basic problems*. *First* we must decide what kind of equation (or curve) is to be used, and this question is usually answered by plotting the data points $(x, y)$, which correspond to the given pairs of measurements or observations, on a piece of graph paper—hopefully, this will enable us to discern some definite pattern.* So far as the work of this chapter is concerned, this question is really academic, for linear equations are the only ones we shall discuss. This is not as confining as it may seem. *Linear equations are useful not only because many of the relationships which we meet in actual practice are of this form, but also because they often provide excellent approximations to relationships which would otherwise be difficult to describe in mathematical terms.*

* There exist methods for putting this kind of decision on a more *objective* basis, but they are fairly advanced and will not be treated in this text.

280   THE ANALYSIS OF PAIRED DATA                                    CHAP. 10

Once we have decided to use a straight line, we are faced with the *second* kind of problem, namely, that of finding the equation of the particular line which in some sense provides the *best possible fit* to the **data points** (paired data) with which we are concerned. To illustrate, let us refer to the following data (based on information supplied by the mail order business for 12 cities), which actually led to the line of Figure 10.1:

| Number of Catalogs Distributed (in thousands) $x$ | Number of Mail Orders Received (in thousands) $y$ |
|---|---|
| 6 | 20 |
| 2 | 14 |
| 5 | 20 |
| 1 | 14 |
| 10 | 28 |
| 7 | 23 |
| 15 | 36 |
| 3 | 16 |
| 11 | 32 |
| 13 | 33 |
| 2 | 13 |
| 12 | 30 |

If we plot the points which correspond to these 12 pairs of $x$'s and $y$'s as in Figure 10.2, it is apparent that although the points do not actually fall on a straight line, the over-all pattern of the relationship is pretty well described by the dashed line. Consequently, it would seem *reasonable*, or *justifiable*, to represent the relationship between the number of catalogs distributed and the number of mail orders received by means of a linear equation.

Logically speaking, there is no end to the number of different lines which we could draw on the diagram of Figure 10.2. Some of these obviously would not fit the given data and they can immediately be ruled out, but we can draw many different lines which come fairly close to the 12 points and, hence, must be taken into account. In order to single out *one* line as *the line* which "best" fits the given set of paired data, we shall have to state explicitly what we mean here by "best"—in other words, we shall have to give a criterion, a "yardstick," which will enable us to choose the one line which is thus considered "best." (If all of the points fell on a straight line, *this would be it*, but this kind of situation is seldom, if ever, met in actual practice.)

The criterion which is most widely used for fitting straight lines to paired data is based on the **method of least squares,** which we shall study in the next section; it dates back to the early part of the nineteenth century and it has

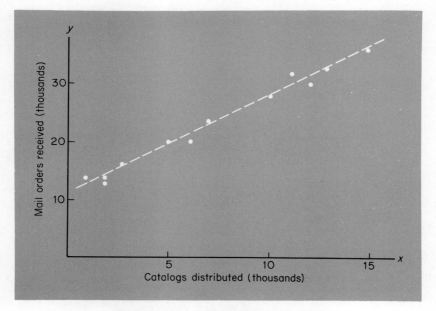

**FIGURE 10.2** Graph showing given data.

found many different important applications in various branches of statistics. Before we go into this, however, let us point out that after we have decided to fit a straight line and after we have decided to use the method of least squares, we still face a *third* kind of question. We must ask ourselves "How well does this line actually fit?" After all, if the data points are scattered all over the place as in Figure 10.3, even the best-fitting line cannot provide a good fit. This third kind of question about the analysis of paired data will be discussed later in the section beginning on page 290.

## THE METHOD OF LEAST SQUARES

If the reader has ever had the occasion to work with paired data which were plotted as points on a piece of graph paper, he has probably felt the urge to take a ruler, move it around, and decide *by eye* on a line which presents a fairly good fit. There is no rule in the book which says that this *cannot* be done, but it would certainly not be very "scientific." More objective is the **least squares criterion** which requires that

> The line which we fit to a set of data points must be such that the sum of the squares of the vertical deviations (distances) from the points to the line is as small as possible.

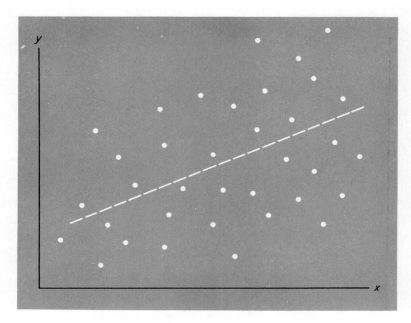

**FIGURE 10.3** Poor fit.

With reference to the illustration of the preceding section, the least-squares criterion requires that the sum of the squares of the distances represented by the heavily-drawn line segments of Figure 10.4 be a minimum.

To explain why this is done, let us refer to the data on page 280, and take one of the cities, say, the one in which 6 thousand catalogs were distributed and 20 thousand mail orders were received. If we substitute $x = 6$ into the equation (as we did on page 279), we find that the corresponding number of mail orders is roughly 21 thousand, so that the *error* of this "prediction," *represented by the vertical distance from the point to the line*, is $20 - 21 = -1$. Altogether, there are 12 such errors in our example, one for each of the 12 cities, and the least-squares criterion requires that we *minimize the sum of their squares*. To explain why we minimize the sum of the *squares* of the deviations and not just the sum of the deviations, themselves, let us point out that the deviations (namely, the differences between the observed values of $y$ and the corresponding values on the line) are *positive* for points which fall above the line, *negative* for points which fall below the line, so that their sum can be very small, even zero, in spite of the fact that the points are scattered all over the page. By working with the squares of the deviations which *cannot be negative*, we are, in fact, making sure that we are paying attention to the *magnitude* of the deviations and not to their signs.

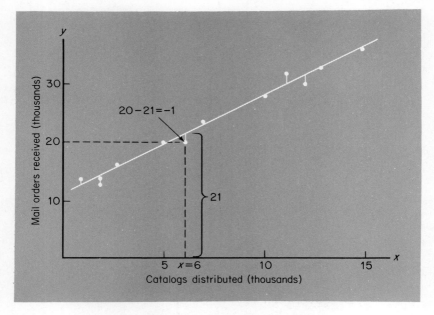

**FIGURE 10.4**  Line fit to given data.

To demonstrate how a **least-squares line** is actually fit to a set of paired data, let us consider $n$ pairs of numbers $(x_1, y_1)$, $(x_2, y_2)$, ..., and $(x_n, y_n)$, which might represent the ages and incomes of $n$ persons, the size of the population and the size of the police department of $n$ cities, the prices and weights of $n$ kinds of woolen blankets, the mid-term and final examination grades of $n$ law students, and so on. If we write the equation of the line which we want to fit to the corresponding $n$ points as

$$y = a + bx$$

we shall have to deal with *two values of $y$ for each of the given values of $x$: the corresponding value of $y$ which was actually observed and the $y$-coordinate of the corresponding point on the least-squares line* (which is obtained by substitution as in the numerical example on page 279). Thus, for $x_1$ the observed value of $y$ is $y_1$ and the $y$-coordinate of the point on the line is $a + bx_1$, for $x_2$ the observed value of $y$ is $y_2$ and the $y$-coordinate of the point on the line is $a + bx_2$, ..., and for $x_n$ the observed value of $y$ is $y_n$ and the $y$-coordinate of the point on the line is $a + bx_n$. As can be seen from Figure 10.5, the corresponding differences, namely, the vertical deviations from the points representing the observed data to the least-squares line, are $y_1 - (a + bx_1)$,

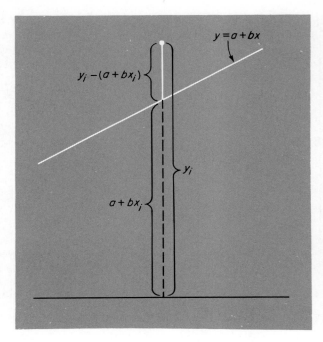

**FIGURE 10.5** Vertical deviation from the point $(x_i, y_i)$ to the line, where $i$ can be 1, 2, 3, ..., or $n$.

$y_2 - (a + bx_2)$, ..., and $y_n - (a + bx_n)$, and the least-squares criterion requires that we minimize the sum of their squares, namely,*

$$\Sigma\,[y - (a + bx)]^2$$

As it takes calculus (or a fairly tedious algebraic process called "completing the square") to find the values of $a$ and $b$ which will minimize this sum, let us merely state the result that these values are given by the solution of the following system of linear equations in the two unknowns $a$ and $b$:

$$\Sigma\,y = n \cdot a + b(\Sigma\,x)$$
$$\Sigma\,xy = a(\Sigma\,x) + b(\Sigma\,x^2)$$

*In the notation of Exercise 17 on page 38, this sum of squares can be written more explicitly as

$$\sum_{i=1}^{n} [y_i - (a + bx_i)]^2$$

Here $n$ is the number of pairs of observations, $\Sigma x$ and $\Sigma y$ are, respectively, the sum of the given $x$'s and the sum of the given $y$'s, $\Sigma x^2$ is the sum of the *squares* of the given $x$'s, and $\Sigma xy$ is the sum of the *products* obtained by multiplying each of the given $x$'s by the corresponding observed value of $y$. It is customary to refer to the above system of equations as the **normal equations,** whose solution gives the desired least-squares values of $a$ and $b$.

Returning now to our numerical example and copying the first two columns from page 280, we obtain the sums needed for substitution into the normal equations by means of the calculations shown in the following table:

|  | Number of Catalogs Distributed (in thousands) | Number of Mail Orders Received (in thousands) |  |  |
|---|---|---|---|---|
|  | $x$ | $y$ | $x^2$ | $xy$ |
|  | 6 | 20 | 36 | 120 |
|  | 2 | 14 | 4 | 28 |
|  | 5 | 20 | 25 | 100 |
|  | 1 | 14 | 1 | 14 |
|  | 10 | 28 | 100 | 280 |
|  | 7 | 23 | 49 | 161 |
|  | 15 | 36 | 225 | 540 |
|  | 3 | 16 | 9 | 48 |
|  | 11 | 32 | 121 | 352 |
|  | 13 | 33 | 169 | 429 |
|  | 2 | 13 | 4 | 26 |
|  | 12 | 30 | 144 | 360 |
| Totals | 87 | 279 | 887 | 2,458 |

(If this work is done with a desk calculator or even fancier equipment, the sum of the $x^2$'s and the sum of the $xy$'s can be accumulated directly, and there is no need to fill in all the details.) Then, if we substitute $n = 12$ and the appropriate column totals into the two normal equations, we get

$$279 = 12a + 87b$$
$$2{,}458 = 87a + 887b$$

and all that remains to be done is to solve this system of equations for $a$ and $b$. There are several ways in which this can be done—simplest, perhaps, is the **method of elimination,** which the reader may recall from elementary algebra. Using this method, let us eliminate $a$ by dividing the expressions on both sides of the first equation by 12, dividing the expressions on both sides of the second equation by 87, and then subtracting "equals from equals." We thus get

$$23.25 = a + 7.25b$$
$$28.25 = a + 10.20b$$

and by subtraction

$$5.00 = 0 + 2.95b$$

Hence, $b = \dfrac{5.00}{2.95} = 1.695$ (or 1.7 rounded to one decimal), and if we substitute this value of $b$ into the first of the original equations, we get $279 = 12a + 87(1.695)$, $279 = 12a + 147.5$, $12a = 131.5$, and finally $a = \dfrac{131.5}{12} = 10.96$ (or 11.0 rounded to one decimal). Thus, the equation of the least squares line is

$$y = 11.0 + 1.7x$$

If the reader is not familiar with the method of elimination, he can instead use the following formulas for $a$ and $b$, which were actually obtained by *symbolically* solving the two normal equations on page 284 by the method of elimination (see Exercise 5 on page 289):

$$a = \frac{(\Sigma\, y)(\Sigma\, x^2) - (\Sigma\, x)(\Sigma\, xy)}{n(\Sigma\, x^2) - (\Sigma\, x)^2}$$

$$b = \frac{n(\Sigma\, xy) - (\Sigma\, x)(\Sigma\, y)}{n(\Sigma\, x^2) - (\Sigma\, x)^2}$$

Had we used these formulas for $a$ and $b$ in our numerical example, we would have obtained

$$a = \frac{(279)(887) - (87)(2{,}458)}{12(887) - (87)^2} = \frac{33{,}627}{3{,}075} = 10.94$$

$$b = \frac{12(2{,}458) - (87)(279)}{3{,}075} = \frac{5{,}223}{3{,}075} = 1.70$$

where we did not have to calculate the denominator in the formula for $b$ because it is the same as that in the formula for $a$. Actually, the results should have been the same as before, and the difference between $a = 10.94$ and $a = 10.96$ (obtained earlier by the method of elimination) is entirely due to rounding.

Now that we have determined the equation of the least-squares line for the data on page 280, suppose we want to predict, say, the number of mail orders they can expect to receive from a city in which they distribute 14,000 of the catalogs. Substituting $x = 14$ into the equation, we get

$$y = 11.0 + 1.7(14) = 34.8$$

(or 35 rounded to the nearest whole number), and we can thus predict that they can expect to get 35,000 mail orders. Although the prediction is for 35,000 mail orders, and not for 34,000 or 37,000, *we should be delighted if it turned out even that close.* After all, it would be very unreasonable to suppose that they will get exactly 35,000 mail orders from *each* city in which they distribute 14,000 catalogs—in fact, it can be seen from Figure 10.2 that the original data did not all fall on the straight line, and that different numbers of mail orders were received from two cities in which they distributed the same number of catalogs (see Exercise 1 below). Indeed, we said originally that they can *expect* to get 35,000 mail orders, and *to make predictions based on least-squares lines meaningful, they will always have to be interpreted as averages, namely, as mathematical expectations.* Interpreted in this way, namely, as a line which enables us to read off (or calculate) the expected value of $y$ for any given value of $x$, a least-squares line is also referred to as a **regression line** or, better, as an **estimated regression line.** This term is due to Sir Francis Galton, the nineteenth century English scientist who first employed it in a study of the relationship between the heights of fathers and sons, in which he observed a "regression," or turning back from the heights of sons to the heights of their fathers.

The equation $y = 11.0 + 1.7x$ is only an *estimated* regression line because it is based on sample data, and it must be evident that if we repeated the whole study with data for 12 different cities, we would probably get different values for $a$ and $b$. Thus, if the *true* relationship is given by the equation $y = \alpha + \beta x$, we must look upon the value we get for $a$ as an estimate of $\alpha$ (*alpha*) and we must look upon the value we get for $b$ as an estimate of $\beta$ (*beta*), and this raises a virtual *Pandora's box* of questions: "*How good* are these estimates?" "*How good* are predictions based on the equations of estimated regression lines?" "*How good* are corresponding estimates of the *true* average value of $y$ for a given value of $x$?" All these questions are answered in a fairly difficult branch of statistics called **regression analysis;** so far as the work of this book is concerned, a partial answer to the first of these questions will be given in Exercise 10 on page 299.

## EXERCISES

1. Among the data on page 280, there were two cities in which 2 thousand catalogs were distributed, and from which, respectively, 13 thousand and 14 thousand mail orders were received. If we use the equation of the least-

squares line on page 286, what number of mail orders would we predict for a city in which 2 thousand of the catalogs are distributed?

2. The following data show the improvement (gain in reading speed) of six students in a speed-reading program, and the number of weeks they have been in the program:

| Number of Weeks | Speed Gain (words per minute) |
|---|---|
| 4 | 81 |
| 2 | 30 |
| 8 | 210 |
| 6 | 164 |
| 9 | 241 |
| 3 | 69 |

(a) Use the formulas on page 286 to calculate $a$ and $b$ for the least-squares line which will enable us to predict the reading-speed gain of a person who has been in the program for a given length of time.
(b) Repeat part (a) by solving the two normal equations directly, and compare the results.
(c) Use the equation of the least-squares line to predict the increase in reading speed which a person can expect if he is in the program for five weeks.

3. The following table shows 15 weeks' sales of a department store in Boston, Massachusetts, and of its suburban branch store:

| Sales of Boston Store (in $1,000) | Sales of Suburban Store (in $1,000) |
|---|---|
| 64 | 32 |
| 60 | 22 |
| 71 | 51 |
| 71 | 47 |
| 66 | 25 |
| 63 | 24 |
| 66 | 35 |
| 70 | 51 |
| 63 | 30 |
| 60 | 18 |
| 71 | 50 |
| 64 | 32 |
| 69 | 40 |
| 64 | 38 |
| 68 | 45 |

(a) Find the equation of the least-squares line which will enable us to predict the sales of the suburban store from the sales of the Boston store.

(b) Plot the least-squares line together with the original data on one diagram.

(c) Predict the sales of the suburban store for a week in which the sales of the Boston store are $67 thousand, by substituting $x = 67$ into the equation obtained in part (a), and also by reading the value off the line plotted in part (b).

4. The following table shows the number of weeks that nine waitresses have been employed in a certain restaurant, and the number of customers each can serve on the average in one hour:

| Weeks Employed $x$ | Customers Served per Hour $y$ |
|---|---|
| 4 | 13.3 |
| 7 | 18.5 |
| 2 | 10.1 |
| 14 | 32.4 |
| 10 | 26.6 |
| 1 | 8.3 |
| 9 | 21.1 |
| 5 | 14.6 |
| 12 | 28.2 |

(a) Fit a least-squares line to these data, and plot the line together with the actual data on one diagram.

(b) If a waitress has been employed in this restaurant for eight weeks, how many customers "should" she be able to serve in one hour?

(c) Would it be reasonable to use the equation obtained in part (a) to predict how many customers a waitress should be able to serve, if she has been employed by the restaurant for five *years*? Explain your answer.

5. Multiply the expressions on both sides of the first of the two *normal* equations on page 284 by $\Sigma x$, multiply those of the second of the two normal equations by $n$, eliminate $a$ by subtraction, and then solve for $b$. Then, substitute the expression obtained for $b$ into the first of the two normal equations, and solve for $a$.

## THE COEFFICIENT OF CORRELATION

Now that we have learned how to fit a least-squares line to paired data, let us see how we might answer the third question raised on page 281—*how well does the line actually fit the given data?* Of course, we can get a fair idea by inspection, say, by looking at a diagram like that of Figure 10.2, but to be more objective let us analyze the *total variation* of the observed $y$'s (in our example, the variation among the mail orders received) and look for possible causes, or explanations. As can be seen from the second column of the table on page 280, there are considerable differences among the $y$'s, with the smallest being 13 and the largest being 36. However, it can also be seen from this table as well as from Figure 10.2 that these differences must be due partly to the fact that the mail order business did not distribute the same number of catalogs in each of the 12 cities. For instance, 36 thousand mail orders were received from a city in which 15 thousand catalogs were distributed, while only 16 thousand mail orders were received from a city in which 3 thousand catalogs were distributed. This raises the following question:

**How much of the total variation of the observed $y$'s can be attributed to the relationship between the two variables $x$ and $y$ (namely, the extent to which $y$ depends on $x$), and how much of it can be attributed to all other factors, including chance?**

With reference to our example, we would thus want to know what part of the differences among the numbers of mail orders received can be accounted for by the differences in the number of catalogs distributed in the respective cities, and what part can be attributed to all other factors (including the amount of competition received from stores and other mail order businesses, the average income of customers in the different cities, . . ., and, last but not least, *chance*).

A convenient measure of the *total variation* of the observed $y$'s is the *sum of squares* $\Sigma (y - \bar{y})^2$, which is simply their *variance* (see page 190) multiplied by $n - 1$. Since $\Sigma y = 279$ in our example (according to the table on page 285), we find that $\bar{y} = \frac{279}{12} = 23.25$, and, hence, that

$$\begin{aligned}\Sigma (y - \bar{y})^2 &= (20 - 23.25)^2 + (14 - 23.25)^2 + (20 - 23.25)^2 \\ &+ (14 - 23.25)^2 + (28 - 23.25)^2 + (23 - 23.25)^2 \\ &+ (36 - 23.25)^2 + (16 - 23.25)^2 + (32 - 23.25)^2 \\ &+ (33 - 23.25)^2 + (13 - 23.25)^2 + (30 - 23.25)^2 \\ &= 752.25\end{aligned}$$

If the number of catalogs distributed in the various cities were the *only* thing that affects the number of mail orders received, the data points of our example all would have fallen on a straight line (assuming, of course, that the relationship is linear); however, the fact that they did *not* fall on a straight line, as is apparent from Figures 10.2 and 10.4, indicates that there are other factors at play. The extent to which the points fluctuate above and below the line, namely, *the variation caused by the other factors*, is usually measured by the sum of the squared deviations from the points to the line (see Figure 10.4 on page 283), that is, by the quantity

$$\Sigma \, [y - (a + bx)]^2$$

on which we based the least squares criterion on page 284.

To calculate this sum of squares for our numerical example, we shall first have to determine $a + bx$ (namely, the $y$-coordinate of the corresponding point on the line) for *each* of the given values of $x$. Thus, we obtain $11.0 + 1.7(6) = 21.2$, $11.0 + 1.7(2) = 14.4$, $11.0 + 1.7(5) = 19.5$, $11.0 + 1.7(1) = 12.7$, $11.0 + 1.7(10) = 28.0$, $11.0 + 1.7(7) = 22.9$, $11.0 + 1.7(15) = 36.5$, $11.0 + 1.7(3) = 16.1$, $11.0 + 1.7(11) = 29.7$, $11.0 + 1.7(13) = 33.1$, $11.0 + 1.7(2) = 14.4$, and $11.0 + 1.7(12) = 31.4$, and it follows that

$$\begin{aligned}\Sigma \, [y - (a + bx)]^2 &= (20 - 21.2)^2 + (14 - 14.4)^2 + (20 - 19.5)^2 \\ &+ (14 - 12.7)^2 + (28 - 28.0)^2 + (23 - 22.9)^2 \\ &+ (36 - 36.5)^2 + (16 - 16.1)^2 + (32 - 29.7)^2 \\ &+ (33 - 33.1)^2 + (13 - 14.4)^2 + (30 - 31.4)^2 \\ &= 13.03\end{aligned}$$

Thus, we can claim that

$$\frac{\Sigma \, [y - (a + bx)]^2}{\Sigma \, (y - \bar{y})^2} = \frac{13.03}{752.25} = 0.0173$$

is the proportion of the total variation of the number of mail orders received (in the various cities) which can be attributed to *all other factors*, while $1 - 0.0173 = 0.9827$ is the proportion of this variation which can be attributed to *differences in the number of catalogs distributed by the mail order business.*

If we take the square root of this last proportion (namely, the proportion of the total variation of the $y$'s that is accounted for by the relationship with $x$), we obtain the statistical measure which is called the **coefficient of correlation**. It is generally denoted by the letter $r$, and its *sign* is always chosen so that it is the same as that of $b$ in the equation of the least-square line. Thus, for our example we get

$$r = \sqrt{0.9827} = 0.99$$

rounded to two decimals. Note that

> If the dependence between $x$ and $y$ is strong, most of the variation of the $y$'s can be attributed to the relationship with $x$, and $r$ will be close to 1 or $-1$.

Indeed, $r = 1$ or $r = -1$ is indicative of a *perfect fit*, namely, it is indicative of the fact that the observed points actually fall on a straight line. On the other hand,

> If the dependence between $x$ and $y$ is weak, very little of the variation of the $y$'s can be attributed to the relationship with $x$, and $r$ will be close to 0.

When $r = 0$, we say that there is *no correlation*, which means that none of the variation of the $y$'s can be attributed to the relationship with $x$. (Of course, $r$ cannot take on a value greater than 1 or less than $-1$, as it was defined as *plus or minus the square root of a proportion*.)

It should also be observed that $r$ is *positive* when the least-squares line has an *upward slope*, that is, when the relationship between $x$ and $y$ is such that *small* values of $y$ tend to go with *small* values of $x$ and *large* values of $y$ tend to go with *large* values of $x$. Correspondingly, $r$ is *negative* when the least-squares line has a *downward slope*, that is, when the relationship between $x$ and $y$ is such that *small* values of $y$ tend to go with *large* values of $x$ and *large* values of $y$ tend to go with *small* values of $x$. Depending on the sign of $r$, we thus say that there is a **positive correlation** or a **negative correlation**, as is illustrated in Figure 10.6.

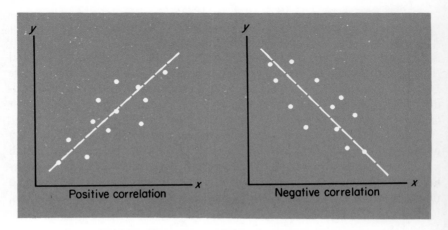

**FIGURE 10.6** Positive and negative correlation.

The correlation coefficient is a very widely used measure of the correlation (relationship, association, or dependence) between two variables. In actual practice, we seldom, if ever, determine $r$ as we did in our example; instead we use the *short-cut computing formula*

$$r = \frac{n(\Sigma\, xy) - (\Sigma\, x)(\Sigma\, y)}{\sqrt{n(\Sigma\, x^2) - (\Sigma\, x)^2}\ \sqrt{n(\Sigma\, y^2) - (\Sigma\, y)^2}}$$

which has the added advantage that it automatically gives $r$ the correct sign. Although this short-cut formula may look rather forbidding, it is actually quite easy to use. All we have to do, really, is substitute the values of $n$, $\Sigma\, x$, $\Sigma\, y$, $\Sigma\, x^2$, $\Sigma\, y^2$, and $\Sigma\, xy$ for the paired data with which we happen to be concerned, and it should be observed that, except for $\Sigma\, y^2$, these are exactly the same quantities which we needed to fit a straight line by the method of least squares.

Returning to our numerical example, if we square the $y$'s in the second column of the table on page 285, we get $\Sigma\, y^2 = 7{,}239$, and if we substitute this quantity together with $n = 12$ and the column totals of the table into the above formula for $r$, we get

$$r = \frac{12(2{,}458) - (87)(279)}{\sqrt{12(887) - (87)^2}\ \sqrt{12(7{,}239) - (279)^2}} = 0.99$$

which agrees, as it should, with the result obtained on page 291.

As we have defined the coefficient of correlation, $r^2$ gives the *proportion* (or $100r^2$ the *percentage*) of the total variation of the $y$'s which is accounted for by the relationship with $x$. For instance, if $r = 0.80$ in a given example, then $100(0.80)^2 = 64$ percent of the total variation of the $y$'s is accounted for by the relationship with $x$, and if $r = 0.40$, only $100(0.40)^2 = 16$ percent of the variation of the $y$'s is accounted for by the relationship with $x$. Thus, in the sense of "percentage of the variation of the $y$'s which is accounted for by the relationship with $x$" we can say that $r = 0.80$ is *four times as strong a correlation* as $r = 0.40$ since $(0.80)^2 = 4(0.40)^2$; by the same token we can say that $r = 0.75$ is *nine times as strong a correlation* as $r = 0.25$ since $(0.75)^2 = 9(0.25)^2$.

As a word of caution let us add that the coefficient of correlation is not only a very widely used statistical measure, but that it is also very widely *abused*. For instance, it is often abused in the sense that we forget that our whole discussion has been about *linear relationships* and, hence, *linear correlation*. In other words, $r$ should be calculated only when it is reasonable to assume that the relationship between the two variables under consideration can be described fairly well by means of a straight line. The correlation coefficient is also abused in the sense that values of $r$ close to $-1$ or 1 are (regrettably often) interpreted as *cause–effect* relationships. For instance, it is easy to find data on the annual consumption of liquor in the United States and teachers' salaries in the same years which will yield a very high value of $r$ (namely, a value of $r$ close to 1). Obviously, though, this is not indicative of a cause–effect relationship; it results from the fact that both variables are *effects of a common cause*—the over-all standard of living. (If it is high, greater amounts of money are available for teachers' salaries as well as liquor than when it is low.) Another classical example is the strong positive correlation which has been obtained for the number of storks seen nesting in English villages and the number of childbirths recorded in the same communities. (*In the large villages there are lots of houses with chimneys for storks to nest in, and hence lots of families, storks, and babies; in the smaller villages there are fewer houses with chimneys for storks to nest in, and hence fewer families, fewer storks, and fewer babies.*)

Finally, it is sometimes overlooked that when $r$ is calculated on the basis of sample data, we may get a fairly strong (positive or negative) correlation *purely by chance*, even though there is no relationship whatsoever between the two variables under consideration. To illustrate this point, suppose we take a spinner (similar to the one pictured on page 179) which has the numbers from 1 through 10. Rotating the spinner clockwise and counterclockwise on seven alternate trials, we obtain the following results:

| Clockwise | Counterclockwise |
| --- | --- |
| $x$ | $y$ |
| 1 | 2 |
| 10 | 7 |
| 5 | 4 |
| 2 | 3 |
| 6 | 3 |
| 9 | 5 |
| 4 | 6 |

If we calculate $r$ for these data, we get the surprisingly high value $r = 0.73$, and this raises the obvious question whether anything is wrong with the idea that there should be no relationship at all—*after all, the outcome of each spin should not depend on the outcome of the preceding spin*. To answer this question, we shall have to see whether this relatively high value of $r$ might, perhaps, be attributed to chance.

To test the null hypothesis of *no correlation* (namely, the hypothesis that there is no linear relationship between two given variables), we shall refer to a special table, Table VI at the end of the book, which is based on the assumption that the $x$'s and $y$'s are values of *independent* (and, hence, *uncorrelated*) random variables having at least approximately normal distributions. With reference to this table

> We say that the value of $r$ which we get for a set of paired data is significant at the level of significance $\alpha = 0.05$ if it exceeds $r_{.025}$ or is less than $-r_{.025}$.

If the value we get for $r$ falls between $-r_{.025}$ and $r_{.025}$, we say that there is *no significant correlation*, or that the value of $r$ is *not statistically significant*. Incidentally, Table VI also contains values of $r_{.005}$, which we would use instead of $r_{.025}$ when the level of significance is to be $\alpha = 0.01$.

If we apply this test to the results which we obtained for the catalogs-distributed-and-mail-orders-received example, we find that $r = 0.99$ *is significant* as it exceeds $r_{.025} = 0.576$ as well as $r_{.005} = 0.708$ for $n = 12$. We are justified, therefore, in concluding that there *is* a (linear) relationship between the number of catalogs distributed and the number of mail orders received. If we apply the same test to the example where we operated the spinner clockwise and counterclockwise, we find that the result, namely, $r = 0.73$, *is not significant;* it is less than $r_{.025} = 0.754$, the tabular value for $n = 7$. Thus, we conclude that the apparent relationship between the numbers obtained with clockwise and counterclockwise spins can be attributed to chance. In case the reader may suspect that this example with the spinner was "rigged," let us assure him that this is not true. Even before this "experiment" was performed we could have given odds of 1 to 19 that the value of $r$ which we get will actually be significant. Looking at it differently, if many students tried this "experiment," about 5 percent of them should get significant results even though there is no relationship—this is the risk which is inherent in an $\alpha = 0.05$ level of significance.

## EXERCISES

1. State in each case whether you would expect to obtain a positive correlation, a negative correlation, or no correlation:
   (a) the ages of husbands and wives;
   (b) a woman's dress size and her ability to sing;
   (c) the number of hours a bowler practices and his (or her) scores;
   (d) income and education;
   (e) shoe size and I.Q.;
   (f) the amount of life left in a battery and the number of hours it has been used;

(g) the number of sunny days in Detroit and the attendance at the Detroit Zoo;
(h) the number of influenza inoculations and the incidence of influenza;
(i) shirt size and sense of humor;
(j) pollen count and the sale of anti-allergy drugs.

**2.** Calculate $r$ for the data of Exercise 2 on page 288 and test its significance at the level of significance $\alpha = 0.05$.

**3.** Calculate $r$ for the sales of the Boston store and the suburban store of Exercise 3 on page 288 and test for significance at $\alpha = 0.01$.

**4.** Calculate $r$ for the data of Exercise 4 on page 289 and test its significance at the level of significance $\alpha = 0.05$.

**5.** The following are the speeds (in words per minute) at which eight secretaries can type at 9 A.M. and at 10 A.M.:

| 9 A.M. | 10 A.M. |
|---|---|
| 68 | 71 |
| 72 | 76 |
| 73 | 71 |
| 70 | 72 |
| 75 | 79 |
| 65 | 67 |
| 77 | 82 |
| 74 | 76 |

Calculate $r$ and test its significance at the level of significance $\alpha = 0.05$.

**6.** The following are the amounts of meat (in thousands of pounds) sold by two food markets in five consecutive weeks:

| Black's Market | Green's Market |
|---|---|
| 4.7 | 6.8 |
| 6.4 | 5.6 |
| 8.8 | 3.4 |
| 9.9 | 2.0 |
| 10.3 | 1.9 |

Calculate $r$ and try to *interpret* the result.

**7.** The following are the grades which 16 students received in final examinations in accounting and statistics:

| Accounting | Statistics | Accounting | Statistics |
|---|---|---|---|
| 47 | 65 | 53 | 71 |
| 71 | 84 | 70 | 67 |
| 95 | 86 | 74 | 85 |
| 68 | 70 | 99 | 90 |
| 22 | 30 | 57 | 64 |
| 93 | 88 | 75 | 70 |
| 76 | 77 | 97 | 96 |
| 82 | 83 | 76 | 70 |

Calculate $r$ and check whether the correlation is significant at the level of significance $\alpha = 0.05$.

**8.** Since $r$ does not depend on the scales of measurement used for $x$ and $y$, its calculation can often be simplified by adding a suitable positive or negative number to each $x$, to each $y$, or to both, or by multiplying each $x$, each $y$, or both by a suitable positive constant.
   (a) Recalculate $r$ for the illustration in the text, namely, the data on page 280, after having subtracted 7 from each $x$ and 20 from each $y$.
   (b) Rework Exercise 6 after having multiplied each $x$ by 10 and then subtracted 80, and having multiplied each $y$ by 10 and then subtracted 40.

**9. RANK CORRELATION** Since the calculation of $r$ for large sets of paired data can be fairly tedious, it is sometimes convenient to base $r$ on the ranks of the observations instead of their actual numerical values. Thus, we first rank the $x$'s among themselves, giving Rank 1 to the largest value, Rank 2 to the second largest value, and so forth; then we similarly rank the $y$'s among themselves and calculate the **coefficient of rank correlation** by means of the formula

$$r' = 1 - \frac{6(\Sigma\, d^2)}{n(n^2 - 1)}$$

Here $n$ is the number of pairs of observations and the $d$'s are the *differences* between the respective ranks. If there are **ties,** we assign to each of the tied observations the *mean* of the ranks which they jointly occupy; thus, if the third and fourth largest values are identical we assign each the rank $\frac{3+4}{2} = 3.5$, and if the fifth, sixth, and seventh largest values are identical we assign each the rank $\frac{5+6+7}{3} = 6$. To illustrate this technique, let us refer again to the example in the text which dealt with the number of catalogs distributed and the corresponding number of mail orders received. Copying the first two columns from page 280, we get

| $x$ | $y$ | Rank of $x$ | Rank of $y$ | $d$ | $d^2$ |
|---|---|---|---|---|---|
| 6 | 20 | 7 | 7.5 | 0.5 | 0.25 |
| 2 | 14 | 10.5 | 10.5 | 0 | 0 |
| 5 | 20 | 8 | 7.5 | 0.5 | 0.25 |
| 1 | 14 | 12 | 10.5 | 1.5 | 2.25 |
| 10 | 28 | 5 | 5 | 0 | 0 |
| 7 | 23 | 6 | 6 | 0 | 0 |
| 15 | 36 | 1 | 1 | 0 | 0 |
| 3 | 16 | 9 | 9 | 0 | 0 |
| 11 | 32 | 4 | 3 | 1 | 1 |
| 13 | 33 | 2 | 2 | 0 | 0 |
| 2 | 13 | 10.5 | 12 | 1.5 | 2.25 |
| 12 | 30 | 3 | 4 | 1 | 1 |

$\Sigma d^2 = 7.00$

and, hence,

$$r' = 1 - \frac{6(7.00)}{12 \cdot 143} = 1 - \frac{42}{1716} = 0.98$$

This is very close to the value of $r$ obtained on page 293.

(a) Calculate $r'$ for the data of Exercise 2 on page 288 and compare it with the value of $r$ obtained in Exercise 2 above.
(b) Calculate $r'$ for the data of Exercise 3 on page 288 and compare it with the value of $r$ obtained in Exercise 3 above.
(c) Calculate $r'$ for the data of Exercise 5 above and compare it with the value obtained for $r$.
(d) The following are the rankings of the sales of 12 different styles of women's blouses in women's apparel shops in Los Angeles and San Francisco:

|  | Los Angeles | San Francisco |
|---|---|---|
| Style A | 8 | 11 |
| Style B | 1 | 3 |
| Style C | 12 | 8 |
| Style D | 4 | 1 |
| Style E | 7 | 7 |
| Style F | 11 | 10 |
| Style G | 6 | 5 |
| Style H | 3 | 9 |
| Style I | 5 | 2 |
| Style J | 10 | 12 |
| Style K | 9 | 6 |
| Style L | 2 | 4 |

Calculate $r'$ as a measure of the *consistency* (or *inconsistency*) of these two rankings.

10. **CONFIDENCE LIMITS FOR $\beta$** As we indicated on page 287, the value we obtain for $b$ by the method of least squares is only an *estimate* of $\beta$, the slope of the *true* regression line $y = \alpha + \beta x$. Using $r$, we can now write 0.95 confidence limits for $\beta$ as

$$b\left[1 \pm t_{.025} \cdot \frac{\sqrt{1-r^2}}{r\sqrt{n-2}}\right]$$

where the number of degrees of freedom for $t_{.025}$ is $n - 2$. For instance, for the example which we used in the text we have $n = 12$, $b = 1.7$ (from page 286), $r = 0.99$ (from page 293), and $t_{.025} = 2.228$ for $12 - 2 = 10$ degrees of freedom (according to Table II). Thus, we get

$$1.7\left[1 \pm 2.228 \cdot \frac{\sqrt{1-(0.99)^2}}{0.99\sqrt{12-2}}\right]$$

which reduces to $1.7(1 \pm 0.100)$, namely, 1.53 and 1.87, and we can write the 0.95 confidence interval for the *slope* of the regression line as

$$1.53 < \beta < 1.87$$

(a) Use this method to construct a 0.95 confidence interval for $\beta$ on the basis of the data of Exercise 2 on page 288, for which $r$ was obtained in Exercise 2 above.

(b) Use this method to construct a 0.95 confidence interval for $\beta$ on the basis of the data of Exercise 4 on page 289, for which $r$ was obtained in Exercise 4 above.

# 11

# THE ANALYSIS OF TIME SERIES

## INTRODUCTION

Statistical data that are collected, observed, or recorded at regular intervals of time are called **time series,** and it was in this sense that we spoke of *index number series* in Chapter 2. Thus dropping the word "time" and referring to the data simply as a "series," we also speak of such things as a *series* of weekly freight car loadings, a *series* of monthly department store sales, or a *series* of annual employment data. Although the terms "time series" and "series" will be used here mainly in connection with business data, it should be understood that they apply also to periodic temperature readings, a laboratory animal's daily intake of food, or, say, a recording of a person's heart beat.

In business and economics, the analysis of time series is of great importance, since the *patterns* which are thus revealed form the basis of all forecasts. To predict the future, we must know what happened in the past, and thus a great deal of time is spent trying to summarize and analyze time series, that is, statistical information about the past.

The methods which we shall present in this chapter are those of *traditional* time series analysis. Although new and highly-sophisticated methods are constantly being developed, the traditional methods continue to be widely employed because of their relative simplicity, and also because they often produce quite adequate results. Most of the more modern methods require a great deal of advanced mathematics and the availability of the most up-to-date computers.

In *traditional time series analysis*, the movements, or fluctuations, of a time series are classified into four basic types of variation which, superimposed and acting in concert, account for the changes in the series over a period of time, and give the series its irregular appearance. These four components are

1. Secular Trend
2. Seasonal Variation
3. Cyclical Variation
4. Irregular Variation

and it is assumed that there is a *multiplicative* relationship between them, namely, that *any particular value in a series is the product of factors which can be attributed to these components.*

By the **secular** (*or* **long-term**) **trend** *of a time series we mean the smooth or regular movement of a series over a fairly long period of time.* In other words, we might say that the trend of a time series reflects the over-all sweep of its development, or that it characterizes the gradual and persistent pattern of its changes. For instance, Figure 11.1 exhibits the persistent upward trend in total civilian employment in the United States, while Figure 11.2 shows the downward trend in the reserve assets of the United States. In both of these examples, the over-all trend is indicated by means of a dashed line, but it should be understood that trends need not be linear. Although we shall limit our discussion in this book to **linear trends,** Figure 11.3 makes it apparent that the trends of some series can adequately be described only by means of more complicated kinds of mathematical curves.

The kind of variation which is, perhaps, easiest to understand is the **seasonal variation,** which consists of regularly recurring patterns like those of Figure 11.4. Although its name implies a connection with the seasons of the year, as in Figure 11.4, it includes any kind of variation that is of a *periodic* nature, provided that the length of its repeating cycles is not more than one year. Simple examples of seasonal patterns may be found in the hourly traffic on freeways, in the daily attendance at a movie theater, and in the monthly sales

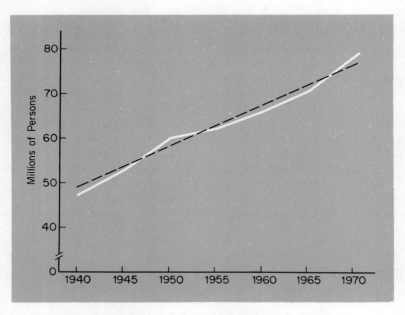

**FIGURE 11.1** Total civilian employment in the United States.

302 THE ANALYSIS OF TIME SERIES

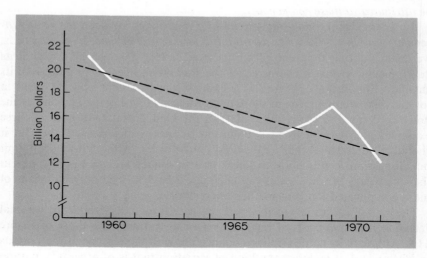

**FIGURE 11.2** Reserve assets of the United States.

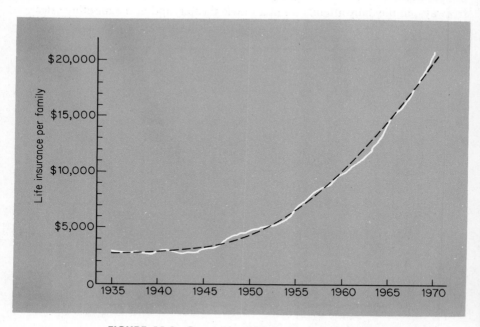

**FIGURE 11.3** Ownership of life insurance in the United States.

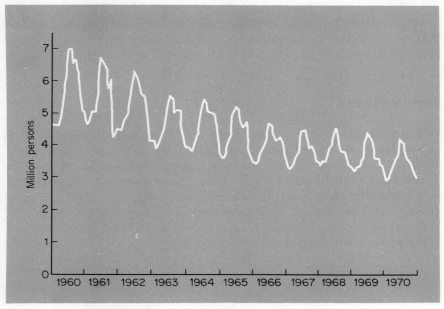

**FIGURE 11.4** Farm employment in the United States.

of ski supplies; indeed, almost every type of retail trade, manufacturing, agriculture, fishing, and other economic activity can be expected to have seasonal variations because of human customs (holidays, for example) as well as natural phenomena (such as the weather).

**Business cycles** are sometimes defined as *the variation which remains in a time series after the trend, seasonal, and irregular variations (yet to be defined) have been eliminated.* Actually, there is more to it than that, but as we shall explain in some detail later, on page 320, the *traditional* way of measuring business cycles is precisely through such a process of elimination. Generally speaking, a business cycle consists of a cyclical variation, an up and down movement of a time series, which differs from a seasonal variation in that it extends over a longer period of time *and*, supposedly, is due to general economic conditions. The periods of prosperity, recession, depression, and recovery, which are sometimes viewed as the *phases* of a complete cycle, are due to factors other than the weather or social customs. Theoretical explanations of business cycles vary from such scientific theories as the *changes-in-income theory* or the *fluctuation-in-discount theory*, to a theory which attributes business cycles to spots on the surface of the sun.

**Irregular variations** of time series are those kind of fluctuations which are *either random or caused by such special events as elections, bank failures, wars, strikes, floods, and so forth.* If they are random (that is, due to chance), there is little to be said except that *in the long run* such fluctuations will tend to average out; if they are due to special events, they can often be *recognized and eliminated* before the other three time series components are investigated.

## LINEAR TRENDS

The most widely-used method for fitting trends is the *method of least squares*, which we studied in the preceding chapter. As we saw, it leads to the normal equations on page 284 and, hence, to the two formulas on page 286 for determining the values of $a$ and $b$ in the equation of the least-squares line $y = a + bx$. When we deal with time series, where the $x$'s are successive years (days, weeks, or months), these formulas can be simplified appreciably by *coding* the $x$'s so that *their sum is equal to zero*. As can easily be verified, the formulas for $a$ and $b$ will thus become

$$a = \frac{\Sigma y}{n} \quad \text{and} \quad b = \frac{\Sigma xy}{\Sigma x^2}$$

When dealing with *annual data* it is convenient to accomplish this by coding the $x$'s by assigning them the values ..., $-3, -2, -1, 0, 1, 2, 3, ...$, when $n$ is *odd*, or the values ..., $-5, -3, -1, 1, 3, 5, ...$, when $n$ is *even*, making sure that this will make $\Sigma x = 0$. Of course, the resulting equation of the trend line will express $y$ in terms of the coded $x$'s, and we will have to take this into account when using the equation to make a prediction. For instance, if we are given a store's sales for the years 1966, 1967, 1968, 1969, and 1970, and we code these years as $x = -2, -1, 0, 1$, and 2, we would have to substitute $x = 7$ to predict the store's sales for the year 1975. Clearly, 1975 is *seven* years after 1968, which corresponds to $x = 0$.

To illustrate how this simplification works, let us consider the following data on mobile home shipments in the United States:

| Year | Mobile Home Shipments (thousands of units) |
|------|--------------------------------------------|
| 1962 | 118 |
| 1963 | 151 |
| 1964 | 191 |
| 1965 | 216 |
| 1966 | 217 |
| 1967 | 240 |
| 1968 | 318 |
| 1969 | 413 |
| 1970 | 401 |

Since there are *nine* (an odd number of) years, we code them $-4, -3, -2, -1, 0, 1, 2, 3,$ and $4$, and the quantities needed for substitution into the formulas for $a$ and $b$ are obtained in the following table:

| Year | x | Actual data<br>y | $x^2$ | xy |
|------|----|-----|----|------|
| 1962 | −4 | 118 | 16 | −472 |
| 1963 | −3 | 151 | 9  | −453 |
| 1964 | −2 | 191 | 4  | −382 |
| 1965 | −1 | 216 | 1  | −216 |
| 1966 | 0  | 217 | 0  | 0 |
| 1967 | 1  | 240 | 1  | 240 |
| 1968 | 2  | 318 | 4  | 636 |
| 1969 | 3  | 413 | 9  | 1,239 |
| 1970 | 4  | 401 | 16 | 1,604 |
|      |    | 2,265 | 60 | 2,196 |

Then, substituting $n = 9$, $\Sigma y = 2{,}265$, $\Sigma x^2 = 60$, and $\Sigma xy = 2{,}196$ into the new formulas for $a$ and $b$, we get

$$a = \frac{2{,}265}{9} = 251.7$$

$$b = \frac{2{,}196}{60} = 36.6$$

and the equation of the least-squares trend line can be written as

$$y = 251.7 + 36.6x$$

To facilitate the use of equations like this, it is customary to supplement them with *legends* which explain the origin of $x$ (namely, the point of the time scale which corresponds to $x = 0$) and the units of both $x$ and $y$. For our example we might, thus, write

$$y = 251.7 + 36.6x$$

(*origin, 1966; x units, 1 year; y, annual shipments in thousands*)

where "origin, 1966" means that $x = 0$ corresponds to the middle of the year 1966.

In this equation, $a = 251.7$ is the **trend value** for 1966, as it is the value of $y$ which corresponds to $x = 0$, and $b = 36.6$ is the **annual trend increment,** namely, the *average annual increase* (in thousands of units) of mobile home shipments in the United States. To find other trend values, we simply substitute the corresponding values of $x$, and to find those for 1962 and 1970, for example, we substitute $x = -4$ and $x = 4$, getting $y = 251.7 + 36.6(-4) = 105.3$ and $y = 251.7 + 36.6(4) = 398.1$. These are the two values through which we drew the least-squares trend line, shown in Figure 11.5 together with the original data. As we already indicated on page 300, by substituting the appropriate values of $x$ we can also *project* a trend to the future, namely, to make predictions. To predict mobile home shipments for the year 1975, for example, we substitute $x = 9$ and get $y = 251.7 + 36.6(9) = 581.1$ (thousand units).

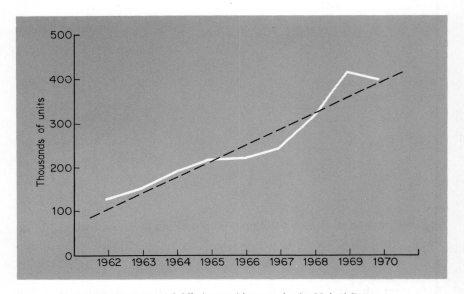

**FIGURE 11.5** Mobile home shipments in the United States.

When the number of years is *even* and we code the years from 1963 through 1970, for example, as $x = -7, -5, -3, -1, 1, 3, 5$, and 7, the work that is required to find the equation of a least-squares line is the same. It must be remembered, however, that now $a$ is the trend value halfway between the two middle years, and $b$ is a **semi-annual trend increment** (since $x$ increases by *two* from year to year). For instance, if we fitted a least-squares trend line to the data of Exercise 4 on page 310, which pertains to the number of one-family housing units started in the United States in the years 1963 through 1970, we would arrive at the equation $y = 887.9 - 14.2x$ (where $y$ is in thousands of units), and this means that the *average annual decrease* in new one-family housing units started is $2(14.2) = 28.4$ thousand units.

## MOVING AVERAGES

If we are interested mainly in the general "behavior" of a series, be it a trend, a business cycle, or possibly both, the general pattern of its movements can often be described quite adequately by what is called a **moving average.** *A moving average is an artificially constructed time series in which each value of a given series is replaced by the mean of itself and some of the values directly preceding it and directly following it.* For instance, in a *three-year moving average* calculated for annual data, each annual figure is replaced by the mean of itself and the figures corresponding to the immediately preceding year and the immediately succeeding year; similarly, in a *five-year moving average* calculated for annual data, each annual figure is replaced by the mean of itself and the figures corresponding to the two immediately preceding years and the two immediately succeeding years. If the averaging is done over an *even* number of years or months, say, 4 years or 12 months, the values of the moving average will fall *between* successive years or months, and it is customary to take care of this by calculating a subsequent two-year or two-month moving average. This will be illustrated on page 314, where we shall calculate a *four-quarter moving average* for quarterly data, which is subsequently "centered" by calculating a *two-quarter moving average.*

To illustrate the construction of a moving average, let us calculate a five-year moving average to *smooth* the time series of crude gypsum production in the United States from 1941 through 1970. In the table which follows, the second column contains the actual production figures and the third column contains the **five-year moving totals,** which for any given year consists of the *sum* of that year's production plus the corresponding figures for the two preceding years and the two succeeding years. The column on the right, which contains the moving average, is obtained by dividing each of the corresponding moving totals by five.

The practical effect of a moving average can readily be seen from Figure 11.6, in which we have plotted the five-year moving average together with the original data. The moving average has appreciably reduced the fluctuations in the series, and thus *smoothed* its over-all appearance. A minor disadvantage of a moving average is that we lose a few values at each end of the series, but this is generally of no consequence.

308  THE ANALYSIS OF TIME SERIES  CHAP. 11

| Year | Crude Gypsum Production (millions of short tons) | Five-Year Moving Totals | Five-Year Moving Average |
|---|---|---|---|
| 1941 | 4.7 | | |
| 1942 | 4.6 | | |
| 1943 | 3.9 | 20.8 | 4.2 |
| 1944 | 3.8 | 21.7 | 4.3 |
| 1945 | 3.8 | 23.3 | 4.7 |
| 1946 | 5.6 | 26.4 | 5.3 |
| 1947 | 6.2 | 29.1 | 5.8 |
| 1948 | 7.0 | 33.4 | 6.7 |
| 1949 | 6.5 | 36.5 | 7.3 |
| 1950 | 8.1 | 38.4 | 7.7 |
| 1951 | 8.7 | 39.6 | 7.9 |
| 1952 | 8.1 | 42.2 | 8.4 |
| 1953 | 8.2 | 44.7 | 8.9 |
| 1954 | 9.1 | 46.3 | 9.3 |
| 1955 | 10.6 | 47.4 | 9.5 |
| 1956 | 10.3 | 48.8 | 9.8 |
| 1957 | 9.2 | 50.6 | 10.1 |
| 1958 | 9.6 | 49.8 | 10.0 |
| 1959 | 10.9 | 49.0 | 9.8 |
| 1960 | 9.8 | 49.8 | 10.0 |
| 1961 | 9.5 | 50.5 | 10.1 |
| 1962 | 10.0 | 50.3 | 10.1 |
| 1963 | 10.3 | 50.5 | 10.1 |
| 1964 | 10.7 | 50.6 | 10.1 |
| 1965 | 10.0 | 50.0 | 10.0 |
| 1966 | 9.6 | 49.7 | 9.9 |
| 1967 | 9.4 | 48.9 | 9.8 |
| 1968 | 10.0 | 48.3 | 9.7 |
| 1969 | 9.9 | | |
| 1970 | 9.4 | | |

## EXERCISES

1. The number of sheep on farms in the United States was 26.7, 25.1, 23.5, 21.8, 21.5, 20.7, 19.1, 18.3, 17.4, 17.0, and 15.8 millions, respectively, in the years 1962 through 1972.
    (a) Find the equation of the least-squares trend line and explain the significance of the values obtained for $a$ and $b$.

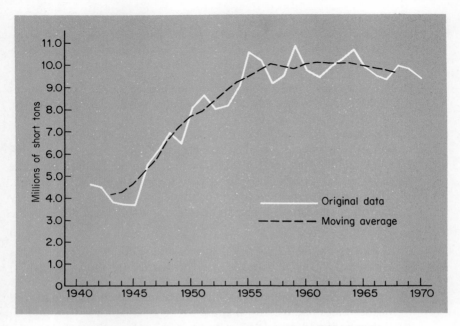

**FIGURE 11.6** Crude gypsum production in the United States.

(b) Calculate the 1962 and 1972 trend values, and use them to draw the least-squares trend line in a diagram showing also the original data.
(c) Assuming that the trend will continue, use the equation obtained in part (a) to predict how many sheep will be on U.S. farms in the year 1976.

2. The following are the U.S. government's appropriations for the fiscal years 1964 to 1971 (in billions of dollars): 102.0, 107.6, 126.0, 140.9, 195.9, 203.0, 222.2, and 247.6.
  (a) Fit a least-squares trend line to these data and explain the significance of the values obtained for $a$ and $b$.
  (b) Draw a diagram showing the trend line obtained in part (a) as well as the original data.
  (c) Based on the given data [that is, the trend line of part (a)], what would have been the prediction for the 1973 fiscal year appropriations of the U.S. government?

3. The following are the 1958 through 1971 values of the *Industrial Production Index* (with 1967 = 100) as published in the *Federal Reserve Bulletin*: 57.9, 64.8, 66.2, 66.7, 72.2, 76.5, 81.7, 89.2, 97.9, 100.0, 105.7, 110.7, 106.7, and 106.8.

(a) Fit a least-squares trend line to these data, and draw a diagram showing the trend line as well as the original data.
(b) Assuming that the trend will continue, use the equation of the least-squares line to predict the value of this index for the year 1977.

**4.** The number of new one-family housing units started in the United States was 1,021, 972, 964, 779, 844, 900, 810, and 813 thousands, respectively, in the years 1963 through 1970.
  (a) Verify that the equation of the least-squares line fit to these data is, in fact, the one given in the text on page 306.
  (b) Calculate the 1964 and 1969 trend values, and use them to draw the least-squares trend line in a diagram showing also the original data.
  (c) If we assume that the trend will continue, how many new one-family housing units can we expect to be started in the United States in the year 1976?

**5.** The following are the 1961 to 1971 figures on the number of immigrants admitted to the United States from all countries (in thousands): 271, 284, 306, 292, 297, 323, 362, 454, 359, 373, and 370.
  (a) Fit a least-squares trend line to these data and explain the significance of the values obtained for $a$ and $b$.
  (b) Calculate the 1961 and 1971 trend values, and use them to draw the least-squares trend line in a diagram showing also the original data.
  (c) Assuming that the trend will continue, how many immigrants can we expect to be admitted to the United States in the year 1978?

**6.** Construct a three-year moving average for the following series of data representing the production of bituminous coal and lignite in the United States (in millions of short tons) for the years 1940 through 1970: 460, 514, 583, 590, 620, 578, 534, 631, 600, 438, 516, 534, 467, 457, 392, 465, 501, 493, 410, 412, 416, 403, 422, 459, 487, 512, 534, 553, 545, 561, and 603. Also draw a diagram like that of Figure 11.6, showing the moving average as well as the original data.

**7.** Calculate a five-year moving average for the data of Exercise 6, and draw a diagram showing the moving average as well as the original data.

**8.** The following figures represent the number of employees (in thousands) of the contract construction industry in the United States for the years 1940 through 1971: 1,294, 1,790, 2,170, 1,567, 1,094, 1,132, 1,661, 1,982, 2,169, 2,165, 2,333, 2,603, 2,634, 2,623, 2,612, 2,802, 2,999, 2,923, 2,778, 2,960, 2,885, 2,816, 2,902, 2,963, 3,050, 3,186, 3,275, 3,203, 3,259, 3,411, 3,345, and 3,259. Construct a three-year moving average and draw a diagram like that of Figure 11.6, which shows the moving average as well as the original data.

9. Calculate a five-year moving average for the data of Exercise 8, and draw a diagram showing the moving average as well as original data.

10. **MODIFYING TREND EQUATIONS** Sometimes it is necessary to modify trend equations by *changing the origin of x*, by *changing the units of x*, or by *changing the units of y*. To illustrate how this is done, let us refer to the mobile-home-shipments example on page 305, and begin by changing the *y* units from *annual shipments* to *average monthly shipments*. Since each *y* will, thus, have to be divided by 12, the values of *a* and *b* in the equation of the least-squares line must be divided by 12, and we get

$$y = 20.98 + 3.05x$$

(*origin, 1966; x units, 1 year; y,
average monthly shipments in thousands*)

where the legend was added in accordance with what we said on page 305.

To continue, let us change the *x*'s so that they refer to successive *months*. Since *b* measures the *trend increment* (that is, the increase or decrease in *y*) corresponding to one unit of *x*, it will have to be changed from an *annual trend increment* to a *monthly trend increment*, and this is accomplished by dividing its value by 12. Thus, we get

$$y = 20.98 + 0.25x$$

(*origin, 1966; x units, 1 month; y,
average monthly shipments in thousands*)

Finally, let us change the *origin* of *x* from the middle of 1966, where it is now, to, say, the middle of January 1966. Since the middle of January 1966 is $5\frac{1}{2}$ months earlier than the origin in the above equation (the middle of 1966), we can perform the desired change of origin by subtracting 5.5 monthly trend increments from 20.98. The new value of *a* is, thus, $20.98 - 5.5(0.25) = 19.60$, and we get

$$y = 19.60 + 0.25x$$

(*origin, January 1966; x units, 1 month; y,
average monthly shipments in thousands*)

(a) Modify the trend equation given for new one-family housing units on page 307 by first changing the *y* units to average monthly data,

then changing the $x$ units to 1 month, and finally changing the origin to (the middle of) January, 1967.
(b) Modify the least-squares trend line obtained in Exercise 1 by changing the origin to 1970.
(c) Modify the least-squares trend line obtained in Exercise 2 by changing the figures to average monthly appropriations, the $x$ units to 1 month, and the origin to January, 1970.
(d) Modify the least-squares trend line obtained in Exercise 5 by first changing the figures to average monthly data, then changing the $x$ units to 1 month, and finally changing the origin to December, 1965.

## SEASONAL VARIATION

Let us now consider the problem of measuring, or describing, seasonal variations, namely, the problem of constructing an **index of seasonal variation.** For monthly data, such an index consists of 12 numbers, one for each month, which are expressed as *percentages of their average.* For instance, if the March seasonal index of a manufacturer's sales is 89, this means that his typical March sales are 89 percent of those of the average month; similarly, if the September seasonal index of a company's production is 114, this means that their typical September production is 114 percent of that of the average month. Correspondingly, for daily data a seasonal index might consist of seven numbers, one for each day of the week, and for quarterly data a seasonal index would consist of four numbers, one for each quarter of the year.

As might be expected, there are many different ways in which an index of seasonal variation can be constructed. The one we shall give here is called the **ratio-to-moving-average method.** Although high-speed computers make the use of more sophisticated methods feasible, the ratio-to-moving-average method is widely used and generally produces quite satisfactory results. In this method we first calculate a moving average (for instance, a 12-month moving average for monthly data) and then we *divide each value of the original series by the corresponding value of the moving average, and multiply by 100.* Note that the resulting percentages are free of any trend or long-range cycles, and the only thing that remains to be done is to eliminate irregular variations (the fourth component mentioned on page 300) by averaging the percentages obtained for each month (day, or quarter).

To illustrate this technique, let us refer to the data given in Column 1 of the table on page 314 and in Figure 11.7, which pertain to the quarterly shipment

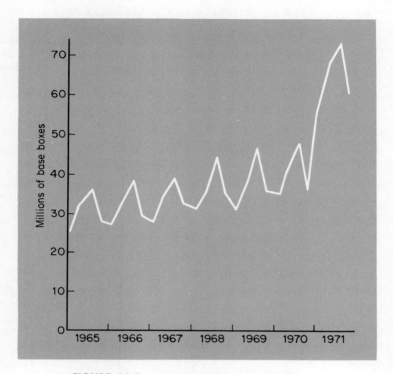

FIGURE 11.7   Quarterly shipments of metal cans.

of metal cans in the United States during the years 1965 through 1971.*
Columns 2, 3, and 4 show the calculations needed to find a **centered** 4-quarter
moving average. First we determine the 4-quarter moving totals shown in
Column 2: 121.1, the first value, is the sum of 25.3, 31.2, 36.6, and 28.0;
122.9, the second value, is the sum of 31.2, 36.6, 28.0, and 27.1; and so forth.
Since these totals are centered *between* the shipment data of Column 1, we
calculate 2-quarter moving totals of the figures in Column 2, getting 121.1 +
122.9 = 244.0, 122.9 + 125.0 = 247.9, . . ., as shown in Column 3. Since
each of the totals in Column 3 is actually the sum of *eight* of the original
figures, we divide each of the entries of Column 3 by eight, and thus get the
moving average shown in Column 4.

* The figures are in millions of *base boxes*, where a base box is an area of 31,360 square inches of metal equivalent to 112 sheets (14 by 20 inches) of metal.

## QUARTERLY SHIPMENTS OF METAL CANS

|  |  | Shipments (1) | 4-Quarter Moving Total (2) | 2-Quarter Moving Total (3) | Centered 4-Quarter Moving Average (4) | Percentages of 4-Quarter Moving Average (5) |
|---|---|---|---|---|---|---|
| 1965 | Quarter 1 | 25.3 | | | | |
| | Quarter 2 | 31.2 | 121.1 | | | |
| | Quarter 3 | 36.6 | 122.9 | 244.0 | 30.5 | 120.0 |
| | Quarter 4 | 28.0 | 125.0 | 247.9 | 31.0 | 90.3 |
| 1966 | Quarter 1 | 27.1 | 127.6 | 252.6 | 31.6 | 85.8 |
| | Quarter 2 | 33.3 | 129.2 | 256.8 | 32.1 | 103.7 |
| | Quarter 3 | 39.2 | 130.0 | 259.2 | 32.4 | 121.0 |
| | Quarter 4 | 29.6 | 131.5 | 261.5 | 32.7 | 90.5 |
| 1967 | Quarter 1 | 27.9 | 131.4 | 262.9 | 32.9 | 84.8 |
| | Quarter 2 | 34.8 | 134.0 | 265.4 | 33.2 | 104.8 |
| | Quarter 3 | 39.1 | 137.1 | 271.1 | 33.9 | 115.3 |
| | Quarter 4 | 32.2 | 137.1 | 274.2 | 34.3 | 93.9 |
| 1968 | Quarter 1 | 31.0 | 143.0 | 280.1 | 35.0 | 88.6 |
| | Quarter 2 | 34.8 | 145.8 | 288.8 | 36.1 | 96.4 |
| | Quarter 3 | 45.0 | 146.1 | 291.9 | 36.5 | 123.3 |
| | Quarter 4 | 35.0 | 150.1 | 296.2 | 37.0 | 94.6 |
| 1969 | Quarter 1 | 31.3 | 152.1 | 302.2 | 37.8 | 82.8 |
| | Quarter 2 | 38.8 | 152.6 | 304.7 | 38.1 | 101.8 |
| | Quarter 3 | 47.0 | 155.8 | 308.4 | 38.6 | 121.8 |
| | Quarter 4 | 35.5 | 157.5 | 313.3 | 39.2 | 90.6 |
| 1970 | Quarter 1 | 34.5 | 159.1 | 316.6 | 39.6 | 87.1 |
| | Quarter 2 | 40.5 | 159.4 | 318.5 | 39.8 | 101.8 |
| | Quarter 3 | 48.6 | 180.7 | 340.1 | 42.5 | 114.4 |
| | Quarter 4 | 35.8 | 208.1 | 388.8 | 48.6 | 73.7 |
| 1971 | Quarter 1 | 55.8 | 232.8 | 440.9 | 55.1 | 101.3 |
| | Quarter 2 | 67.9 | 257.2 | 490.0 | 61.2 | 110.9 |
| | Quarter 3 | 73.3 | | | | |
| | Quarter 4 | 60.2 | | | | |

Finally, dividing each of the original figures in Column 1 by the corresponding entry in Column 4 and multiplying by 100, we obtain the **percentages of moving average** shown in Column 5. *This step is designed to eliminate the* *trend and cyclical components from the data, thus leaving seasonal and irregular variations as the only kinds of variation.*

All that remains to be done is to eliminate, so far as this is possible, the irregular variations by averaging in some way the six values obtained for each quarter, and to facilitate this we rearrange the entries of Column 5 as in the following table:

|           | 1965  | 1966  | 1967  | 1968  | 1969  | 1970  | 1971  | Median | Seasonal Index |
|-----------|-------|-------|-------|-------|-------|-------|-------|--------|----------------|
| Quarter 1 |       | 85.8  | 84.8  | 88.6  | 82.8  | 87.1  | 101.3 | 86.4   | 86.3           |
| Quarter 2 |       | 103.7 | 104.8 | 96.4  | 101.8 | 101.8 | 110.9 | 102.8  | 102.7          |
| Quarter 3 | 120.0 | 121.0 | 115.3 | 123.3 | 121.8 | 114.4 |       | 120.5  | 120.4          |
| Quarter 4 | 90.3  | 90.5  | 93.9  | 94.6  | 90.6  | 73.7  |       | 90.6   | 90.5           |

$$400.3$$

The measure we have chosen to average the six values for each quarter is the *median*, although we could have used the mean or some other measure of central location. These medians are shown in the second column from the right. Since the values of a seasonal index are supposed to be expressed as *percentages of their average*, they should total 400 in this case (or 1,200 for a seasonal index for monthly data), and to correct for this we multiply each of the medians by $\frac{400}{400.3} = 0.99925$. The effect of this modification is very small, as can be seen from the values of the seasonal index shown in the column on the right. (Actually, the four values of a quarterly index should add up to 400; that they do not add up to 400 in this example is due to rounding.) We interpret this seasonal index to mean that the typical Quarter 1 shipments of metal cans are 86.3 percent of those of the average quarter, typical Quarter 2 shipments are 102.7 percent of those of the average quarter, and so on.

Seasonal indexes have many important applications: in *forecasting*, for instance, and in **deseasonalizing data.** To illustrate their use in forecasting, let us use the original data on page 314 and the seasonal index which we have calculated, to predict the quarterly shipments of metal cans for the year 1974. Any such projection consists of *multiplying the predicted trend values by the corresponding values of the seasonal index (written as a proportion)*, and we shall thus make use of the fact that the equation of the least-squares line fit to the total annual shipments in the years 1965 through 1971 (and modified by the method of Exercise 10 on page 311) can be written as

$$y = 37.6 + 1.1x$$

(*origin, Quarter 1, 1968; x units,
1 quarter; y, average quarterly
shipments in millions of base boxes*)

Note that we added the legend explaining the units in accordance with what we said in Exercise 10 on page 311. Since the first quarter of 1974 is 24 quarters after the first quarter of 1968, we substitute $x = 24, 25, 26$, and 27, respectively, into the above equation to obtain the trend values shown in the following table:

|  | 1974 Trend Values | Seasonal Index | Estimated Quarterly Shipments for 1974 |
|---|---|---|---|
| Quarter 1 | 64.0 | 0.863 | 55.2 |
| Quarter 2 | 65.1 | 1.027 | 66.9 |
| Quarter 3 | 66.2 | 1.204 | 79.7 |
| Quarter 4 | 67.3 | 0.905 | 60.9 |

Then, multiplying each of these trend values by the corresponding seasonal index (written as a proportion) and accounting, thus, for trend and seasonal variation, we forecast that in 1974 the first quarter shipments of metal cans will be 55.2 million base boxes, and that the other three quarters' shipments will be, respectively, 66.9, 79.7, and 60.9 million base boxes.

The purpose of a forecast like the one made above is to plan the operations of an industry or a firm so as to arrange its daily activity in the best way possible to meet customers' needs. Of course, predictions based on trend equations and estimated seasonal patterns are not the only factors which must be considered in intelligent planning. The business executive must also consider nationwide, regional, and local forecasts concerning over-all economic activity, employment, income, consumer spending and credit, retail sales, and so on. Any or all of these factors may, and probably will, make it necessary to modify the above forecasts based only on the industry's shipment data for the period 1965–1971.

In the financial pages of newspapers and magazines, we often find that data are "seasonally adjusted," namely, that seasonal variations have been *removed*, so that we are really told what things would have been if there had been no seasonal variations. The process of **deseasonalizing** data, as this process of removing seasonal variations is usually called, is really quite easy. We simply *divide each value of a series by the corresponding value of its seasonal index (written as a proportion)*, and to illustrate this procedure, let us refer again to the shipments-of-metal-cans example, and deseasonalize the figures given for the four quarters of the year 1970. In the table which follows, the actual shipment figures are copied from page 314, the seasonal index is copied from page 315, and the deseasonalized shipment figures in the column on the right are obtained by dividing each value in the first column by the corresponding value of the index:

|  | Actual 1970 Shipments | Seasonal Index | Deseasonalized 1970 Shipments |
|---|---|---|---|
| Quarter 1 | 34.5 | 0.863 | 40.0 |
| Quarter 2 | 40.5 | 1.027 | 39.4 |
| Quarter 3 | 48.6 | 1.204 | 40.4 |
| Quarter 4 | 35.8 | 0.905 | 39.6 |

To understand the significance of such deseasonalized data, let us compare the above figures for Quarters 1 and 2. Comparing the actual data, we find that $\frac{40.5}{34.5} = 1.17$, so that there was an *increase* of 17 percent from Quarter 1 to Quarter 2. However, as can be seen from the index, Quarter 1 is traditionally low, Quarter 2 is traditionally higher than Quarter 1, and some kind of increase should really have been expected. Comparing, similarly, the deseasonalized data, we find that $\frac{39.4}{40.0} = 0.985$, which signifies a *decrease* of 1.5 percent. This shows that the actual increase in shipments from Quarter 1 to Quarter 2, even though it is large, is not as large as we could have expected in accordance with the typical seasonal pattern. Hence, instead of celebrating the increase, it would be more appropriate to investigate why it was less than expected.

## EXERCISES

1. The following data are the 1962–1970 new plant and equipment expenditures (in billions of dollars) for all industries:

|  | 1962 | 1963 | 1964 | 1965 | 1966 | 1967 | 1968 | 1969 | 1970 |
|---|---|---|---|---|---|---|---|---|---|
| Quarter 1 | 8.3 | 8.4 | 10.0 | 11.2 | 13.3 | 14.5 | 15.1 | 16.0 | 17.5 |
| Quarter 2 | 9.8 | 10.1 | 11.8 | 13.6 | 16.0 | 16.7 | 16.8 | 18.8 | 20.3 |
| Quarter 3 | 9.7 | 10.4 | 11.7 | 13.6 | 15.9 | 16.2 | 16.8 | 19.2 | 20.3 |
| Quarter 4 | 10.6 | 11.8 | 13.4 | 15.9 | 18.2 | 18.1 | 19.0 | 21.5 | 21.7 |

(a) Draw a diagram like that of Figure 11.7 to verify that there is, indeed, a consistent seasonal pattern.
(b) Using the median to average the percentages of moving average obtained for the different quarters, use the ratio-to-moving-average method to compute a seasonal index for these data.

(c) Use the index obtained in part (b) to deseasonalize the original 1970 data and discuss the increase from the Quarter 3 to Quarter 4 of that year.

**2.** The following is the equation of the least-squares line fit to the annual totals for the data of Exercise 1 (and modified by the method of Exercise 10 on page 311):

$$y = 14.27 + 0.34x$$

*(origin, Quarter 1, 1966; x units, 1 quarter;
y, average quarterly expenditures in billions
of dollars)*

Use this equation and the seasonal index of part (b) of Exercise 1 to predict the corresponding new plant and equipment expenditures for the four quarters of 1975.

**3.** The following data show the amount of private new construction (in billions of dollars) put in place in the United States during the years 1965 through 1970:

|           | 1965 | 1966 | 1967 | 1968 | 1969 | 1970 |
|-----------|------|------|------|------|------|------|
| January   | 3.3  | 3.7  | 5.0  | 5.6  | 6.2  | 6.1  |
| February  | 3.1  | 3.4  | 4.6  | 5.2  | 5.9  | 5.9  |
| March     | 3.4  | 3.8  | 5.2  | 6.0  | 6.5  | 6.5  |
| April     | 3.8  | 4.2  | 5.8  | 6.8  | 7.3  | 7.1  |
| May       | 4.2  | 4.3  | 6.3  | 7.3  | 8.0  | 7.7  |
| June      | 4.5  | 4.7  | 6.8  | 7.5  | 8.4  | 8.2  |
| July      | 4.6  | 4.8  | 7.1  | 7.7  | 8.5  | 8.3  |
| August    | 4.7  | 4.8  | 7.3  | 8.0  | 8.5  | 8.6  |
| September | 4.7  | 4.8  | 7.4  | 8.1  | 8.6  | 8.5  |
| October   | 4.7  | 4.5  | 7.2  | 7.9  | 8.2  | 8.4  |
| November  | 4.6  | 4.2  | 7.0  | 7.8  | 7.9  | 8.2  |
| December  | 4.4  | 3.9  | 6.4  | 6.8  | 7.0  | 7.7  |

(a) Draw a diagram like that of Figure 11.7, which shows the seasonal pattern in these data.
(b) Using the *mean* to average the percentages of moving average obtained for the different months, use the ratio-to-moving-average method to compute a seasonal index for these data. (*Hint:* To obtain a 12-month moving average calculate first 12-month moving totals, then 2-month moving totals, and then divide by 24.)
(c) Use the index obtained in part (b) to deseasonalize the original 1969 and 1970 data, and discuss the decrease from December 1969 to January 1970.

CHAP. 11  SEASONAL VARIATION  319

4. A motel's revenue was $24,000 in June and $36,000 in July of 1970. If the motel's seasonal index for these two months stands, respectively, at 80 and 120, is the manager right in feeling pleased with the $12,000 increase? Explain your answer.

5. The seasonal index for a bus company's monthly revenue is 69, 89, 100, 98, 109, 117, 131, 141, 111, 91, 71, and 73 for the 12 months of the year. If the company expects its total 1976 revenue to be $1,800,000, prepare a schedule of the company's anticipated monthly revenues (ignoring the possible existence of a trend).

6. Suppose that a finance company determines its trend in monthly loans by means of the equation

$$y = 1,000,000 + 10,000x$$

where the origin, $x = 0$, corresponds to January 1973, the $x$ unit is 1 month, and the $y$'s are the total monthly loans in dollars.
   (a) Use this equation to estimate the trend value for the month of December, 1975.
   (b) If the December value of the seasonal index for the company's total loans is 125, modify the trend value obtained in part (a) to forecast the company's monthly loans for December, 1975.

7. An economist computes the trend in an index of retail food prices in a certain city in the Southwest (with 1967 = 100) by means of the formula

$$y = 110.37 + 0.34x$$

where the origin, $x = 0$, corresponds to January 1970, the $x$ unit is 1 month, and $y$ is the value of the index.
   (a) Verify that the index *is* 100.00 at the middle of the year 1967.
   (b) If the seasonal index for January is 108, predict the value of this retail food price index for January, 1976.

8. **ANNUAL RATES** Deseasonalized quarterly data are often multiplied by *four*, or deseasonalized monthly data by 12, and then referred to as *annual rates*. The advantage of this is that we can thus indicate quarter-to-quarter or month-to-month changes in data which are ordinarily reported on an annual basis, or which are more meaningful on an annual basis. For instance, if we multiply the third of the deseasonalized figures which we obtained on page 317, we could say that in the third quarter of 1970 metal cans were being shipped at an annual rate of 4(40.4) = 161.6 million base boxes.

(a) Referring to Exercise 1, change the deseasonalized data obtained for the four quarters of 1970 into annual rates.
(b) Referring to Exercise 3, change the deseasonalized figure obtained for December 1969 into an annual rate.

## SOME FURTHER CONSIDERATIONS

If we look upon the values of a time series as *products* of the factors $T$, $S$, $C$, and $I$, attributed, respectively, to trend, seasonal variation, cyclical variation, and irregular variation, we have so far learned how to determine $T$ and $S$. ($T$ is a trend value obtained by means of an equation like the one on page 305, and $S$ is a value of a seasonal index.) The product of $T$ and $S$ is sometimes called the **normal** and it represents, hypothetically, what the time series would have been like if trend and seasonal variation had been the only sources of variation present. The practical value of calculating the normal of a time series is that if we *divide* each value of the series by the corresponding value of the normal, we get

$$\frac{T \cdot S \cdot C \cdot I}{T \cdot S} = C \cdot I$$

namely, another hypothetical time series in which only cyclical and irregular variations are at play. The product $C \cdot I$ stands for **cyclical irregulars,** and if we eliminate the irregular variations, perhaps, by means of a moving average, we arrive at the $C$'s which represent whatever cyclical forces there are in the given series. The required calculations are straightforward, though tedious, and we shall not illustrate them in this book.

# BIBLIOGRAPHY

## A. STATISTICS FOR THE LAYMAN

Huff, D., *How to Lie with Statistics.* New York: W. W. Norton, 1954.
Huff, D., and Geis, I., *How to Take a Chance.* New York: W. W. Norton, 1959.
Levinson, H. C., *Chance, Luck, and Statistics.* New York: Dover Publications, Inc., 1963.
Moroney, M. J., *Facts from Figures.* London: Pelican Books, 1956.
Reichman, W. J., *Use and Abuse of Statistics.* New York: Oxford University Press, 1962.
Sielaff, T. J. (ed.), *Statistics in Action.* San Jose, Calif.: Lansford Press, 1963.
Tanur, J. M. (ed.), *Statistics: A Guide to the Unknown.* San Francisco: Holden-Day, Inc., 1972.

## B. SOME BOOKS ON THE THEORY OF PROBABILITY AND STATISTICS

Freund, J. E., *Introduction to Probability.* Encino, Calif.: Dickenson Publishing Co., 1973.
Freund, J. E., *Mathematical Statistics, 2nd ed.* Englewood Cliffs, N.J.: Prentice-Hall, Inc., 1971.
Goldberg, S., *Probability—An Introduction.* Englewood Cliffs, N.J.: Prentice-Hall, Inc., 1960.
Hodges, J. L., and Lehmann, E. L., *Elements of Finite Probability.* San Francisco: Holden-Day, Inc., 1965.
Hoel, P., *Introduction to Mathematical Statistics, 4th ed.* New York: John Wiley & Sons, Inc., 1971.
Mosteller, F., Rourke, R. E. K., and Thomas, G. B., *Probability with Statistical Applications, 2nd ed.* Reading, Mass.: Addison-Wesley, 1970.

## C. SOME GENERAL BOOKS ON BUSINESS STATISTICS

Freund, J. E., and Williams, F. J., *Elementary Business Statistics, 2nd ed.* Englewood Cliffs, N.J.: Prentice-Hall, Inc., 1972.
Hamburg, M., *Statistical Analysis for Decision Making.* New York: Harcourt Brace Jovanovich, Inc., 1970.

Mendenhall, W., and Reinmuth, J. E., *Statistics for Management and Economics.* Belmont, Calif.: Wadsworth Publishing Co., 1971.

Richmond, S. B., *Statistical Analysis*, 2nd ed. New York: Ronald Press, 1964.

Stockton, J. R., *Introduction to Business and Economic Statistics*, 4th ed. Cincinnati: South-Western Publishing Co., 1971.

## D. SOME BOOKS DEALING WITH SPECIAL TOPICS

Cochran, W. G., *Sampling Techniques*, 2nd ed. New York: John Wiley & Sons, Inc., 1963.

Deming, W. E., *Sample Design in Business Research.* New York: John Wiley & Sons, Inc., 1960.

Ferber, R., and Verdoorn, P. J., *Research Methods in Economics and Business.* New York: Macmillan, 1962.

Hicks, C. R., *Fundamental Concepts in the Design of Experiments.* New York: Holt, Rinehart and Winston, 1964.

Mudgett, B. D., *Index Numbers.* New York: John Wiley & Sons, Inc., 1951.

## E. SOME GENERAL REFERENCE WORKS AND TABLES

Buckland, W. R., and Fox, R., *Bibliography of Basic Texts on Statistical Methods.* New York: Hafner, 1963.

Coman, E. T., Jr., *Sources of Business Information*, 2nd ed. Berkeley, Calif.: University of California Press, 1964.

Freund, J. E., and Williams, F. J., *Dictionary/Outline of Basic Statistics.* New York: McGraw-Hill, 1966.

Hauser, F. M., and Leonard, W. R., *Government Statistics for Business Use*, 2nd ed. New York: John Wiley & Sons, Inc., 1956.

Kendall, M. G., and Buckland, W. R., *A Dictionary of Statistical Terms*, 3rd ed. New York: Hafner, 1971.

Owen, D. B., *Handbook of Statistical Tables.* Reading, Mass.: Addison-Wesley, 1962.

RAND Corporation, *A Million Random Digits with 100,000 Normal Deviates.* New York: Free Press, 1955.

Wasserman, P. et al. (eds.), *Statistical Sources*, 3rd ed. Detroit: Gale Research Co., 1965.

# STATISTICAL TABLES

  I. The Standard Normal Distribution
 II. The $t$ Distribution
III. The $\chi^2$ Distribution
 IV. The $F$ Distribution
  V. Confidence Intervals for Proportions
 VI. Critical Values of $r$
VII. Factorials
VIII. Binomial Coefficients
 IX. Random Numbers
  X. Square Roots
 XI. Binomial Probabilities

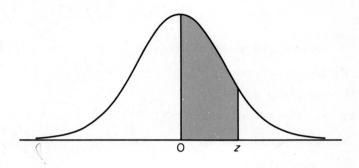

The entries in Table I are the probabilities that a random variable having the standard normal distribution takes on a value between 0 and $z$; they are given by the area under the curve shaded in the above diagram.

## TABLE I
## THE STANDARD NORMAL DISTRIBUTION

| z | .00 | .01 | .02 | .03 | .04 | .05 | .06 | .07 | .08 | .09 |
|---|---|---|---|---|---|---|---|---|---|---|
| 0.0 | .0000 | .0040 | .0080 | .0120 | .0160 | .0199 | .0239 | .0279 | .0319 | .0359 |
| 0.1 | .0398 | .0438 | .0478 | .0517 | .0557 | .0596 | .0636 | .0675 | .0714 | .0753 |
| 0.2 | .0793 | .0832 | .0871 | .0910 | .0948 | .0987 | .1026 | .1064 | .1103 | .1141 |
| 0.3 | .1179 | .1217 | .1255 | .1293 | .1331 | .1368 | .1406 | .1443 | .1480 | .1517 |
| 0.4 | .1554 | .1591 | .1628 | .1664 | .1700 | .1736 | .1772 | .1808 | .1844 | .1879 |
| 0.5 | .1915 | .1950 | .1985 | .2019 | .2054 | .2088 | .2123 | .2157 | .2190 | .2224 |
| 0.6 | .2257 | .2291 | .2324 | .2357 | .2389 | .2422 | .2454 | .2486 | .2517 | .2549 |
| 0.7 | .2580 | .2611 | .2642 | .2673 | .2704 | .2734 | .2764 | .2794 | .2823 | .2852 |
| 0.8 | .2881 | .2910 | .2939 | .2967 | .2995 | .3023 | .3051 | .3078 | .3106 | .3133 |
| 0.9 | .3159 | .3186 | .3212 | .3238 | .3264 | .3289 | .3315 | .3340 | .3365 | .3389 |
| 1.0 | .3413 | .3438 | .3461 | .3485 | .3508 | .3531 | .3554 | .3577 | .3599 | .3621 |
| 1.1 | .3643 | .3665 | .3686 | .3708 | .3729 | .3749 | .3770 | .3790 | .3810 | .3830 |
| 1.2 | .3849 | .3869 | .3888 | .3907 | .3925 | .3944 | .3962 | .3980 | .3997 | .4015 |
| 1.3 | .4032 | .4049 | .4066 | .4082 | .4099 | .4115 | .4131 | .4147 | .4162 | .4177 |
| 1.4 | .4192 | .4207 | .4222 | .4236 | .4251 | .4265 | .4279 | .4292 | .4306 | .4319 |
| 1.5 | .4332 | .4345 | .4357 | .4370 | .4382 | .4394 | .4406 | .4418 | .4429 | .4441 |
| 1.6 | .4452 | .4463 | .4474 | .4484 | .4495 | .4505 | .4515 | .4525 | .4535 | .4545 |
| 1.7 | .4554 | .4564 | .4573 | .4582 | .4591 | .4599 | .4608 | .4616 | .4625 | .4633 |
| 1.8 | .4641 | .4649 | .4656 | .4664 | .4671 | .4678 | .4686 | .4693 | .4699 | .4706 |
| 1.9 | .4713 | .4719 | .4726 | .4732 | .4738 | .4744 | .4750 | .4756 | .4761 | .4767 |
| 2.0 | .4772 | .4778 | .4783 | .4788 | .4793 | .4798 | .4803 | .4808 | .4812 | .4817 |
| 2.1 | .4821 | .4826 | .4830 | .4834 | .4838 | .4842 | .4846 | .4850 | .4854 | .4857 |
| 2.2 | .4861 | .4864 | .4868 | .4871 | .4875 | .4878 | .4881 | .4884 | .4887 | .4890 |
| 2.3 | .4893 | .4896 | .4898 | .4901 | .4904 | .4906 | .4909 | .4911 | .4913 | .4916 |
| 2.4 | .4918 | .4920 | .4922 | .4925 | .4927 | .4929 | .4931 | .4932 | .4934 | .4936 |
| 2.5 | .4938 | .4940 | .4941 | .4943 | .4945 | .4946 | .4948 | .4949 | .4951 | .4952 |
| 2.6 | .4953 | .4955 | .4956 | .4957 | .4959 | .4960 | .4961 | .4962 | .4963 | .4946 |
| 2.7 | .4965 | .4966 | .4967 | .4968 | .4969 | .4970 | .4971 | .4972 | .4973 | .4974 |
| 2.8 | .4974 | .4975 | .4976 | .4977 | .4977 | .4978 | .4979 | .4979 | .4980 | .4981 |
| 2.9 | .4981 | .4982 | .4982 | .4983 | .4984 | .4984 | .4985 | .4985 | .4986 | .4986 |
| 3.0 | .4987 | .4987 | .4987 | .4988 | .4988 | .4989 | .4989 | .4989 | .4990 | .4990 |

## TABLE II

## THE $t$ DISTRIBUTION*

| d.f. | $t_{.050}$ | $t_{.025}$ | $t_{.010}$ | $t_{.005}$ | d.f. |
|---|---|---|---|---|---|
| 1 | 6.314 | 12.706 | 31.821 | 63.657 | 1 |
| 2 | 2.920 | 4.303 | 6.965 | 9.925 | 2 |
| 3 | 2.353 | 3.182 | 4.541 | 5.841 | 3 |
| 4 | 2.132 | 2.776 | 3.747 | 4.604 | 4 |
| 5 | 2.015 | 2.571 | 3.365 | 4.032 | 5 |
| 6 | 1.943 | 2.447 | 3.143 | 3.707 | 6 |
| 7 | 1.895 | 2.365 | 2.998 | 3.499 | 7 |
| 8 | 1.860 | 2.306 | 2.896 | 3.355 | 8 |
| 9 | 1.833 | 2.262 | 2.821 | 3.250 | 9 |
| 10 | 1.812 | 2.228 | 2.764 | 3.169 | 10 |
| 11 | 1.796 | 2.201 | 2.718 | 3.106 | 11 |
| 12 | 1.782 | 2.179 | 2.681 | 3.055 | 12 |
| 13 | 1.771 | 2.160 | 2.650 | 3.012 | 13 |
| 14 | 1.761 | 2.145 | 2.624 | 2.977 | 14 |
| 15 | 1.753 | 2.131 | 2.602 | 2.947 | 15 |
| 16 | 1.746 | 2.120 | 2.583 | 2.921 | 16 |
| 17 | 1.740 | 2.110 | 2.567 | 2.898 | 17 |
| 18 | 1.734 | 2.101 | 2.552 | 2.878 | 18 |
| 19 | 1.729 | 2.093 | 2.539 | 2.861 | 19 |
| 20 | 1.725 | 2.086 | 2.528 | 2.845 | 20 |
| 21 | 1.721 | 2.080 | 2.518 | 2.831 | 21 |
| 22 | 1.717 | 2.074 | 2.508 | 2.819 | 22 |
| 23 | 1.714 | 2.069 | 2.500 | 2.807 | 23 |
| 24 | 1.711 | 2.064 | 2.492 | 2.797 | 24 |
| 25 | 1.708 | 2.060 | 2.485 | 2.787 | 25 |
| 26 | 1.706 | 2.056 | 2.479 | 2.779 | 26 |
| 27 | 1.703 | 2.052 | 2.473 | 2.771 | 27 |
| 28 | 1.701 | 2.048 | 2.467 | 2.763 | 28 |
| 29 | 1.699 | 2.045 | 2.462 | 2.756 | 29 |
| inf. | 1.645 | 1.960 | 2.326 | 2.576 | inf. |

* This table is abridged from Table IV of R. A. Fisher, *Statistical Methods for Research Workers*, published by Oliver and Boyd, Ltd., Edinburgh, by permission of the author's literary executor and publishers.

## TABLE III

## THE $\chi^2$ DISTRIBUTION*

| d.f. | $\chi^2_{.05}$ | $\chi^2_{.01}$ | d.f. |
|---|---|---|---|
| 1 | 3.841 | 6.635 | 1 |
| 2 | 5.991 | 9.210 | 2 |
| 3 | 7.815 | 11.345 | 3 |
| 4 | 9.488 | 13.277 | 4 |
| 5 | 11.070 | 15.086 | 5 |
| 6 | 12.592 | 16.812 | 6 |
| 7 | 14.067 | 18.475 | 7 |
| 8 | 15.507 | 20.090 | 8 |
| 9 | 16.919 | 21.666 | 9 |
| 10 | 18.307 | 23.209 | 10 |
| 11 | 19.675 | 24.725 | 11 |
| 12 | 21.026 | 26.217 | 12 |
| 13 | 22.362 | 27.688 | 13 |
| 14 | 23.685 | 29.141 | 14 |
| 15 | 24.996 | 30.578 | 15 |
| 16 | 26.296 | 32.000 | 16 |
| 17 | 27.587 | 33.409 | 17 |
| 18 | 28.869 | 34.805 | 18 |
| 19 | 30.144 | 36.191 | 19 |
| 20 | 31.410 | 37.566 | 20 |
| 21 | 32.671 | 38.932 | 21 |
| 22 | 33.924 | 40.289 | 22 |
| 23 | 35.172 | 41.638 | 23 |
| 24 | 36.415 | 42.980 | 24 |
| 25 | 37.652 | 44.314 | 25 |
| 26 | 38.885 | 45.642 | 26 |
| 27 | 40.113 | 46.963 | 27 |
| 28 | 41.337 | 48.278 | 28 |
| 29 | 42.557 | 49.588 | 29 |
| 30 | 43.773 | 50.892 | 30 |

* This table is based on Table 8 of *Biometrika Tables for Statisticians, Vol. I* (New York: Cambridge University Press, 1966) by permission of the *Biometrika* trustees.

## TABLE IVa

## THE $F$ DISTRIBUTION (VALUES OF $F_{.05}$)*

Degrees of freedom for numerator

| $v_2 \backslash v_1$ | 1 | 2 | 3 | 4 | 5 | 6 | 7 | 8 | 9 | 10 | 12 | 15 | 20 | 24 | 30 | 40 | 60 | 120 | ∞ |
|---|---|---|---|---|---|---|---|---|---|---|---|---|---|---|---|---|---|---|---|
| 1 | 161 | 200 | 216 | 225 | 230 | 234 | 237 | 239 | 241 | 242 | 244 | 246 | 248 | 249 | 250 | 251 | 252 | 253 | 254 |
| 2 | 18.5 | 19.0 | 19.2 | 19.2 | 19.3 | 19.3 | 19.4 | 19.4 | 19.4 | 19.4 | 19.4 | 19.4 | 19.4 | 19.5 | 19.5 | 19.5 | 19.5 | 19.5 | 19.5 |
| 3 | 10.1 | 9.55 | 9.28 | 9.12 | 9.01 | 8.94 | 8.89 | 8.85 | 8.81 | 8.79 | 8.74 | 8.70 | 8.66 | 8.64 | 8.62 | 8.59 | 8.57 | 8.55 | 8.53 |
| 4 | 7.71 | 6.94 | 6.59 | 6.39 | 6.26 | 6.16 | 6.09 | 6.04 | 6.00 | 5.96 | 5.91 | 5.86 | 5.80 | 5.77 | 5.75 | 5.72 | 5.69 | 5.66 | 5.63 |
| 5 | 6.61 | 5.79 | 5.41 | 5.19 | 5.05 | 4.95 | 4.88 | 4.82 | 4.77 | 4.74 | 4.68 | 4.62 | 4.56 | 4.53 | 4.50 | 4.46 | 4.43 | 4.40 | 4.37 |
| 6 | 5.99 | 5.14 | 4.76 | 4.53 | 4.39 | 4.28 | 4.21 | 4.15 | 4.10 | 4.06 | 4.00 | 3.94 | 3.87 | 3.84 | 3.81 | 3.77 | 3.74 | 3.70 | 3.67 |
| 7 | 5.59 | 4.74 | 4.35 | 4.12 | 3.97 | 3.87 | 3.79 | 3.73 | 3.68 | 3.64 | 3.57 | 3.51 | 3.44 | 3.41 | 3.38 | 3.34 | 3.30 | 3.27 | 3.23 |
| 8 | 5.32 | 4.46 | 4.07 | 3.84 | 3.69 | 3.58 | 3.50 | 3.44 | 3.39 | 3.35 | 3.28 | 3.22 | 3.15 | 3.12 | 3.08 | 3.04 | 3.01 | 2.97 | 2.93 |
| 9 | 5.12 | 4.26 | 3.86 | 3.63 | 3.48 | 3.37 | 3.29 | 3.23 | 3.18 | 3.14 | 3.07 | 3.01 | 2.94 | 2.90 | 2.86 | 2.83 | 2.79 | 2.75 | 2.71 |
| 10 | 4.96 | 4.10 | 3.71 | 3.48 | 3.33 | 3.22 | 3.14 | 3.07 | 3.02 | 2.98 | 2.91 | 2.85 | 2.77 | 2.74 | 2.70 | 2.66 | 2.62 | 2.58 | 2.54 |
| 11 | 4.84 | 3.98 | 3.59 | 3.36 | 3.20 | 3.09 | 3.01 | 2.95 | 2.90 | 2.85 | 2.79 | 2.72 | 2.65 | 2.61 | 2.57 | 2.53 | 2.49 | 2.45 | 2.40 |
| 12 | 4.75 | 3.89 | 3.49 | 3.26 | 3.11 | 3.00 | 2.91 | 2.85 | 2.80 | 2.75 | 2.69 | 2.62 | 2.54 | 2.51 | 2.47 | 2.43 | 2.38 | 2.34 | 2.30 |
| 13 | 4.67 | 3.81 | 3.41 | 3.18 | 3.03 | 2.92 | 2.83 | 2.77 | 2.71 | 2.67 | 2.60 | 2.53 | 2.46 | 2.42 | 2.38 | 2.34 | 2.30 | 2.25 | 2.21 |
| 14 | 4.60 | 3.74 | 3.34 | 3.11 | 2.96 | 2.85 | 2.76 | 2.70 | 2.65 | 2.60 | 2.53 | 2.46 | 2.39 | 2.35 | 2.31 | 2.27 | 2.22 | 2.18 | 2.13 |
| 15 | 4.54 | 3.68 | 3.29 | 3.06 | 2.90 | 2.79 | 2.71 | 2.64 | 2.59 | 2.54 | 2.48 | 2.40 | 2.33 | 2.29 | 2.25 | 2.20 | 2.16 | 2.11 | 2.07 |
| 16 | 4.49 | 3.63 | 3.24 | 3.01 | 2.85 | 2.74 | 2.66 | 2.59 | 2.54 | 2.49 | 2.42 | 2.35 | 2.28 | 2.24 | 2.19 | 2.15 | 2.11 | 2.06 | 2.01 |
| 17 | 4.45 | 3.59 | 3.20 | 2.96 | 2.81 | 2.70 | 2.61 | 2.55 | 2.49 | 2.45 | 2.38 | 2.31 | 2.23 | 2.19 | 2.15 | 2.10 | 2.06 | 2.01 | 1.96 |
| 18 | 4.41 | 3.55 | 3.16 | 2.93 | 2.77 | 2.66 | 2.58 | 2.51 | 2.46 | 2.41 | 2.34 | 2.27 | 2.19 | 2.15 | 2.11 | 2.06 | 2.02 | 1.97 | 1.92 |
| 19 | 4.38 | 3.52 | 3.13 | 2.90 | 2.74 | 2.63 | 2.54 | 2.48 | 2.42 | 2.38 | 2.31 | 2.23 | 2.16 | 2.11 | 2.07 | 2.03 | 1.98 | 1.93 | 1.88 |
| 20 | 4.35 | 3.49 | 3.10 | 2.87 | 2.71 | 2.60 | 2.51 | 2.45 | 2.39 | 2.35 | 2.28 | 2.20 | 2.12 | 2.08 | 2.04 | 1.99 | 1.95 | 1.90 | 1.84 |
| 21 | 4.32 | 3.47 | 3.07 | 2.84 | 2.68 | 2.57 | 2.49 | 2.42 | 2.37 | 2.32 | 2.25 | 2.18 | 2.10 | 2.05 | 2.01 | 1.96 | 1.92 | 1.87 | 1.81 |
| 22 | 4.30 | 3.44 | 3.05 | 2.82 | 2.66 | 2.55 | 2.46 | 2.40 | 2.34 | 2.30 | 2.23 | 2.15 | 2.07 | 2.03 | 1.98 | 1.94 | 1.89 | 1.84 | 1.78 |
| 23 | 4.28 | 3.42 | 3.03 | 2.80 | 2.64 | 2.53 | 2.44 | 2.37 | 2.32 | 2.27 | 2.20 | 2.13 | 2.05 | 2.01 | 1.96 | 1.91 | 1.86 | 1.81 | 1.76 |
| 24 | 4.26 | 3.40 | 3.01 | 2.78 | 2.62 | 2.51 | 2.42 | 2.36 | 2.30 | 2.25 | 2.18 | 2.11 | 2.03 | 1.98 | 1.94 | 1.89 | 1.84 | 1.79 | 1.73 |
| 25 | 4.24 | 3.39 | 2.99 | 2.76 | 2.60 | 2.49 | 2.40 | 2.34 | 2.28 | 2.24 | 2.16 | 2.09 | 2.01 | 1.96 | 1.92 | 1.87 | 1.82 | 1.77 | 1.71 |
| 30 | 4.17 | 3.32 | 2.92 | 2.69 | 2.53 | 2.42 | 2.33 | 2.27 | 2.21 | 2.16 | 2.09 | 2.01 | 1.93 | 1.89 | 1.84 | 1.79 | 1.74 | 1.68 | 1.62 |
| 40 | 4.08 | 3.23 | 2.84 | 2.61 | 2.45 | 2.34 | 2.25 | 2.18 | 2.12 | 2.08 | 2.00 | 1.92 | 1.84 | 1.79 | 1.74 | 1.69 | 1.64 | 1.58 | 1.51 |
| 60 | 4.00 | 3.15 | 2.76 | 2.53 | 2.37 | 2.25 | 2.17 | 2.10 | 2.04 | 1.99 | 1.92 | 1.84 | 1.75 | 1.70 | 1.65 | 1.59 | 1.53 | 1.47 | 1.39 |
| 120 | 3.92 | 3.07 | 2.68 | 2.45 | 2.29 | 2.18 | 2.09 | 2.02 | 1.96 | 1.91 | 1.83 | 1.75 | 1.66 | 1.61 | 1.55 | 1.50 | 1.43 | 1.35 | 1.25 |
| ∞ | 3.84 | 3.00 | 2.60 | 2.37 | 2.21 | 2.10 | 2.01 | 1.94 | 1.88 | 1.83 | 1.75 | 1.67 | 1.57 | 1.52 | 1.46 | 1.39 | 1.32 | 1.22 | 1.00 |

Degrees of freedom for denominator

* This table is reproduced from M. Merrington and C. M. Thompson, "Tables of percentage points of the inverted beta (F) distribution," *Biometrika*, Vol. 33 (1943), by permission of the *Biometrika* trustees.

## TABLE IVb

## THE F DISTRIBUTION (VALUES OF $F_{.01}$)*

| | \multicolumn{12}{c}{Degrees of freedom for numerator} | | | | | | | | | | | | |
|---|---|---|---|---|---|---|---|---|---|---|---|---|---|---|---|---|---|---|
| | 1 | 2 | 3 | 4 | 5 | 6 | 7 | 8 | 9 | 10 | 12 | 15 | 20 | 24 | 30 | 40 | 60 | 120 | ∞ |
| 1 | 4,052 | 5,000 | 5,403 | 5,625 | 5,764 | 5,859 | 5,928 | 5,982 | 6,023 | 6,056 | 6,106 | 6,157 | 6,209 | 6,235 | 6,261 | 6,287 | 6,313 | 6,339 | 6,366 |
| 2 | 98.5 | 99.0 | 99.2 | 99.2 | 99.3 | 99.3 | 99.4 | 99.4 | 99.4 | 99.4 | 99.4 | 99.4 | 99.4 | 99.5 | 99.5 | 99.5 | 99.5 | 99.5 | 99.5 |
| 3 | 34.1 | 30.8 | 29.5 | 28.7 | 28.2 | 27.9 | 27.7 | 27.5 | 27.3 | 27.2 | 27.1 | 26.9 | 26.7 | 26.6 | 26.5 | 26.4 | 26.3 | 26.2 | 26.1 |
| 4 | 21.2 | 18.0 | 16.7 | 16.0 | 15.5 | 15.2 | 15.0 | 14.8 | 14.7 | 14.5 | 14.4 | 14.2 | 14.0 | 13.9 | 13.8 | 13.7 | 13.7 | 13.6 | 13.5 |
| 5 | 16.3 | 13.3 | 12.1 | 11.4 | 11.0 | 10.7 | 10.5 | 10.3 | 10.2 | 10.1 | 9.89 | 9.72 | 9.55 | 9.47 | 9.38 | 9.29 | 9.20 | 9.11 | 9.02 |
| 6 | 13.7 | 10.9 | 9.78 | 9.15 | 8.75 | 8.47 | 8.26 | 8.10 | 7.98 | 7.87 | 7.72 | 7.56 | 7.40 | 7.31 | 7.23 | 7.14 | 7.06 | 6.97 | 6.88 |
| 7 | 12.2 | 9.55 | 8.45 | 7.85 | 7.46 | 7.19 | 6.99 | 6.84 | 6.72 | 6.62 | 6.47 | 6.31 | 6.16 | 6.07 | 5.99 | 5.91 | 5.82 | 5.74 | 5.65 |
| 8 | 11.3 | 8.65 | 7.59 | 7.01 | 6.63 | 6.37 | 6.18 | 6.03 | 5.91 | 5.81 | 5.67 | 5.52 | 5.36 | 5.28 | 5.20 | 5.12 | 5.03 | 4.95 | 4.86 |
| 9 | 10.6 | 8.02 | 6.99 | 6.42 | 6.06 | 5.80 | 5.61 | 5.47 | 5.35 | 5.26 | 5.11 | 4.96 | 4.81 | 4.73 | 4.65 | 4.57 | 4.48 | 4.40 | 4.31 |
| 10 | 10.0 | 7.56 | 6.55 | 5.99 | 5.64 | 5.39 | 5.20 | 5.06 | 4.94 | 4.85 | 4.71 | 4.56 | 4.41 | 4.33 | 4.25 | 4.17 | 4.08 | 4.00 | 3.91 |
| 11 | 9.65 | 7.21 | 6.22 | 5.67 | 5.32 | 5.07 | 4.89 | 4.74 | 4.63 | 4.54 | 4.40 | 4.25 | 4.10 | 4.02 | 3.94 | 3.86 | 3.78 | 3.69 | 3.60 |
| 12 | 9.33 | 6.93 | 5.95 | 5.41 | 5.06 | 4.82 | 4.64 | 4.50 | 4.39 | 4.30 | 4.16 | 4.01 | 3.86 | 3.78 | 3.70 | 3.62 | 3.54 | 3.45 | 3.36 |
| 13 | 9.07 | 6.70 | 5.74 | 5.21 | 4.86 | 4.62 | 4.44 | 4.30 | 4.19 | 4.10 | 3.96 | 3.82 | 3.66 | 3.59 | 3.51 | 3.43 | 3.34 | 3.25 | 3.17 |
| 14 | 8.86 | 6.51 | 5.56 | 5.04 | 4.70 | 4.46 | 4.28 | 4.14 | 4.03 | 3.94 | 3.80 | 3.66 | 3.51 | 3.43 | 3.35 | 3.27 | 3.18 | 3.09 | 3.00 |
| 15 | 8.68 | 6.36 | 5.42 | 4.89 | 4.56 | 4.32 | 4.14 | 4.00 | 3.89 | 3.80 | 3.67 | 3.52 | 3.37 | 3.29 | 3.21 | 3.13 | 3.05 | 2.96 | 2.87 |
| 16 | 8.53 | 6.23 | 5.29 | 4.77 | 4.44 | 4.20 | 4.03 | 3.89 | 3.78 | 3.69 | 3.55 | 3.41 | 3.26 | 3.18 | 3.10 | 3.02 | 2.93 | 2.84 | 2.75 |
| 17 | 8.40 | 6.11 | 5.19 | 4.67 | 4.34 | 4.10 | 3.93 | 3.79 | 3.68 | 3.59 | 3.46 | 3.31 | 3.16 | 3.08 | 3.00 | 2.92 | 2.83 | 2.75 | 2.65 |
| 18 | 8.29 | 6.01 | 5.09 | 4.58 | 4.25 | 4.01 | 3.84 | 3.71 | 3.60 | 3.51 | 3.37 | 3.23 | 3.08 | 3.00 | 2.92 | 2.84 | 2.75 | 2.66 | 2.57 |
| 19 | 8.19 | 5.93 | 5.01 | 4.50 | 4.17 | 3.94 | 3.77 | 3.63 | 3.52 | 3.43 | 3.30 | 3.15 | 3.00 | 2.92 | 2.84 | 2.76 | 2.67 | 2.58 | 2.49 |
| 20 | 8.10 | 5.85 | 4.94 | 4.43 | 4.10 | 3.87 | 3.70 | 3.56 | 3.46 | 3.37 | 3.23 | 3.09 | 2.94 | 2.86 | 2.78 | 2.69 | 2.61 | 2.52 | 2.42 |
| 21 | 8.02 | 5.78 | 4.87 | 4.37 | 4.04 | 3.81 | 3.64 | 3.51 | 3.40 | 3.31 | 3.17 | 3.03 | 2.88 | 2.80 | 2.72 | 2.64 | 2.55 | 2.46 | 2.36 |
| 22 | 7.95 | 5.72 | 4.82 | 4.31 | 3.99 | 3.76 | 3.59 | 3.45 | 3.35 | 3.26 | 3.12 | 2.98 | 2.83 | 2.75 | 2.67 | 2.58 | 2.50 | 2.40 | 2.31 |
| 23 | 7.88 | 5.66 | 4.76 | 4.26 | 3.94 | 3.71 | 3.54 | 3.41 | 3.30 | 3.21 | 3.07 | 2.93 | 2.78 | 2.70 | 2.62 | 2.54 | 2.45 | 2.35 | 2.26 |
| 24 | 7.82 | 5.61 | 4.72 | 4.22 | 3.90 | 3.67 | 3.50 | 3.36 | 3.26 | 3.17 | 3.03 | 2.89 | 2.74 | 2.66 | 2.58 | 2.49 | 2.40 | 2.31 | 2.21 |
| 25 | 7.77 | 5.57 | 4.68 | 4.18 | 3.86 | 3.63 | 3.46 | 3.32 | 3.22 | 3.13 | 2.99 | 2.85 | 2.70 | 2.62 | 2.53 | 2.45 | 2.36 | 2.27 | 2.17 |
| 30 | 7.56 | 5.39 | 4.51 | 4.02 | 3.70 | 3.47 | 3.30 | 3.17 | 3.07 | 2.98 | 2.84 | 2.70 | 2.55 | 2.47 | 2.39 | 2.30 | 2.21 | 2.11 | 2.01 |
| 40 | 7.31 | 5.18 | 4.31 | 3.83 | 3.51 | 3.29 | 3.12 | 2.99 | 2.89 | 2.80 | 2.66 | 2.52 | 2.37 | 2.29 | 2.20 | 2.11 | 2.02 | 1.92 | 1.80 |
| 60 | 7.08 | 4.98 | 4.13 | 3.65 | 3.34 | 3.12 | 2.95 | 2.82 | 2.72 | 2.63 | 2.50 | 2.35 | 2.20 | 2.12 | 2.03 | 1.94 | 1.84 | 1.73 | 1.60 |
| 120 | 6.85 | 4.79 | 3.95 | 3.48 | 3.17 | 2.96 | 2.79 | 2.66 | 2.56 | 2.47 | 2.34 | 2.19 | 2.03 | 1.95 | 1.86 | 1.76 | 1.66 | 1.53 | 1.38 |
| ∞ | 6.63 | 4.61 | 3.78 | 3.32 | 3.02 | 2.80 | 2.64 | 2.51 | 2.41 | 2.32 | 2.18 | 2.04 | 1.88 | 1.79 | 1.70 | 1.59 | 1.47 | 1.32 | 1.00 |

Degrees of freedom for denominator

* This table is reproduced from M. Merrington and C. M. Thompson, "Tables of percentage points of the inverted beta (F) distribution," *Biometrika*, Vol. 33 (1943), by permission of the *Biometrika* trustees.

## TABLE Va

## 0.95 CONFIDENCE INTERVALS FOR PROPORTIONS*

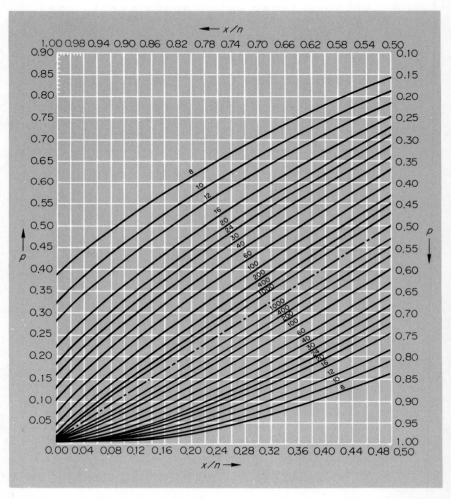

* This table is reproduced from Table 41 of the *Biometrika Tables for Statisticians*, Vol. I (New York: Cambridge University Press, 1966) by permission of the *Biometrika* trustees.

## TABLE Vb

## 0.99 CONFIDENCE INTERVALS FOR PROPORTIONS*

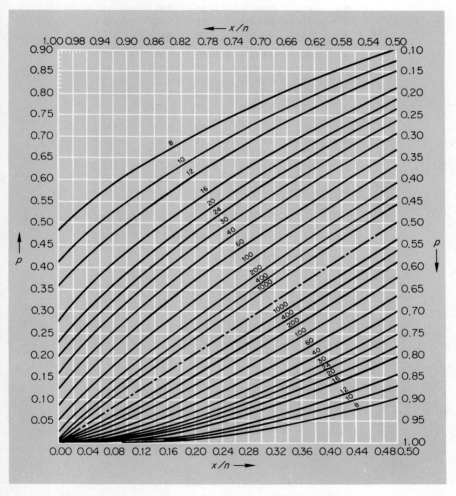

*This table is reproduced from Table 41 of the *Biometrika Tables for Statisticians*, Vol. I (New York: Cambridge University Press, 1966) by permission of the *Biometrika* trustees.

## TABLE VI

## CRITICAL VALUES OF $r^*$

| $n$ | $r_{.025}$ | $r_{.005}$ | $n$ | $r_{.025}$ | $r_{.005}$ |
|---|---|---|---|---|---|
| 3 | 0.997 | | 18 | 0.468 | 0.590 |
| 4 | 0.950 | 0.999 | 19 | 0.456 | 0.575 |
| 5 | 0.878 | 0.959 | 20 | 0.444 | 0.561 |
| 6 | 0.811 | 0.917 | 21 | 0.433 | 0.549 |
| 7 | 0.754 | 0.875 | 22 | 0.423 | 0.537 |
| 8 | 0.707 | 0.834 | 27 | 0.381 | 0.487 |
| 9 | 0.666 | 0.798 | 32 | 0.349 | 0.449 |
| 10 | 0.632 | 0.765 | 37 | 0.325 | 0.418 |
| 11 | 0.602 | 0.735 | 42 | 0.304 | 0.393 |
| 12 | 0.576 | 0.708 | 47 | 0.288 | 0.372 |
| 13 | 0.553 | 0.684 | 52 | 0.273 | 0.354 |
| 14 | 0.532 | 0.661 | 62 | 0.250 | 0.325 |
| 15 | 0.514 | 0.641 | 72 | 0.232 | 0.302 |
| 16 | 0.497 | 0.623 | 82 | 0.217 | 0.283 |
| 17 | 0.482 | 0.606 | 92 | 0.205 | 0.267 |

* This table is abridged from Table VI of R. A. Fisher and F. Yates, *Statistical Tables for Biological, Agricultural, and Medical Research*, published by Oliver and Boyd, Ltd., Edinburgh, by permission of the author's literary executor and publishers.

## TABLE VII  FACTORIALS

| $n$ | $n!$ |
|---|---|
| 0 | 1 |
| 1 | 1 |
| 2 | 2 |
| 3 | 6 |
| 4 | 24 |
| 5 | 120 |
| 6 | 720 |
| 7 | 5,040 |
| 8 | 40,320 |
| 9 | 362,880 |
| 10 | 3,628,800 |
| 11 | 39,916,800 |
| 12 | 479,001,600 |
| 13 | 6,227,020,800 |
| 14 | 87,178,291,200 |
| 15 | 1,307,674,368,000 |

## TABLE VIII  BINOMIAL COEFFICIENTS

| $n$ | $\binom{n}{0}$ | $\binom{n}{1}$ | $\binom{n}{2}$ | $\binom{n}{3}$ | $\binom{n}{4}$ | $\binom{n}{5}$ | $\binom{n}{6}$ | $\binom{n}{7}$ | $\binom{n}{8}$ | $\binom{n}{9}$ | $\binom{n}{10}$ |
|---|---|---|---|---|---|---|---|---|---|---|---|
| 0 | 1 | | | | | | | | | | |
| 1 | 1 | 1 | | | | | | | | | |
| 2 | 1 | 2 | 1 | | | | | | | | |
| 3 | 1 | 3 | 3 | 1 | | | | | | | |
| 4 | 1 | 4 | 6 | 4 | 1 | | | | | | |
| 5 | 1 | 5 | 10 | 10 | 5 | 1 | | | | | |
| 6 | 1 | 6 | 15 | 20 | 15 | 6 | 1 | | | | |
| 7 | 1 | 7 | 21 | 35 | 35 | 21 | 7 | 1 | | | |
| 8 | 1 | 8 | 28 | 56 | 70 | 56 | 28 | 8 | 1 | | |
| 9 | 1 | 9 | 36 | 84 | 126 | 126 | 84 | 36 | 9 | 1 | |
| 10 | 1 | 10 | 45 | 120 | 210 | 252 | 210 | 120 | 45 | 10 | 1 |
| 11 | 1 | 11 | 55 | 165 | 330 | 462 | 462 | 330 | 165 | 55 | 11 |
| 12 | 1 | 12 | 66 | 220 | 495 | 792 | 924 | 792 | 495 | 220 | 66 |
| 13 | 1 | 13 | 78 | 286 | 715 | 1287 | 1716 | 1716 | 1287 | 715 | 286 |
| 14 | 1 | 14 | 91 | 364 | 1001 | 2002 | 3003 | 3432 | 3003 | 2002 | 1001 |
| 15 | 1 | 15 | 105 | 455 | 1365 | 3003 | 5005 | 6435 | 6435 | 5005 | 3003 |
| 16 | 1 | 16 | 120 | 560 | 1820 | 4368 | 8008 | 11440 | 12870 | 11440 | 8008 |
| 17 | 1 | 17 | 136 | 680 | 2380 | 6188 | 12376 | 19448 | 24310 | 24310 | 19448 |
| 18 | 1 | 18 | 153 | 816 | 3060 | 8568 | 18564 | 31824 | 43758 | 48620 | 43758 |
| 19 | 1 | 19 | 171 | 969 | 3876 | 11628 | 27132 | 50388 | 75582 | 92378 | 92378 |
| 20 | 1 | 20 | 190 | 1140 | 4845 | 15504 | 38760 | 77520 | 125970 | 167960 | 184756 |

If necessary, use the identity $\binom{n}{k} = \binom{n}{n-k}$.

## TABLE IX
## RANDOM NUMBERS

| | | | | | | | |
|---|---|---|---|---|---|---|---|
| 02946 | 96520 | 81881 | 56247 | 17623 | 47441 | 27821 | 91845 |
| 85697 | 62000 | 87957 | 07258 | 45054 | 58410 | 92081 | 97624 |
| 26734 | 68426 | 52067 | 23123 | 73700 | 58730 | 06111 | 64486 |
| 47829 | 32353 | 95941 | 72169 | 58374 | 03905 | 06865 | 95353 |
| 76603 | 99339 | 40571 | 41186 | 04981 | 17531 | 97372 | 39558 |
| | | | | | | | |
| 47526 | 26522 | 11045 | 83565 | 45639 | 02485 | 43905 | 01823 |
| 70100 | 85732 | 19741 | 92951 | 98832 | 38188 | 24080 | 24519 |
| 86819 | 50200 | 50889 | 06493 | 66638 | 03619 | 90906 | 95370 |
| 41614 | 30074 | 23403 | 03656 | 77580 | 87772 | 86877 | 57085 |
| 17930 | 26194 | 53836 | 53692 | 67125 | 98175 | 00912 | 11246 |
| | | | | | | | |
| 24649 | 31845 | 25736 | 75231 | 83808 | 98997 | 71829 | 99430 |
| 79899 | 34061 | 54308 | 59358 | 56462 | 58166 | 97302 | 86828 |
| 76801 | 49594 | 81002 | 30397 | 52728 | 15101 | 72070 | 33706 |
| 62567 | 08480 | 61873 | 63162 | 44873 | 35302 | 04511 | 38088 |
| 49723 | 15275 | 09399 | 11211 | 67352 | 41526 | 23497 | 75440 |
| | | | | | | | |
| 42658 | 70183 | 89417 | 57676 | 35370 | 14915 | 16569 | 54945 |
| 65080 | 35569 | 79392 | 14937 | 06081 | 74957 | 87787 | 68849 |
| 02906 | 38119 | 72407 | 71427 | 58478 | 99297 | 43519 | 62410 |
| 75153 | 86376 | 63852 | 60557 | 21211 | 77299 | 74967 | 99038 |
| 14192 | 49525 | 78844 | 13664 | 98964 | 64425 | 33536 | 15079 |
| | | | | | | | |
| 32059 | 11548 | 86264 | 74406 | 81496 | 23996 | 56872 | 71401 |
| 81716 | 80301 | 96704 | 57214 | 71361 | 41989 | 92589 | 69788 |
| 43315 | 50483 | 02950 | 09611 | 36341 | 20326 | 37489 | 34626 |
| 27510 | 10769 | 09921 | 46721 | 34183 | 22856 | 18724 | 60422 |
| 81782 | 04769 | 36716 | 82519 | 98272 | 13969 | 12429 | 03093 |
| | | | | | | | |
| 19975 | 48346 | 91029 | 78902 | 75689 | 70722 | 88553 | 83300 |
| 98356 | 76855 | 18769 | 52843 | 64204 | 95212 | 31320 | 03783 |
| 29708 | 17814 | 31556 | 68610 | 16574 | 42305 | 56300 | 84227 |
| 88014 | 27583 | 78167 | 25057 | 93552 | 74363 | 30951 | 41367 |
| 94491 | 19238 | 17396 | 10592 | 48907 | 79840 | 34607 | 62668 |
| | | | | | | | |
| 56957 | 05072 | 53948 | 07850 | 42569 | 82391 | 20435 | 79306 |
| 50915 | 31924 | 80621 | 17495 | 81618 | 15125 | 48087 | 01250 |
| 49631 | 93771 | 80200 | 84622 | 31413 | 33756 | 15218 | 81976 |
| 99683 | 58162 | 45516 | 39761 | 77600 | 15175 | 67415 | 88801 |
| 86017 | 20264 | 94618 | 85979 | 42009 | 78616 | 45210 | 73186 |
| | | | | | | | |
| 77339 | 64605 | 82583 | 85011 | 02955 | 84348 | 46436 | 77911 |
| 61714 | 57933 | 37342 | 26000 | 93611 | 93346 | 71212 | 24405 |
| 15232 | 48027 | 15832 | 62924 | 11509 | 95853 | 02747 | 61889 |
| 41447 | 34275 | 10779 | 83515 | 63899 | 30932 | 90572 | 98971 |
| 23244 | 43524 | 16382 | 36340 | 73581 | 76780 | 03842 | 64009 |
| | | | | | | | |
| 53460 | 83542 | 25224 | 70378 | 49604 | 14809 | 12317 | 78062 |
| 53442 | 16897 | 61578 | 05032 | 81825 | 76822 | 87170 | 77235 |
| 55548 | 19096 | 04130 | 23104 | 60534 | 44842 | 16954 | 99466 |
| 18185 | 69329 | 02340 | 63111 | 41788 | 74409 | 76177 | 55519 |
| 02372 | 45690 | 38595 | 23121 | 73818 | 74454 | 02371 | 94693 |
| | | | | | | | |
| 51715 | 35492 | 61371 | 87132 | 81585 | 55439 | 98095 | 55578 |
| 24717 | 16786 | 42786 | 86985 | 21858 | 39489 | 39251 | 70450 |
| 78022 | 32604 | 87259 | 93702 | 99438 | 68184 | 62119 | 20229 |
| 35995 | 08275 | 62405 | 43313 | 03249 | 74135 | 43003 | 63132 |
| 29192 | 86922 | 31908 | 42703 | 59638 | 31226 | 89860 | 45191 |

# TABLE IX

## RANDOM NUMBERS (*Continued*)

| | | | | | | | |
|---|---|---|---|---|---|---|---|
| 01654 | 50375 | 23941 | 44848 | 79154 | 30193 | 15271 | 93296 |
| 73750 | 68343 | 40727 | 81203 | 91727 | 06463 | 12248 | 57567 |
| 64163 | 22132 | 22896 | 14305 | 45642 | 71580 | 21558 | 66457 |
| 88445 | 85544 | 23627 | 79176 | 32502 | 34568 | 78777 | 35179 |
| 94180 | 71108 | 19121 | 11958 | 33308 | 75930 | 24865 | 17426 |
| 11433 | 12220 | 36719 | 35435 | 77727 | 78493 | 94580 | 70091 |
| 61838 | 68801 | 49856 | 21739 | 95701 | 11180 | 73936 | 17628 |
| 59908 | 68103 | 36855 | 19127 | 56637 | 06383 | 83182 | 76927 |
| 41018 | 69556 | 06402 | 03436 | 89803 | 95604 | 55735 | 66978 |
| 85095 | 01581 | 92299 | 06166 | 71078 | 31823 | 64316 | 95567 |
| 59705 | 78103 | 66740 | 41743 | 69177 | 59277 | 47629 | 63874 |
| 75094 | 55208 | 77905 | 20705 | 32408 | 16630 | 01242 | 63119 |
| 78425 | 68672 | 79455 | 94334 | 23292 | 90422 | 75540 | 15843 |
| 89088 | 86918 | 20787 | 05691 | 97309 | 96673 | 16389 | 51437 |
| 83345 | 95889 | 39333 | 86027 | 24680 | 51909 | 97230 | 58136 |
| 62896 | 00342 | 66647 | 57096 | 84913 | 67895 | 18804 | 17691 |
| 96498 | 38270 | 80532 | 54307 | 07885 | 38892 | 50990 | 96766 |
| 30974 | 47335 | 04918 | 42974 | 19294 | 72581 | 77377 | 04652 |
| 57901 | 06163 | 99162 | 53285 | 27341 | 02507 | 41858 | 08436 |
| 88494 | 80633 | 47785 | 53996 | 57058 | 04222 | 54488 | 12019 |
| 34883 | 00045 | 89682 | 86664 | 92195 | 42593 | 56488 | 35402 |
| 24373 | 46438 | 28935 | 63903 | 14722 | 10715 | 58795 | 42800 |
| 16828 | 79262 | 23678 | 05509 | 23733 | 95318 | 77730 | 87614 |
| 33723 | 27646 | 92335 | 87136 | 88062 | 21506 | 01750 | 71326 |
| 01542 | 75066 | 73921 | 97188 | 31250 | 41996 | 31680 | 41783 |
| 00100 | 12787 | 74100 | 95536 | 42359 | 01761 | 28842 | 71562 |
| 82697 | 03389 | 19303 | 21646 | 22532 | 81701 | 03425 | 28914 |
| 28137 | 17549 | 22698 | 72955 | 59849 | 02370 | 02784 | 13711 |
| 02248 | 21570 | 33796 | 83789 | 72981 | 96423 | 68791 | 91684 |
| 56175 | 82515 | 23348 | 42207 | 87644 | 57353 | 90349 | 16448 |
| 80020 | 21622 | 67659 | 07878 | 17586 | 65524 | 20162 | 04712 |
| 20271 | 23094 | 48372 | 77621 | 32889 | 19595 | 66500 | 28064 |
| 38734 | 98044 | 02658 | 90698 | 72563 | 15076 | 23780 | 52815 |
| 48183 | 24263 | 49297 | 32923 | 94406 | 63865 | 44336 | 27224 |
| 48163 | 34158 | 03177 | 51696 | 57795 | 31725 | 14403 | 29856 |
| 45658 | 15024 | 66664 | 18730 | 40671 | 92727 | 68626 | 81631 |
| 71128 | 15524 | 55666 | 14763 | 13729 | 51708 | 54104 | 81331 |
| 19041 | 42899 | 49464 | 93965 | 14960 | 88896 | 72784 | 82054 |
| 32672 | 67506 | 93040 | 94527 | 31556 | 80163 | 80203 | 90928 |
| 15823 | 48310 | 04391 | 15521 | 79255 | 69253 | 60254 | 01653 |
| 82810 | 18981 | 62581 | 31642 | 42693 | 78972 | 60322 | 90462 |
| 74772 | 80840 | 05816 | 29023 | 67410 | 12916 | 87933 | 78840 |
| 52931 | 38199 | 85632 | 23761 | 99084 | 48028 | 07184 | 41635 |
| 95395 | 87644 | 09722 | 99251 | 97129 | 70847 | 91864 | 08549 |
| 76695 | 33451 | 57139 | 90612 | 11918 | 90871 | 60965 | 23555 |
| 83560 | 50374 | 04410 | 57272 | 36705 | 51302 | 93147 | 29479 |
| 28355 | 62002 | 85994 | 35807 | 84810 | 14186 | 51153 | 78998 |
| 84684 | 54861 | 41330 | 66808 | 65231 | 14168 | 45193 | 27156 |
| 21135 | 92001 | 43896 | 55887 | 35319 | 03793 | 60344 | 95970 |
| 24236 | 01536 | 43897 | 41294 | 45551 | 46877 | 58631 | 82654 |

To find the square root of any positive number rounded to two digits, use the following rule to decide whether to take the entry of the $\sqrt{n}$ or the $\sqrt{10n}$ column:

> Move the decimal point an even number of places to the right or to the left until a number greater than or equal to 1 but less than 100 is reached. If the resulting number is less than 10 go to the $\sqrt{n}$ column; if it is 10 or more go to the $\sqrt{10n}$ column.

Thus, to find the square root of 14,000 or 0.032 we go to the $\sqrt{n}$ column since the decimal point has to be moved, respectively, four places to the left to give 1.4 or two places to the right to give 3.2. Similarly, to find the square root of 2,200 or 0.000016 we go to the $\sqrt{10n}$ column since the decimal point has to be moved, respectively, two places to the left to give 22 or six places to the right to give 16.

Having found the entry in the appropriate column of Table X, the only thing that remains to be done is to put the decimal point in the right place in the result. To this end we use the following rule:

> Having previously moved the decimal point an even number of places to the left or to the right to get a number greater than or equal to 1 but less than 100, the decimal point of the appropriate entry of Table X is moved half as many places in the opposite direction.

For example, to find the square root of 14,000 we first note that the decimal point has to be moved *four places to the left* to give 1.4. We thus take the entry of the $\sqrt{n}$ column corresponding to 1.4, move its decimal point *two places to the right*, and get $\sqrt{14,000} = 118.322$. Similarly, to find the square root of 0.000016 we note that the decimal point has to be moved *six places to the right* to give 16. We thus take the entry of the $\sqrt{10n}$ column corresponding to 16, move the decimal point *three places to the left*, and get $\sqrt{0.000016} = 0.004$.

## TABLE X
## SQUARE ROOTS

| $n$ | $\sqrt{n}$ | $\sqrt{10n}$ | $n$ | $\sqrt{n}$ | $\sqrt{10n}$ | $n$ | $\sqrt{n}$ | $\sqrt{10n}$ | $n$ | $\sqrt{n}$ | $\sqrt{10n}$ |
|---|---|---|---|---|---|---|---|---|---|---|---|
| 1.0 | 1.00000 | 3.16228 | 3.5 | 1.87083 | 5.91608 | 6.0 | 2.44949 | 7.74597 | 8.5 | 2.91548 | 9.21954 |
| 1.1 | 1.04881 | 3.31662 | 3.6 | 1.89737 | 6.00000 | 6.1 | 2.46982 | 7.81025 | 8.6 | 2.93258 | 9.27362 |
| 1.2 | 1.09545 | 3.46410 | 3.7 | 1.92354 | 6.08276 | 6.2 | 2.48998 | 7.87401 | 8.7 | 2.94958 | 9.32738 |
| 1.3 | 1.14018 | 3.60555 | 3.8 | 1.94936 | 6.16441 | 6.3 | 2.50998 | 7.93725 | 8.8 | 2.96648 | 9.38083 |
| 1.4 | 1.18322 | 3.74166 | 3.9 | 1.97484 | 6.24500 | 6.4 | 2.52982 | 8.00000 | 8.9 | 2.98329 | 9.43398 |
| 1.5 | 1.22474 | 3.87298 | 4.0 | 2.00000 | 6.32456 | 6.5 | 2.54951 | 8.06226 | 9.0 | 3.00000 | 9.48683 |
| 1.6 | 1.26491 | 4.00000 | 4.1 | 2.02485 | 6.40312 | 6.6 | 2.56905 | 8.12404 | 9.1 | 3.01662 | 9.53939 |
| 1.7 | 1.30384 | 4.12311 | 4.2 | 2.04939 | 6.48074 | 6.7 | 2.58844 | 8.18535 | 9.2 | 3.03315 | 9.59166 |
| 1.8 | 1.34164 | 4.24264 | 4.3 | 2.07364 | 6.55744 | 6.8 | 2.60768 | 8.24621 | 9.3 | 3.04959 | 9.64365 |
| 1.9 | 1.37840 | 4.35890 | 4.4 | 2.09762 | 6.63325 | 6.9 | 2.62679 | 8.30662 | 9.4 | 3.06594 | 9.69536 |
| 2.0 | 1.41421 | 4.47214 | 4.5 | 2.12132 | 6.70820 | 7.0 | 2.64575 | 8.36660 | 9.5 | 3.08221 | 9.74679 |
| 2.1 | 1.44914 | 4.58258 | 4.6 | 2.14476 | 6.78233 | 7.1 | 2.66458 | 8.42615 | 9.6 | 3.09839 | 9.79796 |
| 2.2 | 1.48324 | 4.69042 | 4.7 | 2.16795 | 6.85565 | 7.2 | 2.68328 | 8.48528 | 9.7 | 3.11448 | 9.84886 |
| 2.3 | 1.51658 | 4.79583 | 4.8 | 2.19089 | 6.92820 | 7.3 | 2.70185 | 8.54400 | 9.8 | 3.13050 | 9.89949 |
| 2.4 | 1.54919 | 4.89898 | 4.9 | 2.21359 | 7.00000 | 7.4 | 2.72029 | 8.60233 | 9.9 | 3.14643 | 9.94987 |
| 2.5 | 1.58114 | 5.00000 | 5.0 | 2.23607 | 7.07107 | 7.5 | 2.73861 | 8.66025 | | | |
| 2.6 | 1.61245 | 5.09902 | 5.1 | 2.25832 | 7.14143 | 7.6 | 2.75681 | 8.71780 | | | |
| 2.7 | 1.64317 | 5.19615 | 5.2 | 2.28035 | 7.21110 | 7.7 | 2.77489 | 8.77496 | | | |
| 2.8 | 1.67332 | 5.29150 | 5.3 | 2.30217 | 7.28011 | 7.8 | 2.79285 | 8.83176 | | | |
| 2.9 | 1.70294 | 5.38516 | 5.4 | 2.32379 | 7.34847 | 7.9 | 2.81069 | 8.88819 | | | |
| 3.0 | 1.73205 | 5.47723 | 5.5 | 2.34521 | 7.41620 | 8.0 | 2.82843 | 8.94427 | | | |
| 3.1 | 1.76068 | 5.56776 | 5.6 | 2.36643 | 7.48331 | 8.1 | 2.84605 | 9.00000 | | | |
| 3.2 | 1.78885 | 5.65685 | 5.7 | 2.38747 | 7.54983 | 8.2 | 2.86356 | 9.05539 | | | |
| 3.3 | 1.81659 | 5.74456 | 5.8 | 2.40832 | 7.61577 | 8.3 | 2.88097 | 9.11043 | | | |
| 3.4 | 1.84391 | 5.83095 | 5.9 | 2.42899 | 7.68115 | 8.4 | 2.89828 | 9.16515 | | | |

## TABLE XI

## BINOMIAL PROBABILITIES

| n | x | 0.05 | 0.1 | 0.2 | 0.3 | 0.4 | 0.5 | 0.6 | 0.7 | 0.8 | 0.9 | 0.95 |
|---|---|------|-----|-----|-----|-----|-----|-----|-----|-----|-----|------|
| 2 | 0 | 0.902 | 0.810 | 0.640 | 0.490 | 0.360 | 0.250 | 0.160 | 0.090 | 0.040 | 0.010 | 0.002 |
|   | 1 | 0.095 | 0.180 | 0.320 | 0.420 | 0.480 | 0.500 | 0.480 | 0.420 | 0.320 | 0.180 | 0.095 |
|   | 2 | 0.002 | 0.010 | 0.040 | 0.090 | 0.160 | 0.250 | 0.360 | 0.490 | 0.640 | 0.810 | 0.902 |
| 3 | 0 | 0.857 | 0.729 | 0.512 | 0.343 | 0.216 | 0.125 | 0.064 | 0.027 | 0.008 | 0.001 |       |
|   | 1 | 0.135 | 0.243 | 0.384 | 0.441 | 0.432 | 0.375 | 0.288 | 0.189 | 0.096 | 0.027 | 0.007 |
|   | 2 | 0.007 | 0.027 | 0.096 | 0.189 | 0.288 | 0.375 | 0.432 | 0.441 | 0.384 | 0.243 | 0.135 |
|   | 3 |       | 0.001 | 0.008 | 0.027 | 0.064 | 0.125 | 0.216 | 0.343 | 0.512 | 0.729 | 0.857 |
| 4 | 0 | 0.815 | 0.656 | 0.410 | 0.240 | 0.130 | 0.062 | 0.026 | 0.008 | 0.002 |       |       |
|   | 1 | 0.171 | 0.292 | 0.410 | 0.412 | 0.346 | 0.250 | 0.154 | 0.076 | 0.026 | 0.004 |       |
|   | 2 | 0.014 | 0.049 | 0.154 | 0.265 | 0.346 | 0.375 | 0.346 | 0.265 | 0.154 | 0.049 | 0.014 |
|   | 3 |       | 0.004 | 0.026 | 0.076 | 0.154 | 0.250 | 0.346 | 0.412 | 0.410 | 0.292 | 0.171 |
|   | 4 |       |       | 0.002 | 0.008 | 0.026 | 0.062 | 0.130 | 0.240 | 0.410 | 0.656 | 0.815 |
| 5 | 0 | 0.774 | 0.590 | 0.328 | 0.168 | 0.078 | 0.031 | 0.010 | 0.002 |       |       |       |
|   | 1 | 0.204 | 0.328 | 0.410 | 0.360 | 0.259 | 0.156 | 0.077 | 0.028 | 0.006 |       |       |
|   | 2 | 0.021 | 0.073 | 0.205 | 0.309 | 0.346 | 0.312 | 0.230 | 0.132 | 0.051 | 0.008 | 0.001 |
|   | 3 | 0.001 | 0.008 | 0.051 | 0.132 | 0.230 | 0.312 | 0.346 | 0.309 | 0.205 | 0.073 | 0.021 |
|   | 4 |       |       | 0.006 | 0.028 | 0.077 | 0.156 | 0.259 | 0.360 | 0.410 | 0.328 | 0.204 |
|   | 5 |       |       |       | 0.002 | 0.010 | 0.031 | 0.078 | 0.168 | 0.328 | 0.590 | 0.774 |
| 6 | 0 | 0.735 | 0.531 | 0.262 | 0.118 | 0.047 | 0.016 | 0.004 | 0.001 |       |       |       |
|   | 1 | 0.232 | 0.354 | 0.393 | 0.303 | 0.187 | 0.094 | 0.037 | 0.010 | 0.002 |       |       |
|   | 2 | 0.031 | 0.098 | 0.246 | 0.324 | 0.311 | 0.234 | 0.138 | 0.060 | 0.015 | 0.001 |       |
|   | 3 | 0.002 | 0.015 | 0.082 | 0.185 | 0.276 | 0.312 | 0.276 | 0.185 | 0.082 | 0.015 | 0.002 |
|   | 4 |       | 0.001 | 0.015 | 0.060 | 0.138 | 0.234 | 0.311 | 0.324 | 0.246 | 0.098 | 0.031 |
|   | 5 |       |       | 0.002 | 0.010 | 0.037 | 0.094 | 0.187 | 0.303 | 0.393 | 0.354 | 0.232 |
|   | 6 |       |       |       | 0.001 | 0.004 | 0.016 | 0.047 | 0.118 | 0.262 | 0.531 | 0.735 |
| 7 | 0 | 0.698 | 0.478 | 0.210 | 0.082 | 0.028 | 0.008 | 0.002 |       |       |       |       |
|   | 1 | 0.257 | 0.372 | 0.367 | 0.247 | 0.131 | 0.055 | 0.017 | 0.004 |       |       |       |
|   | 2 | 0.041 | 0.124 | 0.275 | 0.318 | 0.261 | 0.164 | 0.077 | 0.025 | 0.004 |       |       |
|   | 3 | 0.004 | 0.023 | 0.115 | 0.227 | 0.290 | 0.273 | 0.194 | 0.097 | 0.029 | 0.003 |       |
|   | 4 |       | 0.003 | 0.029 | 0.097 | 0.194 | 0.273 | 0.290 | 0.227 | 0.115 | 0.023 | 0.004 |

STATISTICAL TABLES 339

## TABLE XI

## BINOMIAL PROBABILITIES (Continued)

| n | x | 0.05 | 0.1 | 0.2 | 0.3 | 0.4 | 0.5 | 0.6 | 0.7 | 0.8 | 0.9 | 0.95 |
|---|---|------|-----|-----|-----|-----|-----|-----|-----|-----|-----|------|
| 7 | 5 |      |     | 0.004 | 0.025 | 0.077 | 0.164 | 0.261 | 0.318 | 0.275 | 0.124 | 0.041 |
|   | 6 |      |     |       | 0.004 | 0.017 | 0.055 | 0.131 | 0.247 | 0.367 | 0.372 | 0.257 |
|   | 7 |      |     |       |       | 0.002 | 0.008 | 0.028 | 0.082 | 0.210 | 0.478 | 0.698 |
| 8 | 0 | 0.663 | 0.430 | 0.168 | 0.058 | 0.017 | 0.004 | 0.001 |       |       |       |       |
|   | 1 | 0.279 | 0.383 | 0.336 | 0.198 | 0.090 | 0.031 | 0.008 | 0.001 |       |       |       |
|   | 2 | 0.051 | 0.149 | 0.294 | 0.296 | 0.209 | 0.109 | 0.041 | 0.010 | 0.001 |       |       |
|   | 3 | 0.005 | 0.033 | 0.147 | 0.254 | 0.279 | 0.219 | 0.124 | 0.047 | 0.009 |       |       |
|   | 4 |       | 0.005 | 0.046 | 0.136 | 0.232 | 0.273 | 0.232 | 0.136 | 0.046 | 0.005 |       |
|   | 5 |       |       | 0.009 | 0.047 | 0.124 | 0.219 | 0.279 | 0.254 | 0.147 | 0.033 | 0.005 |
|   | 6 |       |       | 0.001 | 0.010 | 0.041 | 0.109 | 0.209 | 0.296 | 0.294 | 0.149 | 0.051 |
|   | 7 |       |       |       | 0.001 | 0.008 | 0.031 | 0.090 | 0.198 | 0.336 | 0.383 | 0.279 |
|   | 8 |       |       |       |       | 0.001 | 0.004 | 0.017 | 0.058 | 0.168 | 0.430 | 0.663 |
| 9 | 0 | 0.630 | 0.387 | 0.134 | 0.040 | 0.010 | 0.002 |       |       |       |       |       |
|   | 1 | 0.299 | 0.387 | 0.302 | 0.156 | 0.060 | 0.018 | 0.004 |       |       |       |       |
|   | 2 | 0.063 | 0.172 | 0.302 | 0.267 | 0.161 | 0.070 | 0.021 | 0.004 |       |       |       |
|   | 3 | 0.008 | 0.045 | 0.176 | 0.267 | 0.251 | 0.164 | 0.074 | 0.021 | 0.003 |       |       |
|   | 4 | 0.001 | 0.007 | 0.066 | 0.172 | 0.251 | 0.246 | 0.167 | 0.074 | 0.017 | 0.001 |       |
|   | 5 |       | 0.001 | 0.017 | 0.074 | 0.167 | 0.246 | 0.251 | 0.172 | 0.066 | 0.007 | 0.001 |
|   | 6 |       |       | 0.003 | 0.021 | 0.074 | 0.164 | 0.251 | 0.267 | 0.176 | 0.045 | 0.008 |
|   | 7 |       |       |       | 0.004 | 0.021 | 0.070 | 0.161 | 0.267 | 0.302 | 0.172 | 0.063 |
|   | 8 |       |       |       |       | 0.004 | 0.018 | 0.060 | 0.156 | 0.302 | 0.387 | 0.299 |
|   | 9 |       |       |       |       |       | 0.002 | 0.010 | 0.040 | 0.134 | 0.387 | 0.630 |
| 10 | 0 | 0.599 | 0.349 | 0.107 | 0.028 | 0.006 | 0.001 |       |       |       |       |       |
|    | 1 | 0.315 | 0.387 | 0.268 | 0.121 | 0.040 | 0.010 | 0.002 |       |       |       |       |
|    | 2 | 0.075 | 0.194 | 0.302 | 0.233 | 0.121 | 0.044 | 0.011 | 0.001 |       |       |       |
|    | 3 | 0.010 | 0.057 | 0.201 | 0.267 | 0.215 | 0.117 | 0.042 | 0.009 | 0.001 |       |       |
|    | 4 | 0.001 | 0.011 | 0.088 | 0.200 | 0.251 | 0.205 | 0.111 | 0.037 | 0.006 |       |       |
|    | 5 |       | 0.001 | 0.026 | 0.103 | 0.201 | 0.246 | 0.201 | 0.103 | 0.026 | 0.001 |       |
|    | 6 |       |       | 0.006 | 0.037 | 0.111 | 0.205 | 0.251 | 0.200 | 0.088 | 0.011 | 0.001 |
|    | 7 |       |       | 0.001 | 0.009 | 0.042 | 0.117 | 0.215 | 0.267 | 0.201 | 0.057 | 0.010 |
|    | 8 |       |       |       | 0.001 | 0.011 | 0.044 | 0.121 | 0.233 | 0.302 | 0.194 | 0.075 |
|    | 9 |       |       |       |       | 0.002 | 0.010 | 0.040 | 0.121 | 0.268 | 0.387 | 0.315 |
|    | 10 |      |       |       |       |       | 0.001 | 0.006 | 0.028 | 0.107 | 0.349 | 0.599 |

## TABLE XI

## BINOMIAL PROBABILITIES (*Continued*)

| n | x | 0.05 | 0.1 | 0.2 | 0.3 | 0.4 | 0.5 | 0.6 | 0.7 | 0.8 | 0.9 | 0.95 |
|---|---|------|-----|-----|-----|-----|-----|-----|-----|-----|-----|------|
| 11 | 0 | 0.569 | 0.314 | 0.086 | 0.020 | 0.004 | | | | | | |
| | 1 | 0.329 | 0.384 | 0.236 | 0.093 | 0.027 | 0.005 | 0.001 | | | | |
| | 2 | 0.087 | 0.213 | 0.295 | 0.200 | 0.089 | 0.027 | 0.005 | 0.001 | | | |
| | 3 | 0.014 | 0.071 | 0.221 | 0.257 | 0.177 | 0.081 | 0.023 | 0.004 | | | |
| | 4 | 0.001 | 0.016 | 0.111 | 0.220 | 0.236 | 0.161 | 0.070 | 0.017 | 0.002 | | |
| | 5 | | 0.002 | 0.039 | 0.132 | 0.221 | 0.226 | 0.147 | 0.057 | 0.010 | | |
| | 6 | | | 0.010 | 0.057 | 0.147 | 0.226 | 0.221 | 0.132 | 0.039 | 0.002 | |
| | 7 | | | 0.002 | 0.017 | 0.070 | 0.161 | 0.236 | 0.220 | 0.111 | 0.016 | 0.001 |
| | 8 | | | | 0.004 | 0.023 | 0.081 | 0.177 | 0.257 | 0.221 | 0.071 | 0.014 |
| | 9 | | | | 0.001 | 0.005 | 0.027 | 0.089 | 0.200 | 0.295 | 0.213 | 0.087 |
| | 10 | | | | | 0.001 | 0.005 | 0.027 | 0.093 | 0.236 | 0.384 | 0.329 |
| | 11 | | | | | | | 0.004 | 0.020 | 0.086 | 0.314 | 0.569 |
| 12 | 0 | 0.540 | 0.282 | 0.069 | 0.014 | 0.002 | | | | | | |
| | 1 | 0.341 | 0.377 | 0.206 | 0.071 | 0.017 | 0.003 | | | | | |
| | 2 | 0.099 | 0.230 | 0.283 | 0.168 | 0.064 | 0.016 | 0.002 | | | | |
| | 3 | 0.017 | 0.085 | 0.236 | 0.240 | 0.142 | 0.054 | 0.012 | 0.001 | | | |
| | 4 | 0.002 | 0.021 | 0.133 | 0.231 | 0.213 | 0.121 | 0.042 | 0.008 | 0.001 | | |
| | 5 | | 0.004 | 0.053 | 0.158 | 0.227 | 0.193 | 0.101 | 0.029 | 0.003 | | |
| | 6 | | | 0.016 | 0.079 | 0.177 | 0.226 | 0.177 | 0.079 | 0.016 | | |
| | 7 | | | 0.003 | 0.029 | 0.101 | 0.193 | 0.227 | 0.158 | 0.053 | 0.004 | |
| | 8 | | | 0.001 | 0.008 | 0.042 | 0.121 | 0.213 | 0.231 | 0.133 | 0.021 | 0.002 |
| | 9 | | | | 0.001 | 0.012 | 0.054 | 0.142 | 0.240 | 0.236 | 0.085 | 0.017 |
| | 10 | | | | | 0.002 | 0.016 | 0.064 | 0.168 | 0.283 | 0.230 | 0.099 |
| | 11 | | | | | | 0.003 | 0.017 | 0.071 | 0.206 | 0.377 | 0.341 |
| | 12 | | | | | | | 0.002 | 0.014 | 0.069 | 0.282 | 0.540 |
| 13 | 0 | 0.513 | 0.254 | 0.055 | 0.010 | 0.001 | | | | | | |
| | 1 | 0.351 | 0.367 | 0.179 | 0.054 | 0.011 | 0.002 | | | | | |
| | 2 | 0.111 | 0.245 | 0.268 | 0.139 | 0.045 | 0.010 | 0.001 | | | | |
| | 3 | 0.021 | 0.100 | 0.246 | 0.218 | 0.111 | 0.035 | 0.006 | 0.001 | | | |
| | 4 | 0.003 | 0.028 | 0.154 | 0.234 | 0.184 | 0.087 | 0.024 | 0.003 | | | |
| | 5 | | 0.006 | 0.069 | 0.180 | 0.221 | 0.157 | 0.066 | 0.014 | 0.001 | | |
| | 6 | | 0.001 | 0.023 | 0.103 | 0.197 | 0.209 | 0.131 | 0.044 | 0.006 | | |
| | 7 | | | 0.006 | 0.044 | 0.131 | 0.209 | 0.197 | 0.103 | 0.023 | 0.001 | |

## TABLE XI

## BINOMIAL PROBABILITIES (*Continued*)

| n | x | 0.05 | 0.1 | 0.2 | 0.3 | 0.4 | 0.5 | 0.6 | 0.7 | 0.8 | 0.9 | 0.95 |
|---|---|------|-----|-----|-----|-----|-----|-----|-----|-----|-----|------|
| 13 | 8 |  |  | 0.001 | 0.014 | 0.066 | 0.157 | 0.221 | 0.180 | 0.069 | 0.006 |  |
|  | 9 |  |  |  | 0.003 | 0.024 | 0.087 | 0.184 | 0.234 | 0.154 | 0.028 | 0.003 |
|  | 10 |  |  |  | 0.001 | 0.006 | 0.035 | 0.111 | 0.218 | 0.246 | 0.100 | 0.021 |
|  | 11 |  |  |  |  | 0.001 | 0.010 | 0.045 | 0.139 | 0.268 | 0.245 | 0.111 |
|  | 12 |  |  |  |  |  | 0.002 | 0.011 | 0.054 | 0.179 | 0.367 | 0.351 |
|  | 13 |  |  |  |  |  |  | 0.001 | 0.010 | 0.055 | 0.254 | 0.513 |
| 14 | 0 | 0.488 | 0.229 | 0.044 | 0.007 | 0.001 |  |  |  |  |  |  |
|  | 1 | 0.359 | 0.356 | 0.154 | 0.041 | 0.007 | 0.001 |  |  |  |  |  |
|  | 2 | 0.123 | 0.257 | 0.250 | 0.113 | 0.032 | 0.006 | 0.001 |  |  |  |  |
|  | 3 | 0.026 | 0.114 | 0.250 | 0.194 | 0.085 | 0.022 | 0.003 |  |  |  |  |
|  | 4 | 0.004 | 0.035 | 0.172 | 0.229 | 0.155 | 0.061 | 0.014 | 0.001 |  |  |  |
|  | 5 |  | 0.008 | 0.086 | 0.196 | 0.207 | 0.122 | 0.041 | 0.007 |  |  |  |
|  | 6 |  | 0.001 | 0.032 | 0.126 | 0.207 | 0.183 | 0.092 | 0.023 | 0.002 |  |  |
|  | 7 |  |  | 0.009 | 0.062 | 0.157 | 0.209 | 0.157 | 0.062 | 0.009 |  |  |
|  | 8 |  |  | 0.002 | 0.023 | 0.092 | 0.183 | 0.207 | 0.126 | 0.032 | 0.001 |  |
|  | 9 |  |  |  | 0.007 | 0.041 | 0.122 | 0.207 | 0.196 | 0.086 | 0.008 |  |
|  | 10 |  |  |  | 0.001 | 0.014 | 0.061 | 0.155 | 0.229 | 0.172 | 0.035 | 0.004 |
|  | 11 |  |  |  |  | 0.003 | 0.022 | 0.085 | 0.194 | 0.250 | 0.114 | 0.026 |
|  | 12 |  |  |  |  | 0.001 | 0.006 | 0.032 | 0.113 | 0.250 | 0.257 | 0.123 |
|  | 13 |  |  |  |  |  | 0.001 | 0.007 | 0.041 | 0.154 | 0.356 | 0.359 |
|  | 14 |  |  |  |  |  |  | 0.001 | 0.007 | 0.044 | 0.229 | 0.488 |
| 15 | 0 | 0.463 | 0.206 | 0.035 | 0.005 |  |  |  |  |  |  |  |
|  | 1 | 0.366 | 0.343 | 0.132 | 0.031 | 0.005 |  |  |  |  |  |  |
|  | 2 | 0.135 | 0.267 | 0.231 | 0.092 | 0.022 | 0.003 |  |  |  |  |  |
|  | 3 | 0.031 | 0.129 | 0.250 | 0.170 | 0.063 | 0.014 | 0.002 |  |  |  |  |
|  | 4 | 0.005 | 0.043 | 0.188 | 0.219 | 0.127 | 0.042 | 0.007 | 0.001 |  |  |  |
|  | 5 | 0.001 | 0.010 | 0.103 | 0.206 | 0.186 | 0.092 | 0.024 | 0.003 |  |  |  |
|  | 6 |  | 0.002 | 0.043 | 0.147 | 0.207 | 0.153 | 0.061 | 0.012 | 0.001 |  |  |
|  | 7 |  |  | 0.014 | 0.081 | 0.177 | 0.196 | 0.118 | 0.035 | 0.003 |  |  |
|  | 8 |  |  | 0.003 | 0.035 | 0.118 | 0.196 | 0.177 | 0.081 | 0.014 |  |  |
|  | 9 |  |  | 0.001 | 0.012 | 0.061 | 0.153 | 0.207 | 0.147 | 0.043 | 0.002 |  |
|  | 10 |  |  |  | 0.003 | 0.024 | 0.092 | 0.186 | 0.206 | 0.103 | 0.010 | 0.001 |
|  | 11 |  |  |  | 0.001 | 0.007 | 0.042 | 0.127 | 0.219 | 0.188 | 0.043 | 0.005 |
|  | 12 |  |  |  |  | 0.002 | 0.014 | 0.063 | 0.170 | 0.250 | 0.129 | 0.031 |
|  | 13 |  |  |  |  |  | 0.003 | 0.022 | 0.092 | 0.231 | 0.267 | 0.135 |
|  | 14 |  |  |  |  |  |  | 0.005 | 0.031 | 0.132 | 0.343 | 0.366 |
|  | 15 |  |  |  |  |  |  |  | 0.005 | 0.035 | 0.206 | 0.463 |

# ANSWERS TO ODD-NUMBERED EXERCISES

*In exercises involving extensive calculations, the reader may well get answers differing somewhat from those given here due to rounding at various intermediate stages.*

**Page 6**

1. (a) Descriptive; (b) generalization; (c) generalization; (d) descriptive; (e) descriptive; (f) generalization; (g) descriptive; (h) generalization.
3. (a) For instance, a study conducted in Florida on the per capita expenditures on snow tires in the United States; (b) for instance, a study conducted in December on the annual consumption of ice cream in the United States.
5. (a) Secondary; (b) secondary; (c) secondary; (d) primary; (e) primary; (f) primary; (g) primary.
7. All from U.S. Department of Commerce, Office of Business Economics.
9. (a) 2,213, 1,901, 1,638, 374, and 405 thousand pounds; (b) Hong Kong, 10,864.

**Page 24**

1. (a) 48; (b) cannot be determined; (c) cannot be determined; (d) 57; (e) cannot be determined; (f) cannot be determined.
3. A good choice would be $120.00–$139.99, $140.00–$159.99, $160.00–$179.99, $180.00–$199.99, $200.00–$219.99, $220.00–$239.99.
5. The class limits are 1–5, 6–10, 11–15, 16–20, 21–25, 26–30, and the class interval is 5.
7. (a) The percentages are 14, 26, 33.5, 17, and 9.5; (b) the cumulative "less than" frequencies are 0, 28, 80, 147, 181, and 200; (c) the cumulative "more than" frequencies are 200, 172, 120, 53, 19, and 0; (d) the cumulative "or more" percentages are 100, 86, 60, 26.5, 9.5, and 0.
9. If the class limits are 40–49, 50–59, 60–69, 70–79, 80–89, 90–99, and 100–109, then (a) the class frequencies are 3, 6, 11, 16, 25, 15, and 4; (b) the corresponding percentages are 3.75, 7.5, 13.75, 20.0, 31.25, 18.75, and 5.0; (c) the "or more" percentages are 100, 96.25, 88.75, 75.0, 55.0, 23.75, 5.0, and 0.
15. The pictogram is misleading because we are more likely to compare the *areas* of the certificates than their sides, and, hence, get the impression that the 1970

ANSWERS TO ODD-NUMBERED EXERCISES 343

value of corporation stock held by U.S. life insurance companies is *nine* times the corresponding value for 1960. The ratio of the sides should be $\sqrt{3}$ to 1(or approximately 1.73 to 1), so that the ratio of the areas will be 3 to 1.

## Page 34

1. The mean *is* $21,000, but it is not really indicative of the income of the "average family" as it is inflated by the one very large value of $100,000.
3. (a) $6,484.60; (b) $6.48 to the nearest cent.
5. 0.542%.
7. $45.90.
9. (a) 79.1; (b) 78.88.
11. (a) 45.90; (b) 58.50; (c) 78.88; (d) 37.91.
13. (a) 400 mph, which is the harmonic mean; (b) $35.29, which is the harmonic mean.
15. 6.525%.
17. (a) $\bar{x} = \dfrac{\sum_{i=1}^{n} x_i}{n}$; (b) $\bar{x} = \dfrac{\sum_{i=1}^{k} x_i f_i}{n}$; (c) $\bar{x}_w = \dfrac{\sum_{i=1}^{n} w_i x_i}{\sum_{i=1}^{n} w_i}$;
    (d) $x_1 + x_2 + x_3 + x_4 + x_5$;
    (e) $y_1 + y_2 + y_3 + y_4 + y_5 + y_6$; (f) $x_1 f_1 + x_2 f_2 + x_3 f_3 + x_4 f_4$;
    (g) $x_1^2 + x_2^2 + x_3^2 + x_4^2$; (h) $x_2 + x_3 + x_4 + x_5 + x_6$; (i) $(y_3 - 2) + (y_4 - 2) + (y_5 - 2)$.

## Page 43

1. 62.1 and 37.6.
3. $5,694.
5. 32.9 and 32.65; it would not have affected the median, but the mean would have been 31.65.
7. (a) the medians are 3, 5, 5, 3, 2, 2, 4, 5, 3, 4, and the means are $3\frac{1}{3}$, 4, 4, $3\frac{1}{3}$, 3, 3, $4\frac{2}{3}$, $4\frac{1}{3}$, $2\frac{2}{3}$, 4; (b) for the medians the frequencies are 2, 3, 2, 3, and for the means the frequencies are 0, 5, 4, 1.
9. (a) 81.5; (b) 81.1.
11. (a) 7.88; (b) 27.96, 45.47, and 61.26; (c) 53.17 and 62.00; (d) 69.5 and 89.1.
13. (a) 2.62 and 3.64; (b) 0.93 and 97.39; (c) 46.00 and 79.17; (d) 42.17, 51.17, 99.5, and 107.5.

## Page 47

1. (a) 28; (b) 52 and 58; (c) no mode.

**344** ANSWERS TO ODD-NUMBERED EXERCISES

3. 32.25.
5. The modes are 55 and 57, and the mid-range is 64.
7. West.

## Page 51

1. (a) 90.0; (b) 93.4.
3. 119.5.
5. (a) 95.4 and 96.6; (b) 120.2 and 96.2.

## Page 56

1. (a) 140.1; (b) 137.4; (c) 102.0, 129.0, and 136.7.
5. (a) 793.9, 829.4, 847.3, 839.6, 866.0, and the two percentage increases are, respectively, 12.9% and 2.2%; (b) 14.90, 15.50, 15.61, 15.92, 16.08, so that the prices of these chairs exceeded the general increases in wholesale prices.

## Page 66

1. (a) Both will ship the same way; (b) the typewriter manufacturer will ship by truck; (c) one will ship by rail and the other by air; (d) they do not both ship the same way; (e) the clock manufacturer does not ship by truck and the typewriter manufacturer does not ship by rail; (f) the clock manufacturer does not ship by rail and the typewriter manufacturer does not ship by air.
3. (a) Not mutually exclusive; (b) mutually exclusive; (c) not mutually exclusive; (d) mutually exclusive.
5. (b) A is the event that there are as many check-out clerks as stock boys, B is the event that there are two stock boys, C is the event that altogether there are at most two, and D is the event that there are more check-out clerks than stock boys; (c) (i) there are not two stock boys, (1,0), (1,1), (1,3), (2,0), (2,1), (2,3), (3,0), (3,1), (3,3); (ii) there are more than two altogether, (1,2), (1,3), (2,1), (2,2), (2,3), (3,0), (3,1), (3,2), (3,3); (iii) there are at least as many stock boys as check-out clerks, (1,1), (1,2), (1,3), (2,2), (2,3), (3,3); (iv) there are two of each, (2,2); (v) there are at least as many check-out clerks as stock boys, (1,0), (1,1), (2,0), (2,1), (2,2), (3,0), (3,1), (3,2), (3,3); (vi) ∅; (vii) ∅; (viii) there are three check-out clerks and two stock boys, (3,2); (ix) there are either two of each or three of each, (2,2), (3,3); (d) (i) mutually exclusive; (ii) not mutually exclusive; (iii) not mutually exclusive; (iv) mutually exclusive.
7. (a) 4; (b) 7; (c) 1 and 2; (d) 4 and 6; (e) 3, 5, 6, and 8.
9. (a) Repairman will come today, he can fix the washing machine, and his charges will be fair; (b) the repairman will come today and his charges will be

ANSWERS TO ODD-NUMBERED EXERCISES 345

fair, but he will not be able to fix the washing machine; (c) the repairman will come today, but he will not be able to fix the washing machine and his charges will not be fair; (d) the repairman will be able to fix the washing machine and his charges will be fair; (e) the repairman will come today, but will not be able to fix the washing machine; (f) the repairman will not be able to come today and his charges will not be fair.

## Page 76

3. (b) 6; (c) 2.
5. 30.
7. (a) 8; (b) 16.
9. 30.
11. (a) 27; (b) 81; (c) 243.
13. (a) ∅, General Motors, Ford, Chrysler, General Motors and Ford, General Motors and Chrysler, Ford and Chrysler, General Motors and Ford and Chrysler; (b) ∅, manufacturing, mining, public utilities, transportation, manufacturing and mining, manufacturing and public utilities, manufacturing and transportation, mining and public utilities, mining and transportation, public utilities and transportation, manufacturing and mining and public utilities, manufacturing and mining and transportation, manufacturing and public utilities and transportation, mining and public utilities and transportation, manufacturing and mining and public utilities and transportation; (c) 64; (d) 1,024.
15. 5040.
17. 40,320.
19. (a) 190; (b) 15,504; (c) 125,970.
21. 120.

## Page 83

1. 0.38.
3. 0.76.
5. (a) $\frac{1}{3}$; (b) 2 to 1; (c) the counselor.
7. $\frac{2}{3}$.
9. $\frac{11}{16}$.
11. "At least 0.90."

## Page 89

1. $0.225, No.

346 ANSWERS TO ODD-NUMBERED EXERCISES

3. $4,800, Yes.
5. 424.44.
7. 0.128.
9. $p > \frac{1}{16}$.
11. (b) Continue; (c) do not continue; (d) does not matter.
13. (a) Build the new factory; (b) drive to library; (c) continue.

**Page 100**

1. (a) The sum of the probabilities exceeds 1; (b) the sum of the probabilities is less than 1; (c) the second probability should not exceed the first; (d) the first probability should not exceed the second.
3. (a) 0.73; (b) 0.52; (c) 0; (d) 0.75; (e) 0.25.
5. Not consistent.
7. (a) 0.28; (b) 0.72; (c) 0.66; (d) 0.88.
9. (a) 0.08, 0.35, and 0.57; (b) 0.27, 0.72, and 0.43.
11. (a) $\frac{1}{26}$; (b) $\frac{3}{13}$; (c) $\frac{11}{13}$; (d) $\frac{1}{2}$; (e) $\frac{3}{13}$.
13. $\frac{1}{8}, \frac{3}{8}, \frac{3}{8}$, and $\frac{1}{8}$.
15. 0.45.
17. 0.25.

**Page 108**

1. (a) The probability that a production worker who received a low grade in the manual dexterity test will not meet the production quota; (b) the probability that a production worker who does not meet the production quota will have received a low grade in the manual dexterity test; (c) the probability that a production worker who did not receive a low grade in the manual dexterity test will not meet the production quota; (d) the probability that a production worker who meets the production quota will have received a low grade in the manual dexterity test; (e) the probability that a production worker who meets the production quota will not have received a low grade in the manual dexterity test.
3. (a) The probability that an applicant with prior sales experience has a college degree; (b) the probability that an applicant who has a college degree will also have a car; (c) the probability that an applicant without prior sales experience has a college degree; (d) the probability that an applicant who does not own a car will have prior sales experience; (e) the probability that an applicant without a college degree will not own an automobile either; (f) the probability that an applicant with prior sales experience will have an automobile and a college degree; (g) the probability that an applicant with prior sales experience and a car will also have a college degree.

ANSWERS TO ODD-NUMBERED EXERCISES   347

5. (a) $\frac{23}{40}$; (b) $\frac{17}{40}$; (c) $\frac{1}{2}$; (d) $\frac{1}{2}$; (e) $\frac{9}{40}$; (f) $\frac{3}{20}$; (g) $\frac{9}{20}$; (h) $\frac{9}{23}$; (i) $\frac{3}{10}$; (j) $\frac{6}{17}$.
7. $\frac{5}{8}$ and $\frac{11}{34}$.
9. (a) $\frac{10}{13}$; (b) $\frac{10}{17}$.
13. (a) $\frac{33}{95}$; (b) $\frac{14}{95}$; (c) $\frac{48}{95}$.
15. 0.01.
17. (a) $\frac{1}{8}$; (b) 0.003375; (c) $\frac{1}{256}$; (d) 0.590.
19. (a) 0.048, 0.096, and 0.76; (b) 0.729, 0.0625, and 0.844; (c) 0.18, 0.06, and 0.50.

Page 118

1. $\frac{6}{7}$.
3. $\frac{5}{7}$.
5. (a) $\frac{4}{29}$; (b) $\frac{20}{29}$.

Page 129

1. (a) No, the sum of the probabilities exceeds 1; (b) yes; (c) no, f(1) and f(3) are negative.
3. (a) $\frac{2}{9}$; (b) $\frac{1}{9}$; (c) $\frac{1}{3}$.
5. (a) 0.26; (b) 0.39; (c) 0.25; (d) 0.10.
7. (a) 0.185 and 0.420; (b) 0.185 and 0.421.
9. (a) 0.069; (b) 0.020; (c) 0.794.
13. (a) $\frac{33}{95}$; (b) $\frac{14}{95}$; (c) $\frac{48}{95}$ also, 0.40.
15. (a) 0.092; (b) 0.369; (c) 0.539.
17. 0.655; binomial approximation is off by 0.001.

Page 141

1. $\mu = 1$.
3. $\mu = 5.991$ compared to $np = 6$.
5. $\mu = 3$.
7. $\mu = 2.399$ and $\mu = 2.4$ by means of special formula.
9. (c) $\frac{1}{16}$.
11. 0.998.
13. The probability is at least 0.96.
15. The probability is at least $\frac{63}{64}$ or 0.984.

## Page 154

1. (a) $\frac{1}{4}$; (b) $\frac{1}{8}$, yes; (c) 0.65; (d) 0.20.
3. (a) 0.2764; (b) 0.1844; (c) 0.9382; (d) 0.4522; (e) 0.1093; (f) 0.8186; (g) 0.1593; (h) 0.2569; (i) 0.5549; (j) 0.7888.
5. (a) 1.63; (b) 2.22; (c) $-1.00$; (d) 0.44; (e) $-0.50$; (f) 2.17.
7. (a) 0.9332; (b) 0.2266; (c) 0.3891; (d) 0.9104.
9. $\sigma = 10$

## Page 164

1. (a) 0.0102; (b) 0.2483; (c) 145.2.
3. (a) 0.62%; (b) 6.68%; (c) 0.3830; (d) 4.244.
5. (a) 7.08%; (b) 0.54%; (c) 87.6; (d) 65.1.
7. (a) 0.1210; (b) 0.0735.

## Page 169

1. 0.1937.
3. (a) 0.1846; (b) 0.4522; (c) 0.1020; (d) 0.8090.
5. 0.0838.
7. 0.9535.
9. (a) 0.2358; (b) 0.4908; (c) 0.9556.

## Page 176

1. (a) In a study of the expense accounts filed by the executives of many corporations; (b) In a tax audit of the expense accounts filed by the executives of the given corporation in January, 1974.
3. (a) If one is interested also in the letters mailed to their constituents by other congressmen; (b) if one is interested only in the letters mailed by the particular congressman.
5. (a) Finite, consisting of the 30 applicants; (b) infinite, consisting of the hypothetically infinite number of measurements that could be made of the volume of the rock; (c) infinite, consisting of the hypothetically infinite number of times the actor could rehearse his lines; (d) finite, consisting of the 14 postcards; (e) infinite, consisting of the hypothetically infinite determinations of the mileage.

ANSWERS TO ODD-NUMBERED EXERCISES 349

## Page 183

1. (a) $\frac{1}{220}$; (b) $\frac{1}{1140}$.
3. (a) $\frac{1}{1225}$; (b) $\frac{1}{67525}$.
5. $\frac{2}{5}$; yes.
9. 12782, 12735, 12716, 12729, and 12742.
13. 80, 60, 20, 32, and 8.
15. 15, 10, 11, 7, and 10; 7, 7, 20, 27, and 16; 11, 24, 12, 19, and 55; 7, 17, 9, 12, and 3; 50, 12, 21, 29, and 2; 11, 12, 10, 9 and 12; 5, 10, 14, 12, and 13; 0, 15, 7, 23, and 9; 24, 7, 3, 2, and 17; 15, 5, 4, 11, and 3.

## Page 194

1. $28,000 to the nearest $1,000.
3. 3.8.
5. 51.
7. No changes.
9. $23 to the nearest dollar.
11. (a) $30,500 to the nearest $100; (b) 0.56%; (c) 3.57.
15. (a) 133%; (b) 42%.

## Page 206

1. (a) It is divided by 2; (b) it is divided by 5; (c) it is multiplied by 4.
3. 0.026.
5. 0.0026.
7. (a) 0.3830; (b) 0.5934; (c) 0.8664.
9. Both standard errors are approximately $0.112\sigma$.
11. There is a fifty-fifty chance that the error will be less than $3.88.

## Page 215

1. Error is less than 0.157 hours.
3. Error is less than 2.24 minutes.
5. (a) 84; (b) 259; (c) 154; (d) 145.
7. (a) $5.83 < \mu < 6.57$; (b) $5.72 < \mu < 6.68$.
9. $613.1 < \mu < 626.9$.
11. $36.88 < \mu < 47.42$.
13. (a) $312.7 < \mu < 391.3$; (b) $11,961 < \mu < 13,321$; (c) $14.20 < \mu < 14.44$; (d) $44,782 < \mu < 80,164$.
15. $2.00 < \sigma < 2.70$.

## Page 230

1. (a) $\mu > 230$; (b) $\mu < 230$.
3. $z = 3.03$, reject the null hypothesis $\mu = \$62.00$.
5. $z = 1.14$, cannot reject the hypothesis $\mu = 8.5$.
7. $z = 2.04$; (a) reject the null hypothesis; (b) cannot reject the null hypothesis.
9. (a) $t = -0.43$, cannot reject the claim; (b) $t = -2.14$, reject $\mu = 4.30$; (c) $t = 2.65$, accept the alternative; (d) $t = -0.67$, cannot reject the claim.

## Page 237

1. $z = -4.2$, and the difference is significant.
3. $z = 3.0$, and the difference is significant.
5. (a) $t = -2.84$, and the difference is not significant; (b) $t = -3.4$, and the difference is significant; (c) $t = 0.89$, and the difference is not significant.
7. $t = -2.34$, and the difference is not significant.

## Page 245

1. $F = 1.53$, and the null hypothesis cannot be rejected.
3. $F = 1.14$, which is not significant.
5. $F = 0.44$, which is not significant.

## Page 253

1. $0.33 < p < 0.39$.
3. $39\% - 71\%$.
5. (a) $0.81 < p < 0.87$; (b) $0.808 < p < 0.872$.
7. $0.33 < p < 0.39$.
9. (a) $0.31 < p < 0.39$; (b) error is less than 0.038.
11. Error is less than 0.08.
13. (a) 385 and 196; (b) 1,068 and 1,849; (c) 2,401 and 865.

## Page 259

1. $z = -1.15$, cannot reject the null hypothesis.
3. Fewer than 4 or more than 11.
5. $z = -4.6$, reject the claim.
7. $z = -2.31$; (a) reject the null hypothesis; (b) cannot reject the null hypothesis.
9. $z = 1.22$, cannot reject the claim.

## ANSWERS TO ODD-NUMBERED EXERCISES 351

**Page 265**

1. $x^2 = 2.49$, which is not significant.
3. $x^2 = 5.84$, so that the difference is significant.
5. $x^2 = 35.3$, and the null hypothesis can be rejected.
7. (a) $z = -2.42$, so that the difference is significant; (b) $z = 1.05$, and the null hypothesis cannot be rejected.

**Page 272**

1. $x^2 = 5.6$, and the null hypothesis cannot be rejected.
3. $x^2 = 34.3$, reject the null hypothesis.
5. $x^2 = 4.0$, and the null hypothesis cannot be rejected.

**Page 276**

1. (a) 18, 54, 54, and 18; (b) $x^2 = 7.19$, and the null hypothesis (that the binomial distribution provides a good fit) cannot be rejected.

**Page 287**

1. 14,400.
3. (a) $y = -139.57 + 2.66x$; (c) $38.65 thousand.

**Page 295**

1. (a) Positive; (b) none; (c) positive; (d) positive; (e) none; (f) negative; (g) positive; (h) negative; (i) none; (j) positive.
3. $r = 0.92$, which is significant.
5. $r = 0.91$, which is significant.
7. $r = 0.91$, which is significant.
9. (a) 1; (b) 0.92; (c) 0.88; (d) 0.64.

**Page 308**

1. (a) $y = 20.63 - 1.04x$; (b) 25.83 and 15.43; (c) 11.27.
3. (a) $y = 85.93 + 2.18x$; (c) 140.43.
5. (a) $y = 335.55 + 12.72x$; (c) 488.19 thousands.

7. 553.4, 577, 581, 590.6, 592.6, 556.2, 543.8, 543.8, 511, 482.4, 473.2, 463, 456.4, 461.6, 452.2, 456.2, 446.4, 426.8, 412.6, 422.4, 437.4, 456.6, 482.8, 509, 526.2, 541, 559.2.
9. 1583, 1550.6, 1524.8, 1487.2, 1607.6, 1821.8, 2062, 2250.4, 2380.8, 2471.6, 2561, 2654.8, 2734, 2791.8, 2822.8, 2892.4, 2909, 2872.4, 2868.2, 2905.2, 2923.2, 2983.4, 3075.2, 3135.4, 3194.6, 3266.8, 3298.6, 3295.4.

## Page 317

1. (b) 88.8, 102.1, 99.2, and 109.8; (c) 19.71, 19.88, 20.46, and 19.76.
3. (b) 86.1, 79.8, 88.4, 97.4, 103.4, 109.0, 110.9, 111.7, 111.6, 106.9, 102.5, and 92.4; (c) 1969: 7.2, 7.4, 7.4, 7.5, 7.7, 7.7, 7.7, 7.6, 7.7, 7.7, 7.7, and 7.6; 1970: 7.1, 7.4, 7.4, 7.3, 7.4, 7.5, 7.5, 7.7, 7.6, 7.9, 8.0, and 8.3.
5. 103,500, 133,500, 150,000, 147,000, 163,500, 175,500, 196,500, 211,500, 166,500, 136,500, 106,500, 109,500.
7. (b) 145.64.

# INDEX

## A

Absolute value, 130, 136
Addition rule:
 general, 100
 generalized, 96
 special, 95
Aggregative index number, 51
 weighted, 55
$\alpha$ (alpha), level of significance, 223
$\alpha$ (alpha), regression line, 287
Alternative hypothesis:
 one-sided, 225
 two-sided, 225
Analysis of variance, 240–245
Annual rate, 319
Annual trend increment, 306
Area sampling, 182
Arithmetic mean (*see* Mean)
Arithmetic probability paper, 162
Association, measures of, 29
Average (*see* Mean)

## B

Bar chart, 20
Base year, 49
Bayes' rule, 115, 116
Bayesian analysis, 89
$\beta$ (beta), regression line, 287, 299
Bias, 2
Binomial coefficients, 75, 80
 table, 333
Binomial distribution:
 and hypergeometric distribution, 129
 and normal distribution, 166
 formula, 124
 mean, 134
 table, 338–341
  use of, 126
 variance, 138
Binomial probability function (*see* Binomial distribution)
Boundaries, class, 15
 upper and lower, 15
Business cycles, 303

## C

Categorical distribution (*see* Qualitative distribution)
Central limit theorem, 203
Chance fluctuations, 120
Chance variation, 135, 186
Chebyshev's theorem, 139
 and distribution of means, 203
 and law of large numbers, 140
Chi-square distribution, 263
 degrees of freedom, 264, 271, 275
 table, 327
$\chi^2$ (Chi-square) statistic, 263, 271, 274
Class:
 boundaries, 15
 frequencies, 14
 interval, 16
 limits, 14
  real, 15
 marks, 15
 modal, 46
 open, 13
Cluster sampling, 182
Coding, 33, 35, 196
Coefficient of correlation (*see* Correlation coefficient)
Coefficient of rank correlation, 297
Coefficient of skewness, Pearsonian, 199
Coefficient of variation, 199
Coefficient, binomial, 75, 80
 table, 333
Combinations, 75
Complement, 64
Conditional probability, 103, 105
Confidence, degree of, 213
Confidence interval:
 for mean, 213
  small samples, 218
 for proportion, 249
  large samples, 252
 for slope of regression line, 299
 for standard deviation, 214
Confidence limits, 213
Consistency criterion, 95

## 354 INDEX

Contingency table, 268
   $\chi^2$ statistic, 271
   degrees of freedom, 271
   expected frequencies, 269
Continuity correction, 159, 167
Continuous distribution, 145
   mean and standard deviation, 147
Continuous sample space, 144
Correction factor, finite population, 202
Correlation:
   and causation, 294
   linear, 295
   negative, 292
   positive, 292
   rank, 297
Correlation coefficient:
   computing formula, 293
   definition, 291
   significance test, 295
Count data, 210
Critical values, 226
Cumulative distribution, 16
Cyclical irregulars, 320
Cyclical variation, 300, 303, 320

## D

Data:
   biased, 2
   count, 210
   deseasonalized, 315
   external, 4
   internal, 4
   paired, 278
   primary, 4
   raw, 10
   secondary, 4
Data points, 280
Deciles, 44
Decision theory, 5
Deflating, 57
Degree of confidence, 213
Degrees of freedom:
   Chi-square distribution, 264, 271, 275
   $F$ distribution, 243
   sample standard deviation, 190
   $t$ distribution, 218, 231, 238
Density, probability, 145
Dependent events, 106
Dependent samples, 239
Description, statistical, 29
Descriptive statistics, 1
Deseasonalized data, 315
Deviation from mean, 135
Difference between means:
   dependent samples, 239
   large-sample test, 236

Difference between means (*cont.*)
   small-sample test, 238
   standard error, 236
Differences among means, 240
Differences among proportions, 260, 266
Distribution (*see also* Frequency
      distribution):
   binomial, 124
   Chi-square, 263
   continuous, 145
   cumulative, 16
   $F$, 242
   frequency, 10
   hypergeometric, 127
   normal, 147, 148
   percentage, 16
   probability, 123
   population, 174, 180
   qualitative, 10
   quantitative, 10
   sampling, 201
   skewed, 199
   symmetrical, 199
   $t$, 218, 231, 238
Double summation, 244

## E

Empty set, 64
Equiprobable events, special rule for, 98
Equitable game, 86
Error:
   probable, 208
   standard, 202
   Type I and Type II, 232
Estimate:
   interval, 214
   point, 210
Estimated regression line, 287
Estimation of mean:
   confidence interval, 213, 218
   maximum error, 211
   point estimate, 210
   sample size, determination of, 216
Estimation of proportion:
   confidence interval, 249, 252
   maximum error, 252
   sample size, determination of, 255
Estimation of $\sigma$:
   confidence interval, 214
Events, 62
   dependent, 106
   independent, 106
   mutually exclusive, 63
Expectation, mathematical, 85, 86
Expected frequencies, 269, 274
Expected profit with perfect information, 92

## INDEX

Experiment, 59
  outcome of, 59
External data, 4

### F

$F$ distribution, 242
  degrees of freedom, 243
  table, 328, 329
$F$ statistic, 242
  computing formula, 244
Factorials, 74
  table, 332
Finite population, 172
  correction factor, 202
Fixed-weight aggregative index, 56
Fractiles, 42
Frequency:
  class, 14
  expected, 269, 274
  relative, 248
Frequency distribution, 10
  class boundaries, 15
  class frequencies, 14
  class interval, 16
  class limits, 14
  class marks, 15
  cumulative, 16
  interval, 16
  percentage, 16
  qualitative, 10
  quantitative, 10
Frequency interpretation, probability, 80
Frequency polygon, 20
Function, probability, 121

### G

Game, equitable, 86
General addition rule, 100
General multiplication rule, 107
General purpose index, 49
Geometric mean, 36
Given year, 49
Goodness of fit:
  least-square line, 290, 291
  probability distribution, 274
Grand mean, 244
Graphical presentation:
  bar chart, 20
  frequency polygon, 20
  histogram, 18
  ogive, 20
  pictogram, 21
  pie chart, 22

### H

Harmonic mean, 37
Histogram, 18

Hypergeometric distribution, 127
  and binomial distribution, 129
  mean, 135
Hypothesis:
  null, 228
  one-sided alternative, 225
  two-sided alternative, 225

### I

Independent events, 106
Independent samples, 236
Index:
  Consumer Price, 48
  general purpose, 49
  Industrial Production, 49
  Laspeyres, 55
  Paasche, 55
  quantity, 52
  value, 53
  Wholesale Price, 49
Index numbers:
  base year, 49
  fixed-weight aggregative, 56
  given year, 49
  mean of price relatives, 50
  series, 56
  shifting the base of, 57
  simple aggregative, 51
  units test, 51
  unweighted, 49
  use in deflating, 57
  weighted aggregative, 55
  weighted mean of price relatives, 54
Index of seasonal variation, 312
Inductive statistics, 3
Inequality signs, 212
Inference, statistical, 3
Infinite population, 173
  sampling from, 180
Internal data, 4
Intersection, 63
Interval, class, 16
  of distribution, 16
Interval estimate, 214
Irregular variation, 300, 303, 320

### J

Judgment sample, 182

### L

Large numbers, law of, 140
Large-sample confidence intervals:
  for means, 213
  for proportions, 252
  for standard deviations, 214
Laspeyres index, 55
Law of averages, 140

Law of large numbers, 140
Least-squares line, 283
Least-squares, method of, 281
 criterion for, 284
Level of significance, 223
Limit, class, 14
 upper and lower, 14
Linear correlation, 295
Linear equation, 278
Linear trend, 301, 304
Location, measures of, 29
Logarithmic probability paper, 162
Lower class boundaries, 15
Lower class limits, 14

## M

Mark, class, 15
Markov chain, 112, 114
Mathematical expectation, 85, 86
 and decision making, 87
 formula for, 86
Mean, 29
 arithmetic, 36
 binomial distribution, 134
 combined data, 37
 confidence intervals, 213, 218
 continuous distribution, 147
 deviation from, 135
 error of estimate, 211
 geometric, 36
 grand, 244
 harmonic, 37
 hypergeometric distribution, 135
 of distribution, 32
 population, 173
 probability distribution, 133
 probable error, 208
 properties of, 30
 reliability of, 31, 43
 sampling distribution, 201
 standard error, 202
 tests for, 227–229, 231
 weighted, 33
Means, differences between, 236
Means, differences among, 240
Measure, statistical, 29
Measurements, 209
Measures of association, 201
Measures of location, 29
Measures of skewness, 29
Measures of variation, 29
Median, 39
 of distribution, 42
 properties, 40
 reliability of, 40, 43
 standard error, 207
Method of elimination, 285

Method of least squares, 281
 normal equations, 284, 285
Mid-range, 45
Modal class, 46
Mode, 45
Moving average, 307
 centered, 313
$\mu$ (mu), mean:
 population, 173
 probability distribution, 133
Multiplication rule:
 general, 107
 special, 108
Mutually exclusive events, 63

## N

Negative correlation, 292
Nonparametric tests, 231
Normal distribution, 147
 and binomial distribution, 166
 probability paper, 162
 standard, 148
 table, 149, 325
Normal equations, 284, 285
Normal of time series, 320
Null hypothesis, 228
Number of degrees of freedom (*see* Degrees of freedom)
Numbers, random, 178
 table, 334, 335

## O

Odds, 80
 and probabilities, 82, 83
Ogive, 20
One-sided alternative, 225
One-tail test, 224
Open class, 13
Operating characteristic curve, 234
Opportunity loss, 91
Outcome of an experiment, 59

## P

Paasche index, 55
Paired data, 278
Parameter, 173
Pascal's triangle, 79
Pearsonian coefficient of skewness, 199
Percentage distribution, 16
Percentiles, 45
Perfect information, expected value of, 92
Permutation, 73
Personal probability, 82
 consistency criterion, 95
Pictogram, 21
Pie chart, 22

Point estimate, 210
Population, 172
  distribution, 174, 180
  finite, 172
  infinite, 173
  mean, 173
  size, 177
  standard deviation, 189
Positive correlation, 292
Price relative, 50
Primary data, 4
Probability:
  and area under curve, 145
  and odds, 82, 83
  Bayes' rule, 116
  conditional, 103, 105
  equiprobable events, 98
  frequency interpretation, 80
  general addition rule, 100
  general multiplication rule, 107
  personal, 82
  special addition rule, 95
  special multiplication rule, 108
  subjective, 82
Probability density, 145
Probability distribution, 121, 123
  binomial, 124
  hypergeometric, 127
  mean, 133
  standard deviation, 137
  variance, 136
Probability function (*see* Probability distribution)
Probability of Type I and Type II errors, 233, 235
Probability paper:
  arithmetic, 162
  logarithmic, 162
Probable error of the mean, 208
Proportions:
  confidence interval, 249, 252
  differences among, 260, 266
  maximum error of estimate, 252
  standard error, 252
  tests concerning, 258
Purchasing power of dollar, 57

## Q

Qualitative distribution, 10
Quantitative distribution, 10
Quantity index, 52
Quantity relatives, 52
Quota sampling, 182
Quartiles, 44

## R

*r* (*see* Correlation coefficient)

Random numbers, 178
  table, 334, 335
Random sample, 177
  finite population, 178
  infinite population, 180
  simple, 177
Random variable, 120
Rangle, sample, 197
Rank correlation, 297
Ratio-to-moving-average method, 312
Raw data, 10
Real class limits, 15
Regression analysis, 287
Regression line, 287
Regret, 91
Relative frequency, 248
Relative variation, 199
Relatives:
  price, 50
  quantity, 52
  value, 53
Reliability, 31, 187
Rounding numbers, 15
Rule of Bayes, 115, 116

## S

Sample, 171
  random, 177
  size, 177, 216, 255
  standard deviation, 190
  variance, 190
Sample space, 59
  continuous, 144
  finite, 61
Samples, dependent, 239
Samples, independent, 236
Sampling:
  area, 182
  cluster, 182
  from infinite population, 180
  judgment, 182
  quota, 182
  random, 177
  stratified, 181
  systematic, 182
  with replacement, 173
Sampling distribution:
  difference between means, 236
  mean, 201
  median, 207
  proportion, 253
  standard deviation, 214
Seasonal variation, 300, 301, 312
  index of, 312
Secondary data, 4
Secular trend, 300, 301
Semi-annual trend increment, 306

Series, index numbers, 56
Sets:
  complement of, 64
  empty, 64
  intersection, 63
  mutually exclusive, 63
  number of subsets, 78
  union of, 63
Shifting the base, 57
$\sigma$ (sigma) standard deviation:
  population, 189
  probability distribution, 137
$\Sigma$ (sigma) summation, 38
  double, 244
Significance, level of, 223
Significance test, 223 (*see also* Tests of hypotheses)
Simple aggregative index, 51
Simple random sample, 177
Simulation, 185
Size:
  population, 177
  sample, 177, 216, 255
Skewness, 199
  measures of, 29
  Pearsonian coefficient, 199
Slope of regression line, 287, 299
Small-sample confidence interval for $\mu$, 218
Small-sample tests:
  differences between means, 238
  mean, 231
Special addition rule, 95
Special multiplication rule, 108
Square-root table, 337
  use of, 336
Standard deviation:
  binomial distribution, 138
  confidence interval for, 214
  continuous distribution, 147
  grouped data, 193
  population, 189
  probability distribution, 137
  sample, 190
  standard error, 214
Standard error:
  difference between two means, 236
  mean, 202
  median, 207
  proportion, 252
  standard deviation, 214
Standard normal distribution, 148
  table, 149
Standard units, 148, 198
Statistic, 29, 173
Statistical description, 29
Statistical inference, 3
Statistical measure, 29

Statistical significance, 223, 226
Statistics:
  descriptive, 1
  inductive, 3
Stratified sampling, 181
Subjective probability, 82
  consistency criterion, 95
Subscripts, 30, 38
Summation, 38
  double, 244
Symmetrical distribution, 199
Systematic sampling, 182

## T

$t$ Distribution, 218
  table, 326
$t$ statistic, 231, 238
Table, two-way, 268
Tests of hypotheses:
  analysis of variance, 241
  correlation coefficient, 295
  difference between means, 236
  differences among proportions, 261, 264
  goodness of fit, 274
  means, 227–229, 231
  null hypothesis, 228
  one-sided, 224
  one-tail, 224
  proportions, 258
  significance test, 223
  two-sided, 224
  two-tail, 224
  Type I error, 232
  Type II error, 232
Test of significance, 223 (*see also* Tests of hypotheses)
Ties in rank, 297
Time series, 300
Tree diagram, 70
Trend:
  linear, 301, 304
  secular, 300
Trend increment, 306
Trials, 123
Triangular density, 154
Two-sided alternative, 225
Two-sided test, 224
Two-tail test, 224
Two-way table, 268
Type I error, 232
  probability of, 233, 235
Type II error, 232
  probability of, 233, 235

## U

Uniform density, 154
Uniform distribution, 277

Union, 63
Units, standard, 148, 198
Units test, 51
Unweighted index numbers, 49
Upper class boundaries, 15
Upper class limits, 14

## V

Value index, 53
Value relatives, 53
Variable, random, 120
Variance:
  binomial distribution, 138
  continuous distribution, 147
  grouped data, 193

Variance (*cont.*)
  probability distribution, 136
  sample, 190
Variance, analysis of, 240, 245
Variance ratio, 242
Variation:
  coefficient of, 199
  measures of, 29
  relative, 199
Venn diagram, 64

## W

Weighted aggregative index, 55
Weighted mean, 33
Weighted mean of price relatives, 54
Weights, 33